The air engine

Related titles:

Materials for energy conversion devices
(ISBN 978-1-85573-932-1)
The term electroceramic is used to describe ceramic materials that have been specially formulated with specific electrical, magnetic or optical properties. Electroceramics are of increasing importance in many key technologies including: microelectronics, communications and energy conversion. This innovative book presents current research from leading innovators in the field and is the first comprehensive survey on major new developments in electroceramics for energy conversion devices.

Industrial gas turbines: performance and operability
(ISBN 978-1-84569-205-6)
This important new book provides operators of industrial gas turbines with a review of the principles of gas turbine operation and how they can be used to predict and improve turbine performance. The book is accompanied by a CD which allows readers to simulate various aspects of performance such as emissions, changes in pressure and power augmentation.

HCCI and CAI engines for the automotive industry
(ISBN 978-1-84569-128-8)
Homogeneous charge compression ignition (HCCI) and controlled auto ignition (CAI) has emerged as one of the most promising engine technologies with the potential to combine fuel efficiency and improved emissions performance. Despite the considerable advantages, its operational range is rather limited and controlling the combustion (timing of ignition and rate of energy release) is still an area of ongoing research. However, commercial applications are close to realisation. This book reviews the key international research on optimising its use, including gasoline HCCI/CAI engines; diesel HCCI engines; HCCI/CAI engines with alternative fuels; and advanced modelling and experimental techniques.

Details of these books and a complete list of Woodhead's titles can be obtained by:

- visiting our web site at www.woodheadpublishing.com
- contacting Customer Services (e-mail: sales@woodhead-publishing.com; fax: +44 (0) 1223 893694; tel.: +44 (0) 1223 891358 ext.130; address: Woodhead Publishing Ltd, Abington Hall, Abington, Cambridge CB21 6AH, England)

The air engine

Stirling cycle power for a sustainable future

Allan J. Organ

CRC Press
Boca Raton Boston New York Washington, DC

WOODHEAD PUBLISHING LIMITED
Cambridge England

Published by Woodhead Publishing Limited, Abington Hall, Abington,
Cambridge CB21 6AH, England
www.woodheadpublishing.com

Published in North America by CRC Press LLC, 6000 Broken Sound Parkway, NW,
Boca Raton FL 33487, USA

First published 2007, Woodhead Publishing Limited and CRC Press LLC

British Library Cataloguing in Publication Data
A catalogue record for this book is available from the British Library.

Library of Congress Cataloging in Publication Data
A catalog record for this book is available from the Library of Congress.

Woodhead Publishing ISBN 978-1-84569-231-5 (book)
Woodhead Publishing ISBN 978-1-84569-360-2 (e-book)
CRC Press ISBN 978-1-4200-6672-2
CRC Press order number: WP6672

The publishers' policy is to use permanent paper from mills that operate a sustainable forestry policy, and which has been manufactured from pulp which is processed using acid-free and elementary chlorine-free practices. Furthermore, the publishers ensure that the text paper and cover board used have met acceptable environmental accreditation standards.

Project managed by Macfarlane Production Services, Dunstable, Bedfordshire, England
(e-mail: macfarl@aol.com)
Typeset by Replika Press Pvt Ltd, India
Printed by TJ International Limited, Padstow, Cornwall, England

Contents

Notation

Symbol		Units
A_C	cross-sectional area of compression cylinder	m^2
A_{ff}	free-flow area – cross-sectional area of flow passage, e.g. of regenerative annulus	m^2
a	local isentropic acoustic speed $\sqrt{\gamma RT}$ or $\sqrt{RT}$ as per context	m/s
b	height of flow passage in direction of axis of spiral	m
a, b	coefficients of equation – defined in text	(as required)
c	local acoustic speed, isentropic$\sqrt{\gamma RT}$ or isothermal $\sqrt{RT}$ as per context	m/s
c	radial gap between concentric cylinders	m
c_p, c_v	specific heat at constant pressure/volume	J/kgK
c_w	specific heat of material forming matrix of regenerator	J/kgK
d	diameter	m
D	internal diameter of inlet manifold	m
D	total or substantial derivative $\partial/\partial t + u\partial/\partial x$	s^{-1}
d	diameter	m
d_w	diameter of individual wire	m
F	friction term $\frac{1}{2}\underline{u}^2 C_f \text{sign}(\underline{u})/r_h$	m/s^2
f	cyclic frequency	s^{-1}
g	mean mass velocity ρu based on free-flow area	kg/m^2s
G	mean mass velocity ρu based on frontal area	kg/m^2s
h	coefficient of convective heat transfer	W/m^2K
k	thermal conductivity	W/mK
L	overall length of flow passage	m
L_x	length of flow passage of specified exchanger	m
L_o	piston-face to piston-face linear distance at crank angle datum	m
L_r, L_{reg}	regenerator flow passage length	m
L_{ref}	reference length – equal to $V_{sw}^{1/3}$	m
m	variable mass (of working fluid)	kg

m'	mass rate, dm/dt	kg/s
m_w	mesh number – number of wires/m	m^{-1}
P, Q	coefficients defined in text	(as required)
p	(absolute) pressure	Pa
p_w	wetted perimeter	m
q_s	heat shuttled per cycle	J
q'	heat rate per unit mass	W/kg
Q_C	heat rejected (from compression space) per cycle	J
Q_E	heat input (to expansion space) per cycle	J
R	specific gas constant	J/kgK
r	radial coordinate	m
r	crank-pin offset (piston semi-stroke)	m
r_h	hydraulic radius – (wetted volume)/(wetted area)	m
rpm	revolutions per minute	min^{-1}
s	coordinate in peripheral direction = $r\phi$	m
t	time	s
t	thickness of separating plate, or radial thickness (e.g.) of displacer shell	m
t_c, t_d	radial thickness of wall of displacer/cylinder	m
T	(absolute) temperature	K
T	torque	Nm
T_a	local temperature of air stream	K
T_C	temperature at ambient end of regenerator	K
T_E	temperature at expansion end of regenerator	K
T_{ex}	local temperature of exhaust stream	K
T_g	local, instantaneous absolute temperature at point in fluid/enclosure element	K
T_{su}	Sutherland temperature	K
T_w	local, instantaneous absolute temperature at point in matrix or at enclosure wall	K
T_{wc}	local, instantaneous temperature at given axial location on the cylinder	K
T_{wd}	local, instantaneous temperature at given axial location on the displacer	K
u, v	local, instantaneous particle velocity in x and r coordinate directions respectively	m/s
u	(in the context of the Method of Characteristics) one-dimensional (slab-flow) velocity at a point in physical and state planes	m/s
$\underline{u}$	mean mass velocity or bulk velocity, $m'/\rho A_{ff}$	m/s
$\underline{u}$	(in the context of the Method of Characteristics) mean value of u between two adjacent points on integration mesh	m/s

u_d	instantaneous velocity of displacer	m/s
v	variable volume	m^3
V	variable volume	m^3
V_C	amplitude of volume variation in compression space	m^3
V_E	amplitude of volume variation in expansion space	m^3
V_{sw}	swept volume	m^3
w	width of flow passage measured in radial direction	m
W	work per cycle	J
x	linear distances in x coordinate direction	m
ΔT	local, instantaneous difference $T - T_w$ between temperature of gas and that of immediately adjacent solid element	K
β	angular rotation of fluid filament within flow passage of regenerator gauze aperture	–
δ^*	displacement thickness (of boundary layer)	–
ϕ	angular coordinate	–
μ	coefficient of dynamic viscosity	Pa s
ρ	density	kg/m^3
ρ_w	density of parent material of regenerator matrix	kg/m^3
θ	momentum thickness (of boundary layer)	m
ω	angular speed	s^{-1}

Dimensionless variables (NB: dimensionless *parameters* listed below)

f	pressure coefficient $\Delta p/\frac{1}{2}\rho u^2$	–
F	number of degrees of freedom (of planar mechanism)	–
H	shape factor associated with boundary layer	–
J	number of joints (in planar mechanism)	–
L	number of binary, ternary, etc., links (forming planar mechanism	–
m	momentum thickness variable	–
Ma	Mach number $\underline{u}/\sqrt{\gamma RT}$ (isentropic) or $u/\sqrt{RT}$ (isothermal)	–
Pr	Prandtl number $\mu c_p/k$	–
Re	Reynolds number $4\rho \mid \underline{u} \mid r_h/\mu = 4 \mid g \mid r_h/\mu$	–
Sg	Stirling number $p\underline{u}/\mu r_h$	–
St	Stanton number $h/\rho \mid \underline{u} \mid c_p = h/ \mid g \mid c_p$	–
η_r	thermal recovery ratio of regenerator – ratio of heat recovered per gas pass to maximum potential recovery between cycle temperature limits	–
θ	angular measure – circumferential	–
ψ	normalized pressure p/p_{ref}	–

Dimensionless parameters

a, b	coefficients of heat transfer correlation $StPr^{2/3} = a/Re^b$	–
c, d	coefficients of friction factor correlation $C_f = c/Re + d$	–
C_f	friction factor $\Delta p/(\frac{1}{2}\rho u^2 L/r_h)$	–
n_g	number of gauzes in close-packed stack	–
n_j	number of circumferential subdivisions of single lap	–
n_l	number of laps of duplex flow channel	–
N	abbreviation for the product $\frac{1}{2}\pi m_w d_w$	–
N_{FL}	flush ratio: ratio of mass perfusing matrix to resident mass	–
NFU	number of friction units $C_f L_x/r_h$ (by analogy with NTU below)	–
N_{MA}	Characteristic Mach number or speed parameter $\omega V_{sw}^{1/3}/\sqrt{RT_C}$	–
N_{TCR}	thermal capacity ratio $\rho_w c_w T_C/p_{ref}$	–
NTU	number of transfer units StL/r_h	–
N_{RE}	Characteristic Reynolds number $N_{SG} N_{MA}^2$	–
N_T	temperature ratio T_E/T_C	–
α	thermodynamic phase angle – not to be confused with …	–
α	… aperture ratio or critical area ratio – (minimum free-flow area)/frontal area	–
α_{ffr}	dimensionless free-flow area of the uniformly packed regenerator $A_{ffr}/V_{sw}^{2/3}$	–
κ	thermodynamic volume ratio V_C/V_E	–
γ	specific heat ratio – isentropic index	–
$¶_v$	volume porosity – ratio of void volume to envelope volume	–

Other

$f\{\ \}$	function of { }	–

Subscripts

app	approach value (as in approach velocity)
AIR,air, a	relating to inlet air stream
EX,ex	relating to exhaust stream
C, E	compression, expansion
cr, crit	critical location – point of minimum free-flow area

d	un-swept or 'dead'; also relating to displacer
K&L	as defined by Kays and London (1964)
lam	laminar flow regime
L	measured over gauze stack length L
path	path direction
r, reg	relating to regenerator
ref	reference values
s	separation point
trans	transition flow regime
turb	turbulent flow regime
w	pertaining to wall or to individual (regenerator) wire
0	datum or reference value e.g. p_0
x	relating to heat exchanger
I, II	I and II characteristics directions

Preface

The preface to the author's 2001 text *Air Engines* opened with a question: Why another book on Stirling engines? Two books and two prefaces later the question deserves an updated response.

In the early 1990s the commercial market was limited to niche applications such as auxiliary power generation for yachts. A few years on, the air engine has become the core component of CHP – combined heat and power – a large-scale commercial venture promising a vital contribution towards mitigating the man-made element of global warming and climate change. A potential market of millions of units in Europe alone transforms commercial prospects. But why an academic text rather than a sales catalogue?

At the present time, no single manufacturer is equipped to meet the predicted demand. The Western manufacturing base has retrenched and become specialized; the prospect of open-ended development costs has become increasingly unacceptable; the blank cheque approach which allegedly characterized Fritz Philips' patronage will not be available again until air engines power F-1 cars. For a manufacturer to embrace a new product there has to be a high degree of confidence that the design stage will yield a prototype requiring a minimum of modification and development in the course of achieving performance and service life targets.

But what has advanced most since publication of *Air Engines* is the perception of the thermodynamic design challenge. The internal driving force is convective heat exchange, an *inherently irreversible* process which ensures that the Stirling engine no more aspires to the Carnot efficiency than does the steam engine. Abandoning the mirage of Carnot – which is not an engine anyway but a 'thought experiment' – and discarding the associated 'ideal' cycle avoids unrealistic expectations and the inevitable disenchantment. Preoccupation with optimum volume ratio and phase angle is mercifully giving way to closer scrutiny of the gas processes: an 'optimum' deriving from a defective gas process model is emphatically no optimum at all – and who is to say that it is not the very reverse? Chapter 2 amplifies these considerations.

Increasingly under scrutiny are basic flow data – 'correlations' – of friction factor C_f vs Reynolds number *Re*, and, by association, Stanton number *St* vs *Re*. Data commonly resorted to derive from tests *not* of regenerators, but of stacks of small numbers of screens, some as few as 3 screens per stack! The concept of friction factor C_f is taken arbitrarily from pipe flow where it has physical significance, to flow through matrices where that significance is lost. A far larger body of experimentation on flow through porous media correlates pressure drop to flow without invoking *either* friction factor *or* Reynolds number. The resource is overlooked in regenerator studies, and Chapter 2 probes the alternative correlation equation – the 'Forchheimer law' – further.

Yet another substantial literature recognizes that a gas is a compressible fluid, and accordingly plots correlations between pressure coefficient and Mach number *Ma*. These data might be thought an indispensable resource for an air-charged (or N_2-charged) design, where acoustic speed is relatively low – and *Ma* correspondingly high – but have likewise been bypassed. Perhaps the strongest case for the reassessment undertaken in these pages is that, when the three major sources on flow through matrices are reduced to a common correlation (e.g., C_f vs *Re*) they fail to coincide – *even over the range where all are incompressible* – and fail to do so by a substantial margin. Chapters 10 and 14 explore in some detail, and demonstrate that reacquiring flow correlations as fractional pressure drop $\Delta p/p$ vs Stirling number $pr_h/\mu \underline{u}$ with Mach number *Ma* as parameter not only provides seamless coverage of all flow regimes (creeping, incompressible and compressible) but takes regenerator design a step closer to becoming an intuitive process.

Within the limitations of the 'slab' or 'one-dimensional' view of fluid flow, a priority of the analytical formulations and associated computer code is to achieve relevance to the broadest possible picture of the gas process interactions. This has facilitated an enquiry into the inner workings of the 'thermal lag' engine (Chapter 9) and the pressure wave engine (Chapter 13). Both accounts offer previously unreported insights in the form of quasi-perspective graphical depictions of the gas processes.

It is easy to forget that a Stirling engine is a *system* which includes a source of heat, usually a combustion process. Most forms of external heat provision are irredeemably wasteful. A solution is for the combustion chamber to be enclosed within a wrap-around air pre-heater. The air engine must become a form of *internal* combustion engine, reflecting the reality that the cycle temperature limits are not 600°C and 60°C as commonly cited, but 1500°C and 60°C, thus putting the Carnot criterion into even more realistic perspective. Chapters 3 and 4 describe a heating system which discharges exhaust below 100°C, thereby delivering the higher calorific value (HCV) of the fuel to the engine.

The designer starting afresh in 2007 can expect to capitalize on the 'polymer alloy' technology of the dry seals of nitrogen compressors which now offer lives measured in tens of years. In conjunction with 'solid oil' rolling-element bearings for the drive mechanism, this finally puts oil-free operation within reach. Chapter 16 explores how these technologies might form the basis for a modern successor to Philips' famous MP1002CA engine.

It has been observed (e.g., by O'Sullivan 1953) that '... wisdom comes with age'. Wisdom is relative, but it is true that the earlier impatience of the author to see principle converted to practice now appears to have been unwise: it is not by chance that closed-cycle, regenerative engines have failed to evolve at the pace of the internal combustion engine – things could not have been otherwise. The inscrutable gas process cycle does not reward seat-of-the-pants design by returning competitive performance. The alternative – cut-and-try development – is haphazard and expensive, and offers no guarantee of making a bad design better. History will eventually confirm that the only way to tackle the challenge of the Stirling engine has been the two-pronged approach: parallel programs of experimental development and theoretical enquiry, the one continuously cross-calibrated against the other.

I am grateful to my publisher, Woodhead Publishing Limited for making possible this contribution to the tradition of theoretical enquiry, and in particular to Sheril Leich, my commissioning editor. Project editor Laura Bunney and project manager Stuart Macfarlane held me courteously but firmly to their respective high standards – and to the tightest production schedule I have experienced in an output of five books!

Jimmy James checked some of the numerical calculations, to the benefit particularly of Chapter 16. Ian Larque built and tested the combustion chamber and Spirex air pre-heater. Colleagues who, whenever requested, kindly downed tools to discuss air engines include Peter Feulner, Norris Bomford, John McIntyre, and Geoff Vaisey. Friends and neighbours too numerous to list assisted in one capacity or another with the commissioning and celebration of the 1818 replica.

Allan J. Organ
Cambridge
2007

Introduction: an invention ahead of its time

Those who have acquired this book to further an established interest in Stirling engines will need no prompting as to the significance of the year 2016 – bi-centenary of a unique double invention, the thermal regenerator and the Stirling engine. The regenerator was to have a measurable impact on the Industrial Revolution by offering economies in industrial-scale metal production. The engine, by contrast, made sedate progress towards its commercial potential, and this despite knocking spots off such revered icons of thermodynamic perfection as the Carnot 'engine'.

Reasons for the ascendancy of engines operating on more 'obvious' gas process cycles – steam, internal combustion and gas turbine – are not hard to propose. Conversely, neglect of the more subtle Stirling principle may have cost us a crucial head-start in cutting CO_2 emissions and in addressing the changes in climate attributed to them.

If the assertion sounds like an over-statement, a selective* look at historical milestones since 1816 will put it into perspective:

Full-steam ahead

The nineteeth century saw the Industrial Revolution enter top gear. Steam power displaced water-wheel and windmill, and facilitated explosive growth in mining and manufacturing. In land and marine transport, steam took over from horse and sail. The scientific work of Darwin challenged millennia of traditional religious beliefs: the perception of the origin and destiny of the planet was irrevocably changed – but its future had every appearance of being in safe custody.

At mid-century the air engine attained a technological peak at the hands of the Stirlings – but then entered a hundred years of decline.

The twentieth century was the era of motorized road transport, of aviation, of centralized electricity generation, of nuclear physics and of

*Even more selective than at first sight, being from an essentially UK viewpoint.

telecommunication – most conspicuously, television. Under the impetus of the second of two World Wars came jet engine and rocket. The former led to a dramatic increase in the speed of atmospheric flight; the latter to realization of the ultimate scientific dream – space travel. We had not merely overcome gravity, but gone one step further and exploited it to orbit, and ultimately to set foot on, the moon. Techniques of mass-production made possible the 'consumer society', whose values, shaped by an increasingly pervasive mass-media, became an economic driving force.

Perhaps most spectacularly, the twentieth century saw the development of the digital computer and the accompanying explosion in IT (information technology). The introduction of the 'smokeless zone' brought about the first major change in domestic heating since coal burning began. California's problem with photochemical smog switched attention to the automobile as a polluter, and galvanized the technology of emissions control. A hole was discovered in the ozone layer, the fluorocarbon identified as the culprit, and measures implemented to cut back on industrial use and release into the atmosphere. In an increasingly secular world there stirred a recognition that custody of the Planet entailed certain responsibilities. But between London and New York, Concorde could out-run the sun; forecasting tomorrow's weather had evolved into a science; mechanized farming had secured the food supply, and the overwhelming sense was that, but for a few anomalies, we had shown the environment who was boss.

... after the horse has bolted

So what of the inaugural six years of the twenty-first century? A belated but escalating awareness of the climatic, economic and social consequences of the relentless profligacy with which our natural resources are being plundered. Publicity material for Al Gore's *An Inconvenient Truth* invites us to face up to the difficulty of *getting a man to understand something when his salary depends on his not understanding it*. A decade earlier the Big Screen would not have been the context for throwing down such a challenge.

A respected scientific view is that global warming is a consequence of – or is compounded by – emission into the atmosphere of gases, particularly of carbon dioxide (CO_2), having a 'greenhouse' effect. CO_2 is a product of burning fossil fuels – coal, oil and gas. The nominal quota of CO_2 in the atmosphere is some 0.037 percent, or 370 ppm. Emissions from industry and transport world-wide add millions of tons to the atmosphere daily, so the link with climate change is plausible, even if not yet universally conceded. Those convinced that this is cause-and-effect predict that the result by the end of the 21st Century could be an average temperature rise of 6°C – sufficient to cause widespread flooding, famine and violent storms.

A recent report on climate change by Sir Nicholas Stern for the UK Government warns that the possibility of avoiding a global catastrophe 'is almost out of reach' (*The Independent*, Monday 30th October 2006).

Professor James Lovelock, author of the 'gaia' theory of the equilibrium of the biosphere, paints a scenario which is even more bleak: in a recent speech to the Institution of Chemical Engineers, Lovelock asserted that temperature rises of up to 8°C are already 'built-in', and that efforts to avert them, while morally commendable, are wasted. 'It is a bit like if your kidneys fail you can go on dialysis – and who would refuse dialysis if death is the alternative. We should think of it in that context.' (*The Times*, Wednesday 29 November 2006, p. 15)

Observation by satellite of the area of the Arctic covered by sea ice began in 1978. The steady decline in coverage measured between that date and 2002 is not *per se* the most disturbing feature. The truly worrying aspect is that a reduction in sea-ice cover leads to increased heat absorption by the water thereby exposed – and to an *escalation in the rate* of warming. In other words, the phenomenon is *cumulative* – and potentially unstable. Taking this into account, interim findings of the Intergovernmental Panel on Climate Change (IPCC), published in December 2006, estimate 35 years for the effective disappearance of north-polar summer sea ice – and with it the demise of indigenous peoples, of their way of life, and of animal species such as the polar bear.

Where does the air engine fit in?

In principle, it allows the concept of combined heat and power (CHP) to be turned into a widespread reality – and crucially – into a domestically-acceptable reality. Homes and places of work consume energy for heating, for cooking and for powering electrical appliances. If that energy is supplied in the form of electricity generated centrally by a coal-, gas- or oil-burning facility and distributed via a grid, some 50 percent of the raw hydrocarbon fuel energy has already been discarded at source. Transmission line losses compound the inefficiency.

The domestic CHP unit is essentially an engine-driven generator under microprocessor control and connected to the grid via a metering system. It is also connected to a supply of fossil fuel – conveniently gas. The gas powers the engine which in turn drives the generator. Waste heat from the engine is directed to a heat exchanger which provides hot water for heating, shower, etc. Electric power in excess of instantaneous domestic requirement is fed back into the grid for credit or rebate. A supplementary boiler heated by direct combustion takes care of short-term demand for extra water heating. Any shortfall in electric power generated is compensated by drawing from the grid – in the long term by other CHP installations and/or generators powered by 'renewable' energy sources.

The terms of reference of the CEGB (Central Electricity Generating Board) prior to the 2001 break-up called for the highest practicable efficiency of *electricity generation*. By contrast, CHP aims for the *most efficient overall utilization* of the raw energy supply, with potential for 95 percent utilization.

If in doubt about the principle, compare the cooking of a meal on a gas appliance with cooking the identical meal on an electric appliance supplied via the grid from a gas-burning power station. The latter meal accounts for over twice the gas consumption of the former – and, other things being equal, is responsible for twice the CO_2 emissions.

In principle, any engine (petrol, diesel, micro-turbine) will serve, but the Stirling engine offers the quiet operation expected of a domestic appliance. The technology of the Stirling engine happens to be the technology of the CFC-free refrigerator, the CFC-free heat pump, and therefore the CFC-free air-conditioning unit – all of which now merit urgent re-evaluation.

The report by the Domestic Power Unit CHP Section of the SBGI Trade Association dated 8th September 2003 and entitled *'Micro-CHP – delivering a low-carbon future'* summarizes the potential market for micro-CHP. It affirms that micro-CHP meets all four of HM Government's Policy Objectives outlined in the Energy White Paper:

1. to keep energy costs low
2. to keep energy costs secure
3. to achieve reduction in greenhouse gas emissions
4. to eradicate fuel poverty.

A domestic CHP unit has potential to achieve up to 45 percent reduction in CO_2 emissions – an estimated equivalent 1.5 tonnes annually per installation – and simultaneously to save each household £150 on the annual electricity bill. Some 22 million UK homes have central heating, and of these approximately 18 million are gas systems. The SBGI report considers that 14 million UK homes are suitable for conversion to CHP. This equates to a national CO_2 saving of $1.5 \times 14 \times 10^6 = 21$ million tonnes/year, and according to the report 'would yield around a quarter of the UK's total Kyoto commitment by 2010 and all of the additional carbon savings to 2020 from domestic energy efficiency identified in the (UK Government) White Paper'.

A second Industrial Revolution

The political response to date is in terms of green taxes and the extension of carbon trading. This is worse than merely specious: it creates the false impression that there is time in hand for tackling the climate change threat at source, whereas clearly there is not.

So what are the options? We can accept the Lovelock scenario – that it is already too late, and do nothing. Alternatively, we can decide to do *something*.

Worst of all would be a compromise response, since that would dissipate resources without containing the problem. Richard Lambert is the head of the main employers' body in the UK – the Confederation of British Industry (CBI). Speaking on 28. November 2006 at the end of the two-day CBI conference, he suggested a prime objective for Britain: 'a goal of making the UK a world leader in the great new industries that will certainly emerge in the area of renewables and low-carbon energy'. Elsewhere it is being suggested that only a prompt and full-scale technological response on the scale of the Industrial Revolution will effectively tackle climate change while simultaneously allowing economic growth to be sustained.

A counter-argument is gaining strength: that nothing the West can achieve by way of cutting CO_2 emissions can hope to offset the increase accompanying the industrialization of China and India. Eighty percent of China's electric energy is generated in coal-burning power stations, and as of March 2005 the number of new coal-burning power stations planned stood at 544 (http://news.bbc.co.uk/1/hi/programmes/newsnight/4330469.stm.) According to some estimates, setting these operational will raise atmospheric CO_2 levels to 400 ppm, at which dangerous changes to the climate are anticipated.

Even with the well-being of the biosphere at stake it is morally problematic for the Western World to attempt, from its position of material luxury, to attempt to stifle economic growth elsewhere. But how about an all-out effort to develop an environment-friendly technology *equally acceptable to China's programme of economic development as to Western needs?* China will hardly prejudice its future economic prosperity by contributing to climate change which will gratuitously frustrate realization of the long-term benefits.

It would be difficult to overestimate the magnitude of the technological challenge: the unstoppable underground coal fires which burn in China are already emitting CO_2 at a rate comparable the rate of release by all the automobiles in the USA (website cited above). Addressing controlled mining and burning is a different proposition: the concept of CCT (clean coal technology) is already under development in the West. The coal is gasified into 'syngas' which is burned in a gas turbine. The resulting CO_2 is stored underground. Some of the energy not converted in this first power stage generates steam which drives a second stage of steam turbines. Combining this technology with capture of the methane released during mining (kg for kg methane is more harmful than CO_2) would go a long way to reconciling economic growth with minimization of the accompanying climate impact.

This technology must also be implemented in the coal-fired power stations of the western world. The measure will not circumvent the Second Law of Thermodynamics, so the domestic customer still stands to benefit from only 45 percent or so of the latent energy of the fuel. Domestic CHP thus remains an attractive complement from two points of view (1) resource conservation and (2) the relatively short time-scale required for widespread implementation.

Domestic CHP and the air engine

One thing is beyond doubt – until we apply ourselves we have not the vaguest notion as to what is achievable. Much will have to change: the apparent benefits of domestic CHP cited in this Introduction are almost certainly not the *net* benefit. The latter can be evaluated only in terms of *life-cycle costing* – an 'ore-mine-to-scrapyard' inventory of energy expended (and CO_2 created) during manufacture, generated during service life, and consumed during recycling and/or disposal. The new economic perspective is epitomized in the hypothetical factory manufacturing photo-voltaic solar panels, powered exclusively by solar panels on its own roof, containing its toxic emissions, disposing of its own waste – and making a profit.

Engineering thermodynamics has a long tradition of preoccupation with thermal efficiency – the ratio of work out to heat in. With the new priorities, the criterion of 'best' must now take account of the operating environment – in this case domestic CHP – and of the wider context, the eco-system. The textbook notion of efficiency has to be replaced by a *figure of merit*, η_{CO_2}, possibly along the lines of a ratio of kilowatt-hours (over service life) to kg CO_2 emitted over the product life cycle. The ratio has units of J/kg. This suggests generalizing somewhat: dividing by the calorific value of the fuel (say the higher value, HCV (J/kg)) makes η_{CO_2} fuel-independent:

$$\eta_{CO_2} = \frac{\text{kWh}_{\text{service life}}}{\text{CO}_{2\ \text{life cycle}}\text{HCV}} \quad (-)$$

The figure of merit remains significant even when the fuel is hydrogen – i.e. when there is no carbon to form CO_2. Only when initial manufacture and disposal/recyling involve zero CO_2 production does the denominator vanish and η_{CO_2} (appropriately) become infinite.

It is early days, and the η_{CO_2} criterion will no doubt be superseded. Right now it delineates the magnitude of the immediate task: design of a Stirling engine to specified performance is challenging enough. The new 'chicken-and-egg' dimension makes it even more daunting: without a design there is nothing on which to base a life-cycle cost prediction. Conversely, life-cycle cost target *per se*, even if available, is of no direct help in getting the inevitable bugs out of a prototype engine.

Against the background, this text is the stuff of a 'first-iteration'. On the positive side, its focus, the air engine, has realistic prospects of an important role in a Brave New World: a world where the priorities which have powered economic growth since the Industrial Revolution must now be reconciled with a priority overriding all others – to secure the well-being of a beautiful planet for the prosperity and enjoyment of future generations.

Part I

A long-overdue reappraisal

1

The famous engine that never was

1.1 Status quo

Robert Stirling filed the celebrated patent for his thermal regenerator almost two centuries ago. The invention combined stark constructional simplicity with a subtlety of function which has challenged analysts ever since.

The conventions of patent drafting required Stirling to describe a practical application of his idea. His response was to offer a second invention no less ingenious than the first: an engine operating on a 'principle entirely new'. Nor he did stop at a principle as Carnot was to do eight years later: he produced nothing less than a fully-engineered design supported by detailed technical drawings. Within two years, a full-size engine working on the principle (not, however, the engine of the patent drawings – see Fig. 1.1) was earning a living pumping water from a quarry in Ayrshire (Stirling 1853).

The subsequent technological evolution of the Stirling engine has been less inspired, but has recently led to commercial-scale production as the core component in domestic CHP appliances (Combined Heat and Power). With reliability adequate for the role and set to improve, and with concern for the environment mounting, the potential market in the CHP application alone is some millions.

This long-overdue breakthrough represents not the end of modern development, but the beginning. Markets on this scale will attract competition to supply engines delivering greater electrical (or mechanical) output for given fuel energy input. The first step will be to face up to an unwelcome reality: that the Stirling engine is *not*, after all, simpler than the internal combustion engine. The tangible complexities of valves, gears, contact-breakers and spark-plugs are absent. However, they give way to inscrutable gas process interactions far more intricate than the mechanical timing which they replace.

Preoccupation with determining optimum phase angle and volume ratio has mercifully run its course, and the era of development by intuition and/or trial-and-error is over. The commercial edge will be won by the manufacturer prepared to confront the realities of the gas process cycle against the background

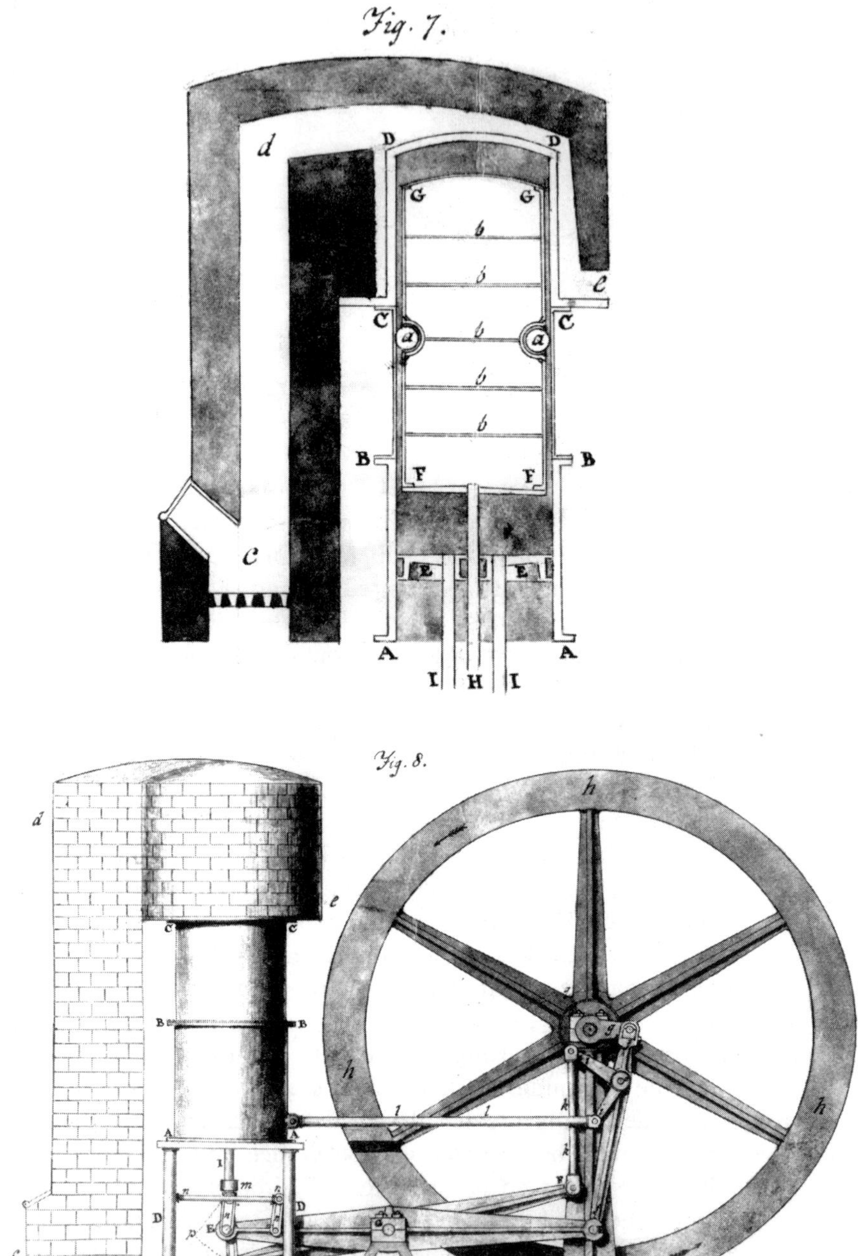

1.1 Drawings accompanying the Edinburgh patent (after Stirling 1816b). Reproduced by kind permission of the National Archives of Scotland under consent reference C20/17/1.

of relevant scientific principles. The Stirling engine is no more capable of approaching the Carnot efficiency than is any other heat engine. Abandoning this spurious criterion in favour of a realistic target is more likely to expedite development than to put a damper on it.

Hard-headed realism might not be a bad guiding principle for a 21st Century text on the air engine. So what better way to start than by addressing the mythology surrounding the first – and most important – air engine of all: Stirling's original of 1818?

If an engine was, indeed, constructed to the drawings of 1816 patents, it may be stated with confidence that it did not develop the two horsepower subsequently attributed to it – nor, indeed, did it produce any horsepower at all. Neither did it pump water from an Ayrshire quarry for two years – nor even for two minutes.

These are the conclusions from firing tests on a furnace, flue and displacer cylinder constructed full-size to dimensions specified in the patent and accompanying drawings. They evidently say much about patent disclosures in relation to practical engineering. But they say even more about the two-centuries-long neglect of heating provision in favour of a focus on phase angle, volume ratio, etc. as the 'obvious' key to performance potential.

1.2 The legend

It has become accepted that an engine, constructed to the drawings of Stirling's London patent (Stirling 1816a) and rated at 2 hp, was in service in 1818 pumping water from an Ayrshire quarry. The notion has been traced back to an article in *The Engineer* for 1917 commemorating the centenary of the original invention – but no further!

On the basis of Stirling's target operating temperatures (hot end 460°F or 255°C above the ambient end), computer simulation suggests that such an engine might, indeed, have functioned – 'though at a more modest half-horsepower (Organ 2001).

Conversely, if the heating provision had been inadequate, one of engineering history's revered icons would not even have turned itself around – not, at least, in the configuration in which it was originally conceived (Fig. 1.1). Now, please indicate by raising a hand if you can remember the last time you succeeded in coaxing power from a Stirling engine with the heat source 10 feet distant from the point of heat delivery! The crucial doubt which arises is worth settling in the interests of historical accuracy alone. A replica furnace and chimney to the drawing of the 1816 patent, fired with a dummy cylinder in position, could shed light.

1.3 History reconstructed

There are grounds (Finkelstein and Organ 2001) for accepting that Stirling intended the prototype to be 10 feet (3.05 m) tall. This allows the dimensions of the side elevation of furnace and chimney to be estimated from drawings accompanying the Edinburgh patent (Stirling 1816b), already introduced as Fig. 1.1. The brickwork shrouding the hot end of the engine appears to be cylindrical. Projecting a plan view from the side elevation on this assumption ought to confirm the line of intersection between cylindrical and plane surfaces, thus revealing the depth of brickwork perpendicular to the plane of the paper. Attempting the projection shows that the drawing is not, after all, orthographic – nor even to scale in all respects. There is no problem, since the information which is material for present purposes is the internal dimension of the flue. This scales from the inset drawing to 7 inches. It is assumed for the moment that the aperture is of square cross-section.

In the Edinburgh patent the vertical location of the fire grate is inconsistent between principal and supplementary side elevations. On the other hand, the side elevation of the London drawing tallies with the principal side elevation of the Edinburgh drawings. These are therefore relied upon for the location of grate and access door.

A common size of brick in use in England in the 19th Century was $9 \times 4\frac{1}{2} \times 3$ in. The modern metric equivalent is not dissimilar, being $210 \times 100 \times 65$ mm. However, achieving both internal and external dimensions of the flue using modern bricks would be possible only with massive wastage. Using uncut bricks for each course allows, for the upper part of the flue, an internal aperture of 250×250 mm (10×10 in.) and an external dimension of 445×445 mm (17.5×17.5 in.). Thus, internal size of the replica is increased somewhat relative to the original, while external size is decreased.

Maximum fuel burn rate is largely determined by grate size. This in turn determines the temperature achievable at the top of the flue. Based on the use of whole bricks, the grate area nearest to the original is 355×250 mm (14×10 in.). A strip metal boot scraper serves the grate itself. It promises short life expectancy, but offers large aperture ratio: for the reconstruction exercise to have a valid outcome it is necessary to avoid restricting the flow of air (and thus the flow of heat) below that which might have been achieved in the original engine.

The combined furnace and flue illustrated in Plate 1 (between pages 134 and 135) was constructed from LBC 'commons' (flettons). Anyone thinking in terms of a similar exercise might wish to note that such bricks have a particularly deep frog, accounting for a disproportionate ratio of mortar to brick. Lime mortar was specified with a view to ready dismantling after testing. The cantilever feature could not be expected to be self-supporting – even if cement mortar had been used. Support was provided by a bolted wooden structure, which doubled as scaffolding during construction.

The cap, of one-to-one cement/sand mixture, was poured over a domed plywood former, allowing incorporation of expanded metal grille reinforcement. It was a two-man job to lift it into position after a 48-hour cure period.

The dummy displacer cylinder was rolled from 16 gauge mild-steel sheet to the original 2 feet (610 mm) diameter. The seam was a 25 mm overlap without a joggle, perfect cylindricity not being paramount. The cap was flame-cut from 3 mm mild steel plate. The seam between end cap and cylinder was not welded. Instead, the cap was a close fit to the internal diameter, and sat on the cylinder rim under its own weight on three mild-steel fingers bolted to the periphery.

The cylinder cap had been cut from a rectangular plate 1000 × 1100 mm, which now served the lower flange. The latter in turn formed the base from which the cantilever portion was built up.

Stirling's engine may have been intended to be coal-fired. Burning coal yields gaseous hydrocarbons. Substituting modern smokeless fuel increases the confidence in estimating heating rate via published calorific values for carbon. However, there are two rates: 9190 kJ/kg for $C \rightarrow CO$; 32,760 kJ/kg for $C \rightarrow CO_2$ (Haywood 1990). Uncertainty as to whether combustion goes to completion suggests basing calculations on a calorific value intermediate to the two cases. The stoking rate calculated settles on the average, namely, CV $\cong$ 20,975 kJ/kg and arbitrarily supposes an overall efficiency (furnace and engine) of 1 percent. Using these values suggests the following stoking rates:

- 1.43 kg at 10-minute intervals if 0.5 kW shaft power is supposed
- 4.29 kg at 10-minute intervals if 2 kW shaft horsepower is supposed
- 5.72 kg at 10-minute intervals if 2 kW shaft power is supposed.

A direct-reading, sheathed thermocouple was available for monitoring cylinder head temperature T_e, and could be applied manually to the under-surface of the hollow displacer cylinder via the open, lower end of the latter. The same instrument was used to note exhaust temperature T_{ex} at the mid-point of the exhaust aperture.

1.4 Exploratory firing tests

The fire kindled readily, due to the healthy draught of the 10 ft-high flue. Stoking was by reference to domestic scales and a kitchen clock attached to supporting structure. With the fire established, 1.66 kg of fuel added at intervals of 10 minutes corresponds to 10 kg/hr, giving slightly in excess of the rate indicated to achieve 0.5 kW brake output.

Three firings, each of some three hours' duration, have yielded the following, interim observations:

- *Operation is highly vulnerable to wind direction.* The exhaust outlet is horizontal, as in the sketch of Fig. 1.2. Optimistically assuming flue–gas temperature to be uniform at 500°C (773K), and using $R = 287$ J/kgK (for air) gives density $\rho_{ex} \approx 0.45$ kg/m^3. This compares with an ambient air value of $\rho_a = 1.15$ kg/m^3. Density difference is evidently 0.7 kg/m^3. Over a 10 ft (3.05 m) height, static pressure difference $(\rho_a - \rho_{ex})gh = 0.7 \times 9.81 \times 3.05 = 20.9$ Pa (0.000209 atm). This is balanced by a dynamic head, $\frac{1}{2}\rho_a u^2$ for $u^2 = 20.9/\frac{1}{2}1.15$, or $u = 6.03$ m/s – a mere 13.4 mph.

  Day-to-day operation of such an engine in 1818 would have called for enclosure in an engine shed.
- There is no indication that continuous stoking with solid fuel would cause the cylinder head to come close to operating temperature (255 + 30 = 285°C): throughout much of the tests it was possible to place the hand in contact with the internal diameter of the displacer cylinder, suggesting a wall temperature below 60°C.
- Peak exhaust temperature noted was 200°C. This was a transient value achieved while stoking briefly with paper, cardboard, small wood chips and plastic waste. These materials can give a substantially higher heat release rate than the solid fuel, and send flames up much of the height of the chimney. Visible flames from the solid fuel, by comparison, are confined to the furnace.
- During the third firing, pork sausages (marketed as 'free-range') were barbecued in the furnace as the fire died down at cessation of fuelling. A pan of potatoes in water previously brought to the boil was placed over a removable aperture on the crown vertically above the flue, and thus with the base of the pan in line of sight of the fire. The sausages took a long time to cook, while the potatoes remained resolutely off-the-boil.

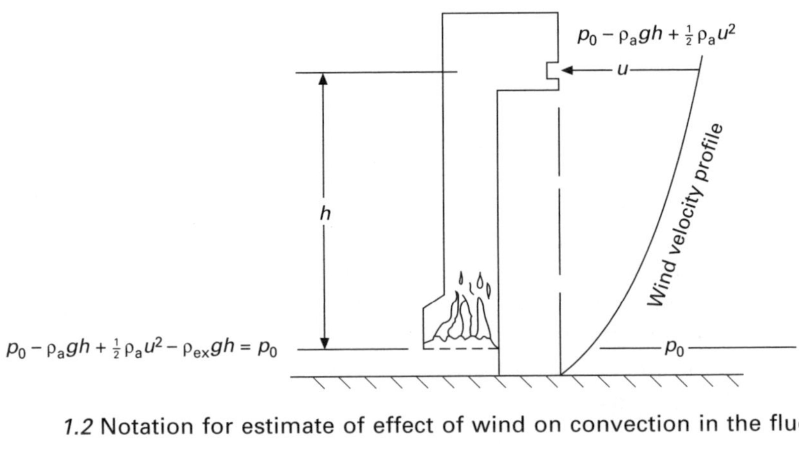

1.2 Notation for estimate of effect of wind on convection in the flue.

- The outer, vertical surface of the furnace attains about 100°C. That of the flue remains barely warm to the touch.
- 0.5 kW output would have called for an estimated 10 kW heat removal from the displacer cylinder, making for even greater difficulty in achieving operating temperature in that member.

1.5 Reassessment

Returning to James Stirling's account at the 1852–3 meeting, this begins:

> ... the engine then described [i.e. in 1816] was nearly similar to that mentioned in the Paper read to the Institution in June 1845 [Stirling 1845]; the only difference was that it had but one air vessel, with a piston working in the open end of it ...

The (twin) displacer cylinders of the 1845 engine to which James refers were heated at the bottom by direct contact with the fire, as the detailed drawing accompanying the inventor's description (Stirling 1845) clearly indicates.

James continues:

> He [Robert] constructed an engine **of this description**, in 1818, for pumping water from a quarry in Ayrshire, which work it performed very well, until, from the carelessness of the engine man, the bottom of the air vessel was overheated, and being made of boiler plate, of a flat conical form, it was crushed down by the pressure of the heated air, and rendered useless. This engine did not work to the power expected ...

Brother James' '*of this description*' implies a cylinder configuration *inverted* relative to that of the patent drawings of 1816. Achieving a viable working temperature by direct contact with the fire would then have been a simple matter, consistent with the account of the *bottom* of the air-vessel becoming overheated. Illustrations of Stirling's engines of 1827 (Galloway 1830, Jenkin 1885) and 1840 (Kennedy 1904–5) show the displacer cylinders heated from below in every case.

Against the new background, other factors now support the view that the 1818 quarry engine was not realized according to the patent drawings:

- Masonry to the left of the vertical surface marked *C–D* in the inset drawing of Fig. 1.1 would have served to impede yet further the limited convection of heat from flue to cylinder.
- The mass of the cantilever section of masonry is estimated (at 2.2 kg/brick) to be at least 350 kg. The patent indicates this weight to be taken on the upper cylinder flange – hardly a promising arrangement, where problems of thermal distortion of the thin cylinder section already threaten, and where the flange at *C–C* is required to seal effectively.

- The two thin-walled (one tenth of an inch) cylinders in series have to sustain the weight of the overhang *plus* the effect of differential thermal expansion between masonry and metal.
- The tall, slender masonry structure would be vulnerable to vibration. The estimated 28 rpm operating speed does not immediately connote high out-of-balance forces. On the other hand, it is necessary only to step inside a working water-mill or windmill to be aware of the level of disturbance which can be produced by the slow rotation of heavy, rustic machinery.

In search of the quarry engine, but on a different tack, Sier (1999) cites a letter received in 1819 by Sir George Cayley from Lord John Campbell, later to become Duke of Argyll, describing an air engine by Stirling. According to the account, a cylinder of '8 inches' raised a weight of 120 lbs through 60 feet per minute, consuming 4 lbs of coal per hour in the process. The work rate equates to 0.22 horsepower, or 160W. This would have made little impact draining a quarry worthy of the name – although it might well have satisfied some local requirement for a modest water supply under a token head.

Even without the handwritten annotation, a sketch attributed to Cayley in 1819, and here reproduced as Fig. 1.3, could be nothing other than the 'business end' of a Stirling engine. In common with all embodiments subsequent to the patent disclosure, heating is by direct contact with the flame. Thus the sketch is consistent with the supposed direction of Stirling's thinking, although the heat source (a lamp) suggests a demonstration model rather than an engine on a scale calling for stoking.

Sier's enquiry fuels our appreciation Stirling as an assiduous experimenter, 'though a definitive description of the pumping engine remains tantalizingly out of grasp.

1.6 Postscript

Confronting a belief which has stood unchallenged for a century is a daunting responsibility, and a certain amount of time has been spent contemplating alternative firing methods by which Stirling might have achieved a viable cylinder temperature. Supplementary trials have been carried out at night to facilitate viewing of the cylinder crown from below for any sign of glowing of the internal surfaces.

Dried ash-tree chippings burn intensely, giving a flame extending part-way up the flue. After an hour's firing the inner surface of the cylinder now becomes too hot to touch, but there is no sign of even the dullest glow. The inconclusive improvement over coal firing requires non-stop chopping and stoking. Coaxing the engine of 1818 into operation in this way would have required the services of a full-time woodsman/stoker, whose energies would possibly have operated a hand pump to better effect.

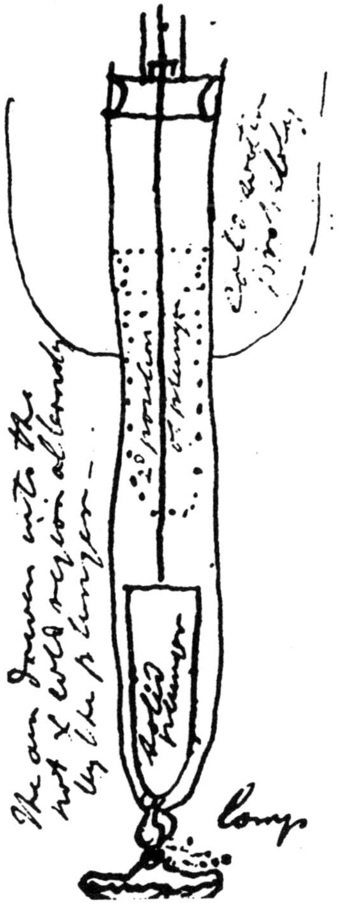

1.3 Sketch of a Stirling engine, said to have been drawn *c.* 1819 by Sir George Cayley. From Sier (1999) with permission.

These findings in no way diminish Stirling's invention and achievements. In particular, they do not bolster Kolin's out-of-character speculation (p. 83 of his 1999 text) that James Stirling, rather than Robert, masterminded subsequent exploitation of the original patent.

The appreciation of the 1818 engine in this author's 2001 text remains intact – but now applies to an engine heated from below, as suggested by Fig. 1.4. The earlier account could, with advantage, be reworked, with T_E of the simulation study increased beyond Stirling's notional 285°C – possibly to 400°C. The increased indicated work potential relaxes somewhat the requirement that regenerator design should have been optimal in order for the engine to function. The case for Stirling having designed his regenerator with some degree of analytical finesse is now less strong.

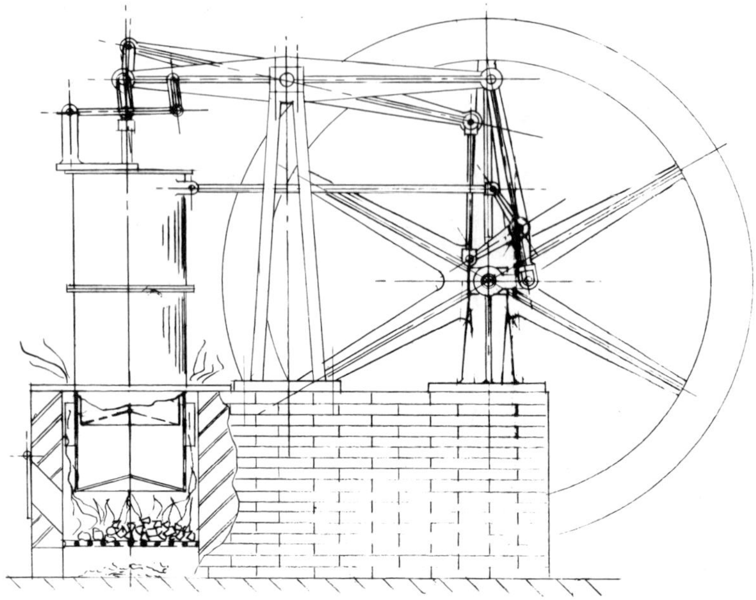

1.4 How the engine of 1818 may have appeared in the quarry. The crank mechanism of the patent descriptions has been retained, but reconnected so as to drive from above. The configuration is now consistent with the account by James Stirling of overheating of the '*bottom* of the air vessel'.

Time, effort and outlay in re-creating the part-engine exceeded expectation. At one stage, the author regretted having failed to capitalize by building a more permanent structure. This would have taken no longer, and would have allowed the option of completing a working engine at leisure. This is now seen to have been a waste of perfectly good regret, since it no longer makes sense to focus on the configuration of the patent drawing.

Without having carried out the calculations, the author would guess that high-quality birch-wood 8-ply or 12-ply, sawn to outline, would serve for many components of the link mechanism. The flywheel could be fabricated in similar fashion. Bearings of wrapped shell type, pressed and/or glued into position, would probably last the length of time required for demonstration runs – and for measurement of the half-horsepower or so which might realistically be expected of the 'real' 1818 engine.

The study has one disappointing outcome: the 1816 embodiment was as elegant as it was inspired. This author's rival concept of Fig. 1.4 is arguably more functional – but eminently unprepossessing!

2
What Carnot efficiency?

2.1 Ideal cycle – or perfect alibi

On the strength of the core invention (the regenerator) the Stirling cycle suffers from a reputation of having the Carnot efficiency as its ultimate limit. This would normally be the cue for a discourse on the 'ideal Stirling cycle'. To follow the predictable course would be to endorse the academic tradition of assessing heat engines on the basis of their respective 'air standard cycles' – but would leave the prospective designer no better-off than before.

It is not generally appreciated outside the university engineering course how much of the academic syllabus is driven by the need for the teacher to produce, annually, a set of examination questions which can be completed by the 'average' student in fixed length of time, and preferably assessed by the examiner in even less time. Perusal of library copies of past papers will show how little room this leaves for 'open-ended' questions, and will suggest what a windfall to engineering thermodynamics the concept of the ideal cycle has been.

The 'ideal Stirling cycle' also serves to indicate just how far technological reality can be perverted to academic priorities. It would be difficult to find a more unideal manufacturing proposition than the 'ideal' discontinuous piston motion. Moreover, basing the criterion for limiting thermal efficiency on exchanger temperatures T_E and T_C, when combustion temperatures are at least twice T_E, is about as fraudulent as the academic approach gets.

Two recent developments change the way one might choose to understand the real Stirling cycle and to assess prospects for the air- (or N_2-) charged variant. One is the 'thermal lag', or 'pulse-tube' engine. The other is the regenerative pressure-wave engine.

At first sight, the thermal lag engine derives from the Stirling, differing only to the extent of dispensing with the displacer. On closer scrutiny, its functioning is seen to depend on the very phenomenon which one attempts (but fails) to eliminate from the Stirling engine – irreversible, time-dependent heat transfer. The imperfections of the Stirling cycle are the driving force of the thermal lag type: optimizing heat exchange will ensure that the thermal lag engine does not function at all.

The pressure wave engine exploits phenomena overlooked in the orthodox analysis of the Stirling engine – and usually ignored in first-principles design.

This book takes the view that to include studies of these two regenerative devices (Chapters 9 and 14) will make better use of resources than would a rehearsal of thread-bare concepts such as ideal cycles. Other chapters will also focus on what is feasible rather than what is unattainable.

2.2 What Carnot efficiency?

A functioning engine is not an isolated thermodynamic cycle – it is part of a *system*. That system comprises the external combustion process in addition to piston and drive linkage. Net efficiency may be thought of as the product of the respective efficiencies of the sub-systems. The major shortfall relative to the Carnot has always been *losses* of one sort or another – hydrodynamic pumping, thermal short – and particularly *exhaust loss*. Starting from Stirling's prototype in 1818, the heat provision for air engines has routinely been the natural-convection flame. The spirit burner which powers desktop models allegedly demonstrating the merits of the Stirling engine in fact sabotages the case by confirming thermal efficiencies close to zero.

The Philips corporation reinvented the Stirling engine in the late 1930s on the basis of 'modern knowledge about heat transfer, flow resistance, etc.'. One outcome was the magnificently-engineered, air-charged MP1002CA. Combustion took place within an enclosed manifold, but exhaust temperature was 600°C (author's measurements (1964)), with heat to exhaust and radiation accounting for 80 percent of fuel energy input. In later designs, incorporation of air pre-heaters has made the heating provision more sophisticated. Nevertheless, exhaust temperatures remain in the region 250–300°C (Hargreaves 2001).

If this situation were inevitable, then the search for improvements to the internal gas process cycle would be misguided. Fortunately it is far from inevitable: in on-going experiments, exhaust temperatures below 100°C are being measured from a novel combustion chamber embodying a 'Spirex' air pre-heater. There is a two-fold gain: (1) the drop in exhaust gas enthalpy between 250 and 100°C is recycled and (2) the difference between the HCV (higher calorific value) and the LCV of the fuel – some 8 percent for propane and 11 percent for methane – is recouped. The only side-effect is a spray of liquid water from the exhaust outlet.

This development justifies a new look at the internal gas processes of the engine. The sole cyclic variable which is readily measured experimentally is pressure, leaving computer simulation as an essential investigative tool. Thus the principal emphasis of this book is identified: in the Stirling engine the gas processes interact in a way which probably has no immediate parallel elsewhere. Meaningful numerical simulation calls for algorithms properly reflecting these complex interactions. Of equal importance is that the flow and heat transfer data used in simulation should derive from appropriately designed experiments.

The two areas offering potential for improved performance may thus be summarized:

1. The heating provision (Chapters 3 and 4): the number of analytical and computer studies of the internal gas processes has reached the hundreds without it becoming clear to the designer how performance enhancement, if any, relates to the numerical machinations. Exhaust loss, on the other hand, is a relatively straightforward steady-flow heat exchange phenomenon, and is readily appreciated and quantified. Despite accounting for a significant fraction of the heat balance it has attracted little interest to date. Improved exhaust heat recuperation now offers potential for substantial economies.
2. The regenerator (Chapters 5 and 11–13) – but not the well-trodden issues of thermal analysis, since the comprehensively-documented options appear adequate. The predictions of any analysis are only as good as the flow and heat transfer data which they invoke. The relevance of the usual sources to flow conditions in the air engine appears never to have been questioned – otherwise it would have been found that data consistent with scientific, first-principles design of air-charged variant are not available, and that reliance on existing data is treacherous.

2.3 Old heat exchanger, new air pre-heater

The pre-eminence of the internal combustion engine owes much to the fact that the heat release phase is totally contained within the working spaces. For the air engine to compete, the combustion process must be similarly contained: the temperature of the outside surfaces and the exhaust temperature must be close to ambient. A step in this direction is for the air pre-heater to surround and enclose the combustion chamber.

A counter-flow exchanger concept which makes this possible is the 'Spirex'. A simple version may be formed by taking a strip of thin conducting material, and winding as indicated in Fig. 2.1. Plane closure plates seal the resulting double spiral at the edges of the strip. Ambient air enters by one of the apertures at the periphery; exhaust leaves by the other. Pre-heated air arrives at one of the central D-shaped manifolds and reacts with the fuel. After imparting heat to the expansion exchanger of the engine the combustion products enter the second D-shaped manifold, exiting via a spiral passage.

As many turns (laps) of the double spiral are specified as achieve the required recovery ratio. Other configurations than the D-shaped manifolds

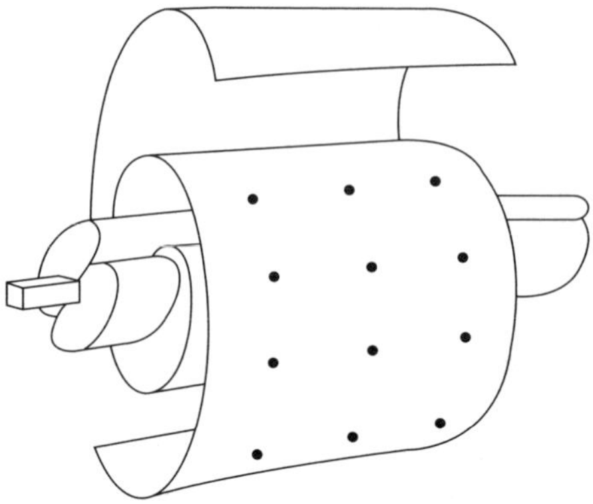

2.1 Schematic representation of wound, sheet metal core of 'Spirex' counterflow exchanger. From Devois and Durastanti (1996) with permission of Taylor and Francis.

are possible. Chapter 3 offers an analysis of the spiral counter-flow type. Chapter 4 is an account of the design and test of a combustion chamber based on the Spirex.

2.4 Resources for first-principles gas path design

Philips' identification of 'modern knowledge . . . ' as the key to releasing the potential of the air engine largely preceded developments which, while pertinent, would today be considered 'old hat'. These include the first appearance of Kays' and London's *Compact Heat Exchangers* (1964); the modern study of *permeability* (steady flow through porous media); a comprehensive literature on steady compressible flow through single gauzes (Pinker and Herbert 1967), solution of the 'regenerator problem' – also called the *conjugate heat exchange problem* (Organ 1997), and linear-wave analysis of unsteady, compressible flow in the interrupted flow passage (Organ 1997).

All of these resources are now freely available, but the Kays and London data uniquely dominate the modelling of Stirling engine and regenerator. What comes to light is that *no fewer than four* modern sciences of flow through porous media give rise to *disparate and incompatible data*. Moreover, with exception of the studies by Su (1986) and by Wirtz *et al.* (2003) one science barely acknowledges the existence of the other!

The implications for air engine analysis and design can be seen on recalling the definitions of Reynolds number *Re*, Mach number *Ma*, friction factor C_f, and pressure coefficient f. With dp/dx (Pa/m) to denote local pressure gradient and r_h (m) for hydraulic radius

$$Re = 4\rho \underline{u} r_h/\mu \qquad 2.1$$

$$Ma = \underline{u}/\sqrt{\gamma RT} \quad 2.2$$

$$C_f = |\,dp/dx\,| r_h / \tfrac{1}{2}\rho \underline{u}^2 \quad 2.3$$

$$f = |\,dp\,| / \tfrac{1}{2}\rho \underline{u}^2 \quad 2.4$$

The compressible-flow case is addressed first.

2.4.1 Steady compressible flow – pressure coefficient f

A substantial literature focuses on flow through the isolated gauze. Experimental results appear in the form of pressure coefficient f vs approach Mach number Ma_{app} with gauze aperture ratio, α as parameter:

$$f = f\,(Ma_{app}, \alpha) \quad 2.5$$

A proper parametric display requires both Re and α to be constant while Ma_{app} is varied. Pinker and Herbert (1967) acknowledge this, citing prior work of Adler (1946) and Morgan (1959) for its failure in this respect. Their attempt to rectify matters amounts to holding constant both α and 'flow factor' $\underline{u}_{app}\rho/\mu$ (m^{-1}) while varying Ma_{app}. Figure 2.2 reproduces Fig. 4 of their account. While α and Re are, indeed, constant for a given gauze (i.e. along any curve) the result falls short of its objective in that 'flow factor' is not Re, and that from one curve to another, *both parameters*, i.e. both α and Re, are changed.

If, for a single value of α, f is plotted against Ma_{app} for a range of values of Re, a family of curves results. Examples will be given later, but for the moment it is sufficient to note that the appropriate form of Eqn 2.5 is:

$$f = f\,(Re, Ma_{app}, \alpha) \quad 2.6$$

Friction factor C_f converts to pressure coefficient f via simple algebra (see Chapter 10, Eqn 10.7). Likewise there is a simple conversion between volume porosity $\P_v$ and aperture ratio α. The notations of incompressible and compressible traditions are thus interchangeable. Accordingly Eqn 2.6 has an equivalent form:

$$C_f = C_f(Re, Ma_{app}, \P_v) \quad 2.7$$

This should cause alarm bells to sound in the regenerator design department:

- at any given volume porosity $\P_v$, friction factor C_f is *more sensitive* to Ma_{app} *than to Re*
- in the vicinity of the 'choking' value of Ma_{app}, friction factor tends to infinity
- at given Ma_{app}, the sensitivity of f to α exceeds the sensitivity of C_f to $\P_v$ at given Re.

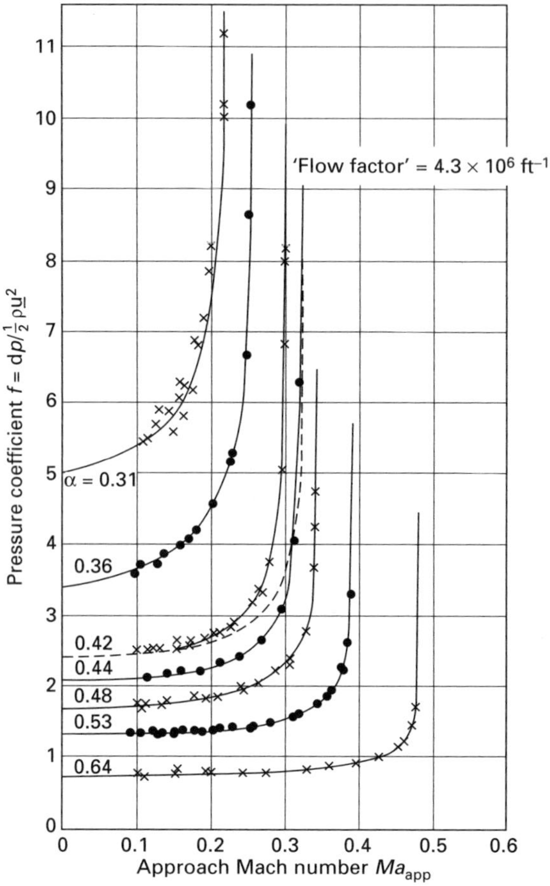

2.2 Pressure coefficient *f* vs approach Mach number Ma_{app}, after Pinker and Herbert (1967). Aperture ratio α is the nominal parameter, but holding flow parameter $\underline{u}_{app}\rho/\mu$ (m^{-1}) constant for all α means that the values of *both* parameters – *Re* and α – change between one curve and the next. Reproduced by permission of the Council of the Institution of Mechanical Engineers.

2.4.2 Steady 'incompressible' flow: Kays and London format – C_f vs *Re*

Figure 2.3(a) needs little introduction. Friction factor C_f is plotted against *Re* to the celebrated Kays and London format, i.e., for steady, 'incompressible' flow, with volume porosity $¶_v$ as parameter:

$$C_f = C_f(Re, ¶_v) \tag{2.8}$$

'Incompressible' is in quotes to reflect the implied assumption that C_f is independent of *Ma*. If, for any given $¶_v$ the variation in *Re* is achieved by

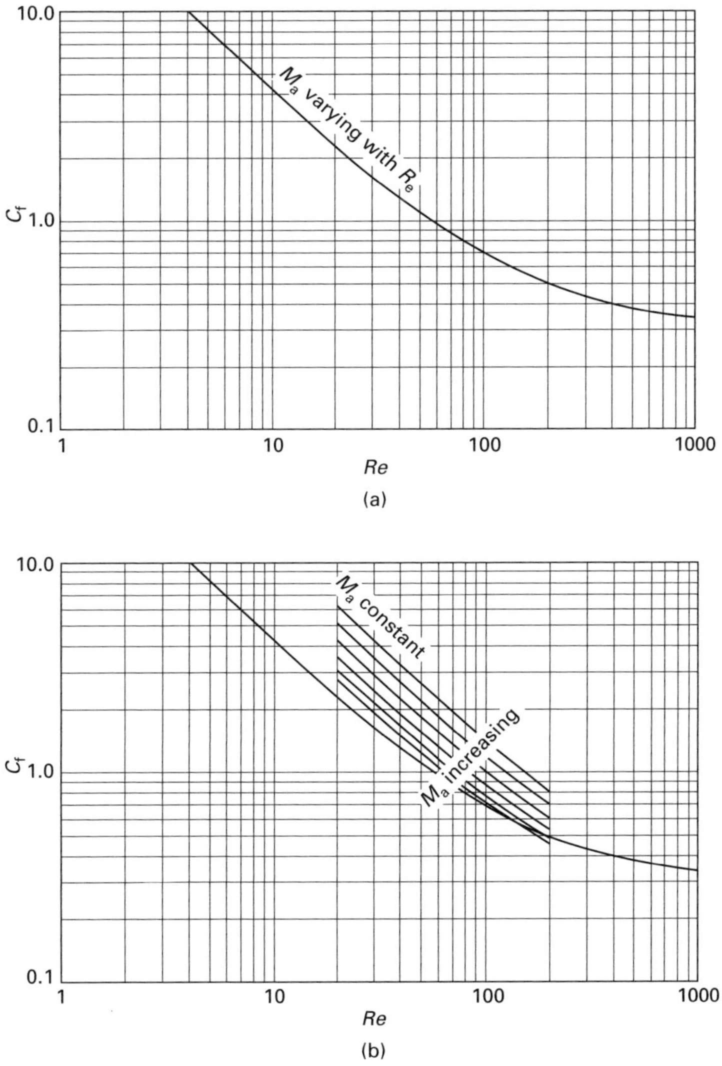

2.3 Experimental determination of C_f–*Re* correlations under two different approaches (a) ignoring dependence on *Ma* (b) taking it into account. (a) 'Incompressible' C_f – *Re* correlations to Kays and London format for a stack of given flow passage geometry (e.g. of given volume porosity $¶_v$). 'Incompressible' because there is no attempt to isolate the respective influences of *Re* and *Ma*. (b) *Re* varied for a range of constant *Ma*.

varying $\underline{u}$, then by definition *Ma* is varying in direct proportion to the variation in *Re*.

If, by contrast, *Ma* is held constant (e.g. by keeping $\underline{u}$ constant and achieving the variation in *Re* by varying overall pressure level and thus ρ) then for any

¶$_v$ a different C_f vs *Re* characteristic results for each invariant value of *Ma* – an outcome consistent with Eqn 2.7. This is illustrated schematically in Fig. 2.3(b), and by reference to experimental measurement in a later chapter.

2.4.3 Permeability studies – the Forchheimer law

A search of the Internet (key in 'Forchheimer law' under Google) will reveal a substantial and growing research interest in flow through porous media – particularly through metal and graphite carbon (GCF) foams. Emphasis is on uniform-density flow, but compressibility effects are under investigation. There is at least one dedicated serial publication (*Transport in Porous Media*, Springer).

The point of departure is a law postulated by Darcy (1856). Where *Re* is sufficiently low that inertia effects are insignificant, pressure gradient relates to flow speed $\underline{u}$:

$$-\mathrm{d}p/\mathrm{d}x = \mu\underline{u}/K \qquad 2.9$$

In Eqn 2.9, *K* is specific permeability (m^2). To extend applicability to higher *Re* a quadratic term is arbitrarily added, resulting in the 'Forchheimer Law':

$$-\mathrm{d}p/\mathrm{d}x = \mu\underline{u}/K + C\rho\underline{u}^2/\sqrt{K} \qquad 2.10$$

In Eqn 2.10, *C* is a 'drag force coefficient' (Khayargoli *et al.* 2004) or 'form drag coefficient' (Thompson (undated)).

Friction factor C_f has been defined (at Eqn 2.3) in terms of pressure gradient d*p*/d*x*, allowing Eqn 2.10 to be rewritten:

$$C_f = 2r_h\mu/\rho\underline{u}K + 2Cr_h/\sqrt{K}$$

With *Re* defined as at Eqn 2.1 (not the sole option) this becomes:

$$C_f = \{8/Re + 2C\sqrt{K}/r_h\}r_h^2/K \qquad 2.11$$

Equation 2.11 is unmistakably the empirical $C_f = c/Re + d$ deployed to fit $C_f - Re$ correlations to Kays and London format.

In permeability studies much effort goes into manipulating values of *K* and *C* to improve the coincidence between experiment and Forchheimer equation. One expedient (Barree and Conway 2004) includes making *K* a function of *Re*. This raises a conflict with the practice of defining *Re* in terms of permeability constant, *K*, namely $Re = \rho\underline{u}\sqrt{K}/\mu$, since the result is the recursive dependency $K = K\{Re[K(Re)]\}$.

The Forchheimer 'law' is founded on a bold assumption: that viscous and inertia effects in the real flow add linearly in precisely the proportions of the two, simple algebraic terms of Eqn 2.10. If the assumption is in any way flawed, then tinkering with coefficients is a lost cause. Efforts by Chen *et al*. (2001) to derive the 'law' by application of the Navier-Stokes equations amount to adjusting reality to fit empiricism and may similarly be expected to fall short of expectation.

2.4.4 Unsteady flow – impedance (or admittance)

The cyclic flow in reciprocating machines is inherently *unsteady*. Unsteady flow proceeds by wave motion. At any point in a closed system, such as the gas path of the Stirling engine, the pressure–time history is in terms of two waves travelling in opposite directions. The interruptions in the regenerator flow passages are sites for wave reflections. Right- and left-travelling waves each comprise a near-infinity of reflected and re-reflected components. Finite time of traverse results in a phase shift between the pressure waveform at one piston face and that at the opposing face. The viscous effects of real flow modify the extent of the phase shift, and cause the pressure amplitude at one piston face to differ from that at the opposite piston face.

The phase shift is frequency-dependent, i.e., it is insignificant at low values of the parameter Lf/a: L is net gas path length, f is frequency (rpm/60) and a is acoustic speed. At high values of Lf/a, phase shift can assume the same order as volume phase angle α. This suggests characterizing the regenerator in terms of an 'impedance' (or 'admittance'), and treating it as a reactive component in an a.c. circuit rather than as a series resistance of a dc circuit.

The author's 1997 text analyses the uniform-temperature case in terms of linear waves, and includes experimental pressure records in support. The approach is not readily modified to handle the severe temperature gradients of the regenerator under service conditions, although attempts to do so (Organ 2001) suggest that the effect of the gradient is to compound asymmetries (e.g. Rix 1984) inherent in the gas processes. Data to C_f – Re format do not convert to admittance form, so to date there is no link to the three sciences of steady flow.

Stirling engine technology draws heavily on Kays and London rather than on Pinker and Herbert, or on Forchheimer, or on any recognition of the capacity of unsteady flow for reactive behaviour. Had the start-point been, say, Pinker and Herbert, then performance predictions would be somewhat different – but not, as has been argued, necessarily more accurate or more relevant. The face-saver is that the regenerators of the 'benchmark' engines – GPU-3, V-160, P-40 etc. – have not, as far as can be ascertained from the literature or from reverse-engineering, been designed at all. To this extent, there is no wrong choice to get worked-up about.

2.5 A further take on friction factor C_f – and not the last

Consider the velocity head term $\frac{1}{2}\rho\underline{u}^2$. For the ideal gas $\rho = p/RT$, and velocity head is:

$$\frac{1}{2}\rho\underline{u}^2 = \frac{1}{2}\gamma p\, M_a^2 \qquad 2.12$$

To the extent that *Re* is an algebraic device, so is *Ma*. The fact that *Re* can be defined for a particular flow does not necessarily make that flow a function of *Re*. Likewise, the fact that the term *Ma* arises in the algebra of a particular flow situation does not mean that compressibility effects necessarily arise.

It appears not to be generally appreciated – indeed, it appears not to be appreciated at all – that the celebrated correlations of Kays and London cannot be reacquired using gauzes in the wire diameters specified for Stirling engines. In the original experiments the high end of the Reynolds number scale was reached by use of large diameters (up to $\frac{3}{8}$ inch, or about 9.5 mm) rather than by elevated particle speed $\underline{u}$. (In fact, straight rods were used, and not wires in the usual sense.) At a pressure of 1 atm (10^5 Pa) and a hydraulic radius of 0.04 mm Kays and London's *Re* of 10^5 converts to hypersonic $\underline{u}$ at Mach number *Ma* = 25!!

Once misgivings arise, others soon follow:

- Kays and London and their contributors did not test Stirling regenerators. Some of the tests from which the famous correlations are compiled involved 'stacks' comprising as few as three screens (Coppage 1952).
- Experimental determination of C_f amounts essentially to dividing experimentally-measured dp/p by Ma^2. Subsequent use of C_f to predict dp/dx involves multiplying by a different Ma^2. If the two values of *Ma* put them in different flow regimes (cf. specimen compressible flow chart of Fig. 2.2) the resulting dp/dx will be meaningless.
- In favour of the Forchheimer approach is that it is explicit in dp/dx – the 'answer' that the analyst/designer is looking for. The counterpart C_f does not provide this same information directly.
- On the consolidated C_f vs *Re* correlation (Kays and London's Figs 7–9 on p. 130, 2nd edn, 1964) the curve for $¶_v = 0.832$ twice crosses those for the other values of $¶_v$. There is no consistent trend with increasing values of the parameter $¶_v$. The disconcerting features are not reflected in the counterpart heat transfer curves.
- The very use of friction factor C_f reflects the thinking of pipe flow, in which context it is unaffected by *Ma* up to *Ma* = unity (Shapiro 1954). Flow through the porous matrix is not pipe flow, evidence of which is that Reynolds' analogy does not apply – a reality obvious from any of

the relevant Kays and London correlations (e.g. their Figs 7, 8, 9, 10, 2nd edn, 1964).

These concerns are less about Kays and London's results *per se* than about the way in which they have been uncritically adopted for Stirling engine use. Nor do the misgivings amount to commendation of the alternative offered by Pinker and Herbert – not at least as it stands: according to Eqn 2.12, pressure coefficient f amounts to dp/p normalized by M_a^2. (In the context the $\frac{1}{2}\gamma$ is incidental.) In other words, plots to the format of Fig. 2.2 amount to taking the dependent variable of primary interest, fractional pressure drop, obscuring its numerical magnitude by dividing by M_a^2 and then displaying against M_a. Moreover, appearances of Ma in the analysis of compressible flow are invariably as M_a^2, so the Pinker and Herbert results cry out to be displayed not against M_a, but against M_a^2. Even if the underlying physical phenomenon is disregarded, there are not too many reputations as analysts to be made by plotting y/x^2 against x.

More of this later, but for now return to the 'fit' commonly applied to $C_f - Re$ correlations in Kays and London format

$$C_f = c/Re + d \qquad 2.13$$

From the definition of friction factor C_f in terms of dp/dx (Eqn 2.3) and from Eqn 2.12:

$$C_f = | dp/dx | r_h / \tfrac{1}{2}\gamma p Ma^2 \qquad 2.14$$

Substitute this and the definition $Re = 4\rho \underline{u} r_h/\mu$ into the empirically-fitted curve (Eqn 2.13):

$$| dp/dx | r_h / \tfrac{1}{2}\gamma p Ma^2 = c/Re + d \qquad 2.15$$

Multiply through by $\frac{1}{2}\gamma Ma^2$:

$$| dp/p | dx/r_h = \tfrac{1}{2} c\gamma Ma^2/Re + \tfrac{1}{2} d\gamma Ma^2 \qquad 2.16$$

Recall the definition Re as $4\rho \underline{u} r_h/\mu$ and substitute $Ma^2 = \underline{u}^2/\gamma RT$ at the first term of the right-hand side:

$$| dp/p | r_h/dx = \tfrac{1}{8} c Sg^{-1} + \tfrac{1}{2} d\gamma Ma^2 \qquad 2.17$$

$$Sg = p r_h/\underline{u}\mu$$

Fractional pressure drop dp/p differs from Forchheimer only by a division by p. So Eqn 2.17, and thus the empirical 'fit' (Eqn 2.13), are the Forschheimer law (or Ergun equation) in dimensionless form. They purport to relate dp/dx to u for Darcy flow, for Newton flow and for the 'transition' region in between.

But why should this be so? Why should the arbitrary linear addition of algebraic groups result in viscous and inertia components combining in precisely the proportions of the physics of this complex flow?

The reality is a term for Stokes flow cobbled together by linear arithmetic addition with a term for Newton flow. Dynamic Similarity principles, endorsed by experiment, confirm that, for given flow passage geometry, as characterized for the wire gauze by r_h/dx for example, dp/p is, indeed, a *function* of Sg and Ma. On the other hand, that function is not a convenient linear sum. Thus a priority of this text is identified: to seek flow correlations giving dp/p in the general form $dp/p\{Sg, Ma, \gamma, r_h/dx\}$.

Friction factor C_f *per se* has never been of any direct interest to the designer, since it conveys no impression of pressure drop or pumping penalty. By contrast with C_f, a numerical value for dp/p says it all at a glance.

To the author's knowledge, only limited data exist – those acquired by Su (1986) – conforming to Dynamic Similarity principles in the spirit of Eqn 2.7. Moreover, only one specimen gauze geometry is represented, corresponding to a single value of volume porosity $¶_v$. This amounts to sufficient information to allow transformation of the specimen $C_f = C_f(Re, Ma, ¶_v)$ plots to maps of fractional pressure drop $dp/p = dp/p(Re, Ma, ¶_v)$ for the specimen values of $¶_v$ and γ. These provide a stencil for the extended experimental coverage of Re and $¶_v$ now urgently required to facilitate exploration of the potential performance envelope of the air engine.

To complete a science of the Stirling engine, heat transfer correlations will be required to the same format and coverage.

2.6 Beyond the sound barrier

The focus to this point has been on the influence of Ma an order of magnitude below the value ($Ma > 0.3$) normally thought of as the threshold of compressibility effects. No theorem known to the writer prohibits the design of closed-cycle, air-charged regenerative engines from operating above that threshold and in the environment of wave phenomena – i.e., from *exploiting* compressibility effects as opposed to being *limited* by them.

Before supersonic flight became a reality there appeared to exist a 'sound barrier'. It is appreciated nowadays that the 'barrier' was imposed by lack of insight rather than by aerodynamics. Part III of this book will present a computer simulation of the *pressure wave engine*, which relies for its very function on pressure wave interaction, and which would not function at all in the absence of substantial compressibility effects.

Practical implementation of any of the concepts so far addressed calls for quantitative definition of matrix geometry. Numerical values for the parameters defined in the section which follows will cover the requirements of Parts I and II.

2.7 Geometric descriptors for wire matrix

The flow characteristics of a stack of square-weave screens of, say, 0.75 volume porosity $¶_v = 0.75$ are not the same as those of Retimet™ metal foam of identical $¶_v$. Because flow patterns differ, respective heat transfer characteristics ($StPr^{2/3}$ vs Re) may be expected to differ also. Chapter 12 will focus on the flow passage geometry of a specific matrix (the wire screen) and use it to predict the pressure-flow characteristic from first principles.

To that point, all that is required is a geometric descriptor which, for a given class of matrix (say, a stack of square-weave gauzes) takes the same numerical value *irrespective of absolute size.* It goes without saying that the descriptor must be dimensionless. Three such parameters are available for the close-packed gauze stack: volume porosity $¶_v$, aperture ratio α and hydraulic radius ratio, r_h/d_w. With reference to Fig. 2.4:

$$¶_v \approx 1 - \tfrac{1}{4}\pi d_w m_w \sqrt{[1 + (d_w m_w)^2]}$$

$$\approx 1 - \tfrac{1}{4}\pi d_w m_w \qquad 2.18$$

$$\alpha \approx (1 - d_w m_w)^2 \qquad 2.19$$

(*or* Pinker and Herbert's definition (1967))

$$r_h/d_w = \tfrac{1}{4}¶_v/(1 - ¶_v) \qquad 2.20$$

Equation 2.20 for hydraulic radius ratio r_h/d_w is valid beyond wire screens to *any matrix of uniformly-packed wires of uniform diameter* provided a value of $¶_v$ is known for the alternative packing (e.g. from calculation or weighing):

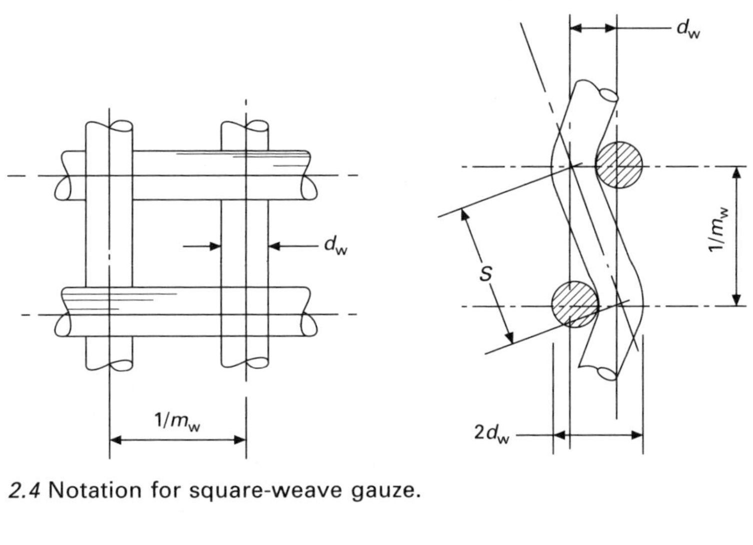

2.4 Notation for square-weave gauze.

Either Eqn 2.19 or the more sophisticated expression of Pinker and Herbert *characterizes* the aperture ratio α of a square-weave gauze completely. However, since both *define* it only approximately, neither is adequate for predicting the *Ma* dependency from first principles. The relevant geometric analysis is carried out in Part III, and is in terms of $d_w m_w$. Indeed, *any geometric description* of the square-weave wire screen is a function of the product $d_w m_w$ – wire diameter (m) *times* pitch (m^{-1}) *alone*. Square-weave screens having a common value of $d_w m_w$ accordingly have the same flow passage geometry notwithstanding different values of d_w and/or m_w separately. For the purposes of flow data acquisition and reduction, a numerical value of $d_w m_w$ is all that is required to characterize each relevant aspect of gauze geometry.

2.8 Inconsistencies uncovered

The flow passage algebra applies equally to the gauzes of Kays and London as to those of Pinker and Herbert. Figure 2.5 uses this, together with the conversion algebra between friction factor C_f and pressure coefficient f to replot Pinker and Herbert's results (Fig. 2.2) as C_f against *Ma* with $¶_v$ and *Re*

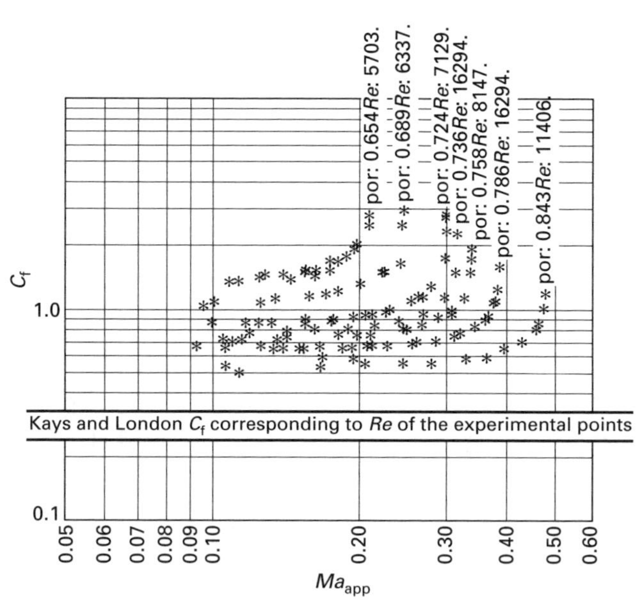

2.5 Pressure coefficient *f* of Pinker and Herbert's Fig. 4 (the 1967 paper) replotted as C_f vs Ma_{app}. All corresponding C_f from the Kays and London correlations (Fig. 7–9 of the 1964 edition) fall between the two bold horizontal lines. On this basis, C_f values in incompressible flow are markedly below those of compressible flow, even at $Ma < 0.1$.

as parameters. The values of *Re* marked against the curves have been back-calculated from data tabulated in the original account.

The Kays and London C_f corresponding to these values of *Re* all lie between the pair of heavy horizontal lines. The discrepancy amounts to an order of magnitude at high values of *Ma*, but remains considerable at *Ma* below 0.09, where compressibility might be thought to have little influence. Increasing $¶_v$ causes a monotonic rightward shift of the Pinker and Herbert curves. The Kays and London counterpart (their Figs 7–9, p. 130) show no such trend.

2.9 Résumé

This chapter opened with two propositions. The first was that reference to ideal equilibrium cycles as performance indicators is of no help in the practical realization of the Stirling engine. A moment's reflection confirms that a prime mover constructed on the isothermal–adiabatic sequence of the Carnot cycle would surpass all others in terms of inefficiency, impotence and ponderousness.

Rather than wondering where the Stirling engine falls short of the specious Carnot criterion we might perhaps celebrate the fact that it is not lumbered with the irreversibilities associated with realizing that iconic concept. Irreversibility is an everyday reality. Equilibrium in reciprocating machinery is not. The theme continues throughout this book to the extent that abstract efficiency goals are pursued no further.

The second proposition was that losses in the external heating provision overwhelm imperfections in regenerator operation. This book divides into four parts, and the remainder of Part I is largely an account of practical measures aimed at reducing thermal losses from the combustion system, particularly in small air engines of below, say, 1 kW target output. This will be recognized as differing fundamentally from a quest for a lossless heating system.

An alternative view is that regenerator performance is not critical, and that there is vastly greater leeway in regenerator design than we have led ourselves to believe. The matter is too urgent – and, indeed, too interesting – to be left to await developments in the areas of combustion and exhaust heat recuperation, so the operation of the regenerator is studied in isolation.

Part II 'Living with Incompressible Data' examines design options which have attracted little attention elsewhere, and which can be probed with the aid of traditional data, namely C_f vs *Re* and $StPr^{2/3}$ vs *Re*. The stacked screen regenerator is inherently inhomogeneous: flow passage geometry perpendicular to the plane of the gauze (axial direction) differs markedly from that in the radial direction. Exploration of the effects of this gross asymmetry does not require the subtleties anticipated in this present chapter, and leads to the valuable insight that flow in the matrix tends naturally to a 'slab' or 'one-dimensional' geometry. This does more than merely justify use of the one-dimensional conservation laws for cycle modelling: it reveals a narrower gap than previously feared between the assumptions under which steady-flow data are acquired and those under which those data are used to model the unsteady gas processes.

Exposing the role of the parameter *Ma* gives continuity between the three regimes – quasi-static, incompressible and compressible. It allows flow data to be formatted for general applicability to the entire sub-sonic flow region. Heat transfer data can be reacquired to the same format – and with the same benefit. Part III 'Working with the realities of compressible flow' examines steady flow through the wire gauze taking account of compressibility. The goal is a format for flow and heat transfer data suitable for modelling the extended performance envelope. The magnitude of the task may be gauged by considering the 'simple' case of compressible, unidirectional, steady flow in the uniform duct. The expression for fractional pressure drop $\mathrm{d}p/p$ over elemental length $\mathrm{d}x$ with account of wall friction cited by White (1979) is:

$$\mathrm{d}p/p = -\tfrac{1}{2}\gamma Ma^2 \frac{1 + \tfrac{1}{2}(\gamma - 1)Ma^2}{1 - Ma^2} C_\mathrm{f}\mathrm{d}x/r_\mathrm{h} \qquad 2.21$$

The value of C_f for use in Eqn 2.21 derives from local *Re* via a Moody diagram. But the Moody diagram is for incompressible flow, and the C_f values come from experimental measurement fed into Eqn 2.3 which, in terms of the algebra of *Ma* is:

$$C_\mathrm{f} = \frac{-\mathrm{d}p/p}{-\tfrac{1}{2}\gamma Ma^2} r_\mathrm{h}/\mathrm{d}x \qquad 2.22$$

Eqns 2.21 and 2.22 have obvious similarities. On the other hand, they do not coincide for any finite *Ma*, and the discrepancy deteriorates as *Ma* increases.

Identifying and rectifying a design defect in a completed Stirling engine can be disproportionately expensive. The best chance of minimizing risks is by use of design data relevant to the gas processes and flow passage geometry involved.

In anticipation of such data becoming available, Part III includes an original, non-linear wave analysis of the pressure-wave engine. This strongly suggests the possibility of extending the performance envelope by achieving high values of work/cycle at unprecedentedly high rpm.

Part IV 'Some design considerations' anticipates how engine design will proceed when a body of appropriate flow and heat transfer data has been accumulated.

3

Counter-flow spiral heat exchanger – Spirex

3.1 Heat provision as an integral part of the engine

Chapter 2 proposed that it makes no sense to scrutinize the internal gas processes for efficiency margins while exhaust temperatures remain in the region 250–300°C. And the only practical hope of reducing the losses associated with these high temperatures is to improve the provision for exhaust gas heat recuperation. Chapter 4 will describe experiments with a 'Spirex' inlet air pre-heater demonstrating a dramatic reduction in exhaust and radiation loss. This chapter introduces the concept and offers a method of first-principles design.

'Spirex' is an acronym for 'spiral counter-flow heat exchanger'. It appears to be due to Alfa-Laval – the inventor (the name happens also to describe proprietary plastic injection equipment, screen savers, fishing reels and a South-Pole InfraRed EXplorer.)

A Spirex can be fabricated by securing a strip of pliable conducting material at mid point, and winding as already indicated in Fig. 2.1. One inlet and one outlet manifold connect near the axis of the spiral. Respective outlet and inlet are at the periphery. Closure plates seal the full length of both edges of the spiral strip. The result is a light-weight, economical construction with favourable heat transfer capacity per unit volume envelope. Temperature recovery ratio is somewhat less than that of a conventional counterflow exchanger of comparable flow passage geometry (as defined in terms of hydraulic radius r_h, free-flow area A_{ff} and length L).

The Spirex appears to offer attractions when compactness and low installation cost are paramount. Applications which come to mind include hospital air-conditioning systems, which are required to meet Code HTM 2025 (Anon., undated, a). This calls for a continuous supply of fresh air, regardless of outside air temperature. Unlike a regenerator, Spirex avoids direct contact between the warm, contaminated stream and the incoming supply.

A further application arises from the recent commercialization of domestic CHP (combined heat and power) powered by the Stirling engine: Powergen offer installations providing supplementary 1 kW(e) while meeting the full domestic heating load. The Microgen unit, based on an alternative implementation of the Stirling cycle, is being readied for market. Making an impact on CO_2 reduction through widespread use of domestic CHP will, however, call for improved conversion of fossil fuel energy to electricity – and a correspondingly smaller fraction to heat. This requires improved engine efficiency, where *engine* now means the system comprising combustion system, exhaust heat recuperator and the Stirling engine *per se*.

Efficiency must now be thought of as the product:

$$(\text{efficiency of combustion system}) \times (\text{efficiency of engine } per\ se)$$

According to Hargreaves (1991), even highly-developed Stirling engines exhaust at 250–300°C. Embodiment of a Spirex air pre-heater into the combustion system of a small (nominal 50W) Stirling engine has permitted Organ and Larque (2004a) to achieve exhaust temperatures below 100°C and thus to extract the higher calorific value (HCV) of the fuel. The low outlet temperature offers a bonus: instead of the combustion system being blown it can instead be drawn by a simple extractor fan downstream of the Spirex. The possibility of exhaust leakage is minimized, although there remains a minor problem of dealing with a fine spray of condensed water expelled by the fan.

While the Spirex has been in commercial use for some decades, particularly in the process industries (Hargis and Beckman 1967), it appears to have attracted little academic study. Devois and Durastanti (1996) obtained predictions of temperature distributions in steady state operation. The investigation covered a range of fluids. The sensitivity of the solution to choice of Nusselt–Reynolds correlation was explored. The authors appear to have made heavy weather of the process of discretization for numerical solution, to the extent that computer CPU capacity was a consideration.

This chapter offers an approach to temperature calculations having no such limitation.

3.2 Thermal analysis

Analysis of the conventional counterflow exchanger can proceed in terms of a new variable $\Delta T = T_g - T_w$. In the Spirex, a given stream exchanges heat with streams of two different temperatures. In this case, there is no unique ΔT.

3.2.1 Simplifying assumptions

The only aspects of the formulation which are in any way taxing are the discretization strategy and the rationale for laying out the matrix of coefficients of the unknowns. The idealizations which follow are largely for convenience in illustrating the set-up procedure. Once the latter is in place, they can be replaced by assumptions more representative of reality:

- the working fluid is an ideal gas
- temperature and particle speed are uniform across the free-flow area of the channel (i.e., 'slab' flow is assumed)
- operation has attained steady-state.

3.2.2 Overall heat transfer coefficient

Between any two adjacent streams, heat is exchanged via a boundary film, a conducting wall of radial thickness t and a further boundary film, as indicated in Fig. 3.1. The 'equivalent' overall heat transfer coefficient h^* between the defining temperatures T_a and T_{ex} of the two streams is the textbook definition:

$$h^* = \frac{1}{1/h_a + t/k + 1/h_{ex}} \tag{3.1}$$

Stanton number St_a on the air side is defined $St_a = h_a/\rho \mid u \mid c_{p(a)}$. Substituting $m' = \rho u A_{ff}$ and transposing:

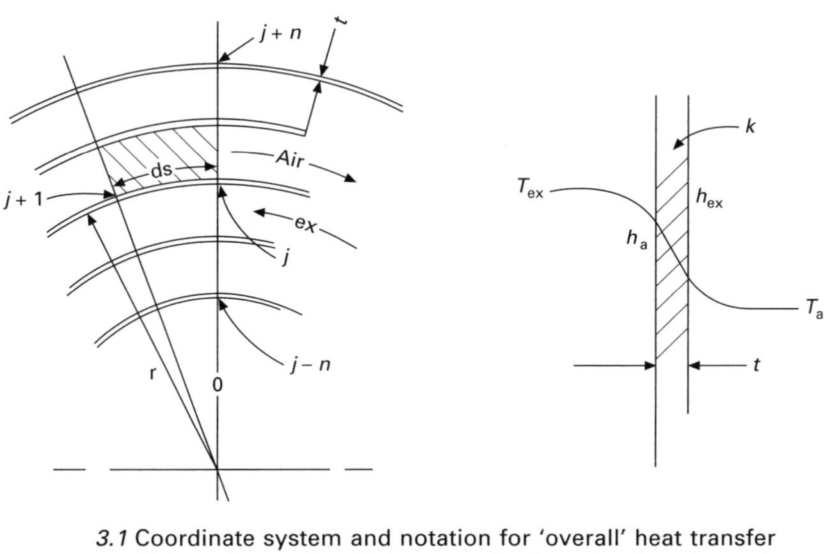

3.1 Coordinate system and notation for 'overall' heat transfer coefficient. A complete Spirex may be thought of as three elements in series: (1) the innermost turn or 'lap'; (2) a number (integral) of internal laps to which Eqns 3.5 and 3.6 apply; (3) the outermost lap. Innermost and outermost laps are atypical and lack the symmetry of the 'internal' laps. Their algebraic treatment varies according to manifold provisions.

$$h_a = St_a \mid m' \mid c_{p(a)}/A_{ff}$$

The corresponding expression for the exhaust side may be written down directly by analogy. Substituting into Eqn 3.1, and assuming free-flow area A_{ff} to be common to both streams:

$$h^* A_{ff}/(\mid m' \mid c_{p(a)}) = St_a^* = \frac{1}{St_a^{-1} + \mid m' \mid c_{p(a)} t/kA_{ff} + St_{ex}^{-1} c_{p(a)}/c_{p(ex)}} \qquad 3.2$$

By similar reasoning for an energy balance referred to the exhaust side:

$$St_{ex}^* = \frac{1}{St_a^{-1} c_{p(ex)}/c_{p(a)} + \mid m' \mid c_{p(ex)} t/kA_{ff} + St_{ex}^{-1}} \qquad 3.3$$

3.2.3 Coordinate system

In Fig. 3.1 flow from the minimum diameter to the maximum follows an anti-clockwise path along which distance s increases. Angle ϕ is positive in the anti-clockwise direction. Radius r is evidently a function of ϕ, namely, $r(\phi)$. This variable identifies the *outer* of the two strips forming a channel pair. The origin of both s and ϕ is the upper vertical axis.

The elemental length $ds = rd\phi$, and thus increases with r (and so with angle ϕ). With w for the width of the channel in the radial direction and t for plate thickness:

$$r = r_i + \{(t + w)/\pi\}\phi \qquad 3.4$$

A representative element of fluid in the exhaust stream exchanges heat with the air streams on either side. The outer air stream belongs to the current lap, i.e., that identified by radius r. The inner air stream belongs to the *previous* lap. Setting up an energy balance calls for focusing on an arc of one lap and taking into consideration the adjacent element of the previous lap as well as the adjacent element of the next lap in the sequence.

With superscript j to identify element location as a function of ϕ, the relevant temperature of the outer air stream is T_a^j, while that for the inner is T_a^{j-n}, n being the number of circumferential subdivisions.

An energy balance for an element of channel of length $ds = rd\phi$ may be written directly in terms of finite differences. Denoting by b the height of the channel in the direction of the axis of the spiral, a heat balance for the element is:

$$h^* b ds\{T_a^{j-n} - T_{ex}^j - (T_{ex}^j - T_a^j)\} = m' c_{p(ex)} (T_{ex}^{j+1} - T_{ex}^j)$$

Temperatures internal to the element have been characterized by respective node temperatures. The alternative – use of averages, $\frac{1}{2}(T_{ex}^{j+1} + T_{ex}^j)$,

$\frac{1}{2}(T_a^{j+1} + T_a^j)$ – has been explored, but found to afford no benefit provided the angular increment $d\phi$ is suitably small.

Overall length of the n_l laps of exhaust channel is denoted L, and NTU_{ex}^* is used to denote $St_{ex}^* L/r_{h(ex)}$:

$$NTU_{ex}^* (ds/L)\{T_a^{j-n} - T_{ex}^j - (T_{ex}^j - T_a^j)\} - T_{ex}^{j+1} + T_{ex}^j = 0 \qquad 3.5$$

For the element of air stream in the channel pair of interest:

$$NTU_a^* (ds/L)\{T_{ex}^j - T_a^j - (T_a^j - T_{ex}^{j+n})\} - T_a^j + T_a^{j+1} = 0 \qquad 3.6$$

Any element, other than the entry element, of the innermost exhaust channel exchanges heat over the inner boundary *not* with fluid at T_a^{j-n} as per Eqn 3.5, but with fluid at T_{EX}. The value of T_{EX} being known, the term moves to the right-hand side of the equation.

Figure 3.2 shows, partly assembled, the Spirex pre-heater of the combustion chamber described by Organ and Larque (2004a). At the innermost radius, pre-heated air enters tangentially to be mixed with gaseous fuel in a ceramic chamber and burned. After heating the cylinder head of the engine, the combustion products, assumed here to be at uniform temperature T_{EX}, are in contact with the full extent of the inner wall, which is therefore taken to be at that same, uniform temperature.

The outer surface (at maximum radius r) is formed by the outermost lap of an air channel exposed to ambient temperature T_{AIR}. For the purposes of this example the outer wall is assumed to be uniformly at that temperature.

3.2 Spirex described by Organ and Larque (2004a) shown during assembly.

Collecting terms:

$$NTU_{ex}^{*}(ds/L)T_{a}^{j} + T_{ex}^{j}\{-2NTU_{ex}^{*}(ds/L) + 1\} - T_{ex}^{j+1}$$

$$= -NTU_{ex}^{*}(ds/L)T_{EX} \qquad 3.5(a)$$

For the first element, entry temperature is the known T_{EX} and:

$$NTU_{ex}^{*}(ds/L)T_{a}^{j} - T_{ex}^{j+1} = \{-NTU_{ex}^{*}(ds/L) - 1\}T_{EX} \qquad 3.5(b)$$

Similarly, for all elements of the outermost lap with the exception of that at which air enters, Eqn 3.6 becomes:

$$\{-2NTU_{a}^{*}(ds/L) - 1\}T_{a}^{j} + T_{a}^{j+1} + NTU_{a}^{*}(ds/L)T_{ex}^{j}$$

$$= -NTU_{a}^{*}(ds/L)T_{AIR} \qquad 3.6(a)$$

For the element at which air enters:

$$\{-2NTU_{a}^{*}(ds/L) - 1\}T_{a}^{j} + NTU_{a}^{*}(ds/L)T_{ex}^{j}$$

$$= -NTU_{a}^{*}(ds/L) + 1\}T_{AIR} \qquad 3.6(b)$$

3.3 Heat transfer and flow friction correlations

Solution of a realistic problem calls for evaluation of Reynolds number Re at each computational node. For the general case Re is defined as $4\rho \mid u \mid r_h/\mu$. The assumption of steady state permits the definition to be rewritten in terms of (steady) mass rate m' by substitution of $m' = \rho u A_{ff}$, in which A_{ff} is free-flow area, assumed uniform. Thus, $Re = 4 \mid m' \mid r_h/A_{ff}\mu$. It then becomes apparent that Re varies with location s only to the extent of the temperature dependence of coefficient of dynamic viscosity, μ. The variation may be accounted for by the Sutherland law:

$$\mu/\mu_0 = f(T) = \left\{\frac{T_0 + T_{su}}{T + T_{su}}\right\}(T/T_{su})^{3/2} \qquad 3.7$$

This permits local Re to be evaluated as the product of a constant term, $Re_0 = 4 \mid m' \mid r_h/A_{ff}\mu_0$ with $f(T)$:

$$Re = \frac{4|m'|r_h}{A_{ff}\mu_0}f(\mathrm{T})^{-1} \qquad 3.8$$

Local Stanton number St follows from Re via the appropriate experimental correlation of $StPr^{2/3}$ vs Re. Graphical correlations, such as those of Kays and London's compendium (1964), may be converted for use in computer-coded solution by fitting laminar, transition and turbulent regions to expressions of the form:

$$Re < 2000: StPr^{2/3} = a_{lam} Re^{b}_{lam} \qquad 3.9(a)$$

$$2000 < Re < 3500: StPr^{2/3} = a_{trans} Re^{b}_{trans} \qquad 3.9(b)$$

$$Re > 3500: StPr^{2/3} = a_{turb} Re^{b}_{turb} \qquad 3.9(c)$$

Corresponding to each value of *St* a value of friction factor C_f is required. For the laminar regime:

$$C_f = c_{lam} Re^{d}_{lam} \qquad 3.10$$

3.4 Special case of high *NTU**

A feature of the eventual numerical solutions may be anticipated on dividing Eqns 3.5 and 3.6 by their respective *NTU** terms. For large values of *NTU** this leaves the pair of equations:

$$T_a^{j-n} - T_{ex}^{j} - (T_{ex}^{j} - T_a^{j}) = 0 \qquad 3.5(c)$$

$$T_{ex}^{j} - T_a^{j} - (T_a^{j} - T_{ex}^{j+n}) = 0 \qquad 3.6(c)$$

The result of adding these may be seen in two ways:

$$^1/_2(T_a^{j-n} + T_{ex}^{j+n}) = {^1/_2}(T_a^{j} + T_{ex}^{j})$$

or

$$T_{ex}^{j} - T_a^{j-n} = T_{ex}^{j+n} - T_a^{j}$$

The first states that the mean of the temperatures of the two streams evaluated across three laps is the same as the mean across a single lap. The physical significance of the second will become apparent when specimen solutions are discussed. For the moment, both are consistent with the unexpected feature: both temperature gradients are linear with uniform inlet-to-outlet temperature difference.

3.5 Numerical integration

Applying the appropriate equation from Eqns 3.5 and 3.6 to each subdivision of the inflow and outflow channels yields a total of $2 \times n_j \times n_l$ simultaneous equations linear in the unknown T_a^j and T_{ex}^j. The matrix of unknowns consists of $n_j \times n_l$ columns of the coefficients of the T_a^j and a similar block of coefficients of the T_{ex}^j.

The sequence of rows of coefficients derives from the sequence of volume elements working anticlockwise from the origin (Fig. 3.2), i.e., in the direction of exhaust flow and decreasing T_{ex}^{j}, but counter to air flow and in the order of T_{a}^{j} decreasing from maximum at air outlet. A row deriving from the coefficients of Eqn 3.5 (for exhaust) is followed by a row from Eqn 3.6 for the inlet air.

The array of coefficients is processed by a proprietary numerical routine SIMQX (Anon, undated, b) which isolates the unknowns by pivotal condensation followed by back-substitution. A single solution with 10 subdivisions ($n_j = 20$) and 6 laps ($n_l = 6$) takes an Intel D845 processor 46 ms from data read to output of screen plot. SIMQX stores and works through all array elements, including zeros. A routine designed for sparse arrays, such as Nag™ F01BRF (decomposition) in conjunction with F04AXF (solution) might be preferred if the solution is to be part of a simulation of a larger system – perhaps a simulation to be subject to mechanised optimization. Either way, problems of CPU capacity experienced by Devois and Durastanti (1996) do not arise.

3.6 Specimen solutions

Devois and Durastanti (1990) display the temperature solutions for the two streams superimposed in spiral form and in perspective. The presentation has its own appeal, but does not allow quantitative interpretation. Indeed, apart from the essential spiral geometry, even qualitative features are obscured.

Figure 3.3 shows a solution obtained for the data of Table 3.1. With slight compromise, these describe the geometry and operating conditions of the air pre-heater described by Organ and Larque (2004a). Temperatures (K) are graphed against cumulative peripheral distance s measured anti-clockwise from the origin and plotted on the diagram in the positive x direction. The horizontal distance between a pair of vertical lines is the circumferential length of the individual lap at current s, and the plot confirms the progressive increase with r (and/or ϕ).

Temperature recovery ratio is not particularly high, but it is reassuring that exhaust outlet (upper curve, right of diagram) is approximately 60°C. above air inlet temperature, and thus below 100°C, consistent with experimental observation recounted earlier.

Within the 'typical' inner laps, temperature difference $\Delta T = T_{ex} - T_a$ tends to a uniform value – i.e., one which is independent of s. In the innermost and outermost laps, by contrast, the variation of ΔT with s is marked, as might be expected from the atypical conditions which apply. Neither of these phenomena is a feature of the solutions offered by Devois and Durastanti (1996).

Experimenting with arbitrary values of *NTU** reveals that the higher the *NTU**, the greater the uniformity of ΔT within the inner laps. If the uniformity is confirmed to be an inherent part of the physics of the problem, a simple analytical solution may exist for this case (high *NTU*).

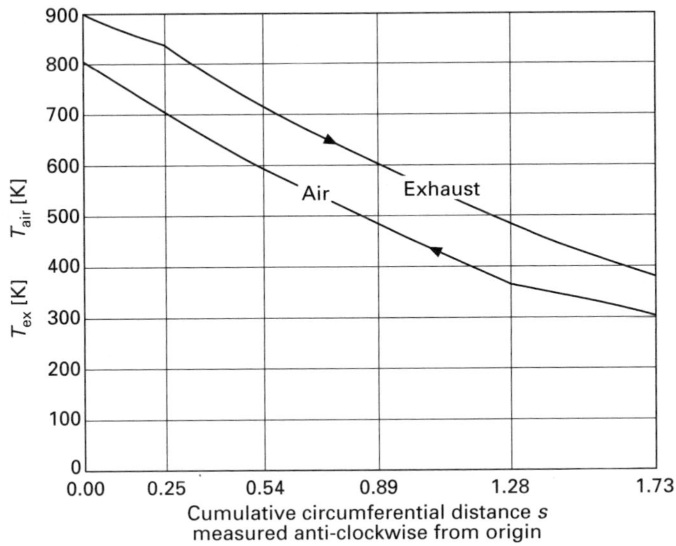

3.3 Temperature solutions corresponding to passage geometry and operating conditions of Table 3.1.

Table 3.1 Geometric parameters and operating conditions of specimen Spirex design

T_{ex}	900.0	K
T_{air}	300.0	K
T_{su}	112.0	K
T_0	300.0	K
μ_0	0.0184E-03	Pas
R	287.0	J/kgK
A_{ff}	75.0E-06	m^2
r_0	35.0E-03	m
m'	10.0E-04	kg/s
w	4.0E-03	m
t	1.5E-03	m
n_j	40	–
n_l	5	–

3.7 Discussion

As in many problems in this category, a deterrent to probing the 'solution' in too great detail is lack of directly-applicable heat transfer and flow friction data: the correlations embodied in the present numerical algorithm may no doubt be relied upon in the context in which Kays and London acquired them. However, that context was not the curved flow passage, where secondary flows are likely to arise.

On the other hand, a tool essential to the acquisition of appropriate experimental data is an analytical or numerical solution to a heat exchange phenomenon in a well-defined flow situation. The required correlations then follow by iterative cross calibration of solution and experiment – a process which the present analysis of the Spirex now makes possible.

On the strength of tentative computation to this point, the Spirex offers unique attractions for specialist applications: in the role of combustion air pre-heater for a Stirling engine, for example, the outermost laps operate near ambient temperature. The feature inhibits convection loss while simultaneously affording effective recuperation of exhaust gas energy.

4
A high-recovery-ratio combustion chamber*

4.1 The design problem

Recent developments, such as flameless oxidation ('Flox', Wünnung, 1996), the 'Swiss-roll' combustor concept, (Ochoa *et al.*, 2003), possibly in combination with catalytic combustion, offer potential for reducing emissions to levels for which the Stirling engine is already optimistically famous. On the other hand, the technical literature does not yet add up to a resource for achieving this benefit by first-principles design. Indeed, no code of practice exists for *any* aspect of Stirling engine design – no British Standard; no ISO; no ASME code. Yet the performance penalty of *ad hoc* combustion chamber design is no less than that of arbitrary regenerator provision.

Little commercial design is undertaken under these handicaps. Much 'new' design is, in reality, evolutionary, and in any case benefits from in-house know-how and skills. Other designs are a synthesis of established concepts – sometimes of proven, off-the-shelf components. A team with adequate resources can tackle a completely new concept through parallel activities, iterating between back-of-envelope calculation, laboratory rigs, materials tests, computer modelling and graphical work. A more modest operation is obliged to carry out design 'serially', confronting the unhappy chicken-and-egg conflict of priorities: to invest in calculation on the basis of estimated data, or to build and test with a view to acquiring the necessary data.

This will be an account of an undertaking in the latter category – and is already somewhat out of chronological sequence. At the outset, a usable analysis of the Spirex exchanger had not been located, and the independent analysis recounted in Chapter 3 had not been undertaken. An elementary simulation of the combustion reaction was available (Organ 1997). This took account of air/fuel ratio and of the temperature dependence of the specific heats of reactants and products, but not of dissociation. Exhaust heat

*In collaboration with Ian Larque, who manufactured and tested the apparatus.

recuperation was dealt with in terms of the arbitrary *NTU* (number of transfer units) in 'generic' counter-flow exchange. As the Spirex is not a simple counter-flow phenomenon, data were not available to justify reliance on the simulation in its existing form.

The status quo is a combustion chamber which lights up readily on a 35 percent/65 percent propane/butane mixture supplied under low pressure from a proprietary cannister. The combustion reaction peaks at about 1200°C, and heat release rate capability is about $1\frac{1}{4}$ kW. At face value this is on target for the proposed application – an unpressurised air engine of some 50–60W brake output for propelling a Shell Eco-Marathon vehicle. Exhaust discharges from the Spirex outlet at below 100°C. Provision for estimating fuel and air flow rates has been incorporated. The numerical model of the Spirex recuperator has been integrated with the combustion model, and the combined simulation is undergoing development and verification.

The work to be described thus amounts to a 'first iteration' of the design. The account is followed by suggestions for a second iteration based on data acquired directly and inferred from simulation and retrospective calculation.

4.2 Principle

Heat loss can be minimized only if all external surfaces of the system operate at, or close to, ambient temperature. Measures which can be taken include (a) use of machinable insulating materials for structural components (b) recovering the maximum possible heat from the outgoing exhaust by using it to pre-heat the inlet air.

Candidate insulating materials for use under (a) are discussed in Section 4.4. Item (b) calls for incorporation of a recuperative heat exchanger. A configuration which fits with both (a) and (b) is a counter-flow exchanger wrapped around the outside of the combustion chamber in the form of a double-spiral bandage. Alternate spiral passages carry outgoing exhaust and incoming combustion air. The configuration is an embodiment of the Spirex concept analyzed in Chapter 3.

A heating system on this principle has been constructed and tested in conjunction with the air engine described by Organ and Larque (2004b). The engine operates inverted, i.e., with cylinder axis vertical, but with the expansion (hot) space lowermost. Figure 4.1 is a cross-section through the cylinder (1), with the heating system assembled into position (Fig. 4.2 shows selected components). Item (2) is the ceramic inner body with integral mixing tube – essentially a Bunsen burner (3). The gas nozzle sits in a flanged, cylindrical ceramic tube (4), whose vertical height is adjustable relative to ceramic sleeve (5), which in turn is adjustable vertically relative to the outer housing (6). The arrangement permits the vertical location of the nozzle to be set relative to the burner venturi independently of the (variable) location of the

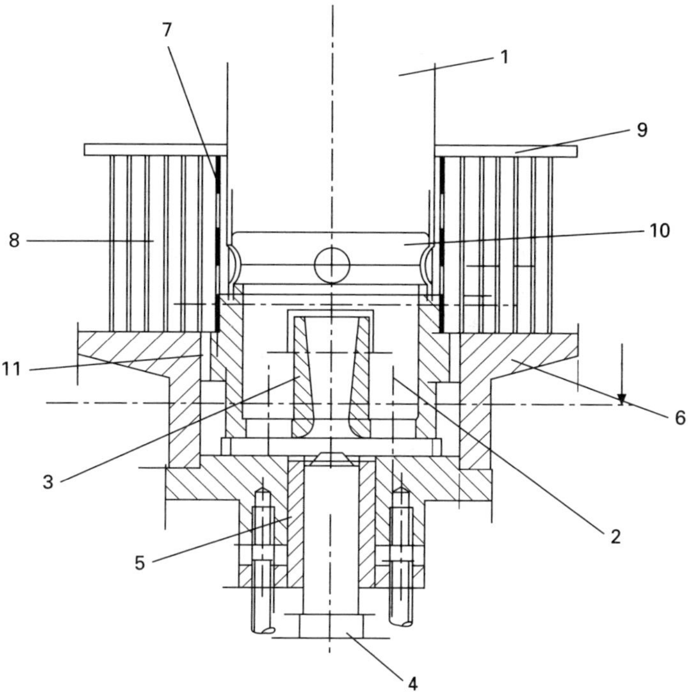

4.1 Cross-section through combustion chamber and double-scroll pre-heater.

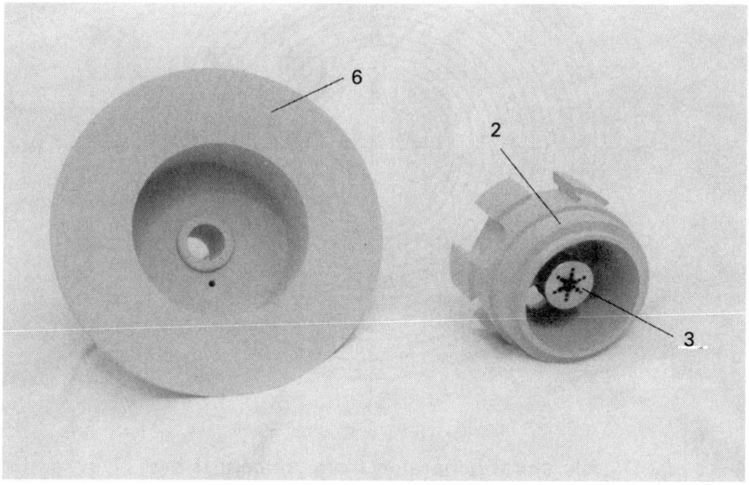

4.2 Inner and outer combustion chamber bodies produced by turning and milling from the machinable natural ceramic *Pyrophyllite*.

(variable) primary air aperture formed between sleeve and body (2). A cylindrical metal sleeve (7) guides the exhaust gases upwards past the outer surface of the engine cylinder and into the spiral recuperator (8). The latter is wound as a double scroll onto the outer diameter of the sleeve between the horizontal faces of outer housing (6) and upper, ceramic closure plate (9). With the assistance of high-temperature sealant, contact under spring pressure between the end faces of the bandage and the closure plates prevents exhaust or inlet air short-circuiting adjacent channels of the scroll.

A miniature fan driven from a 24V d.c. source extracts the exhaust, and causes ambient air to be drawn in for combustion. The combustion chamber operates marginally below atmospheric pressure, thereby discouraging leaks of high-temperature products.

The recuperator double-bandage (8) is wound from corrugated stainless-steel strip of a type used in flame traps. Uniform radial spacing is achieved by locating the full peripheral length of one edge of the strip in a continuous spiral groove machined into a plate of heat resistant material. The double-scroll feature calls for two such grooves per 360° lap.

Without access to computer-controlled machining facilities, a mathematical spiral is difficult to cut. On the other hand, such geometric perfection is not required. Figure 4.3 illustrates an approximation readily achieved from semi-cylindrical grooves turned in a lathe. This calls for two halves of a plate to

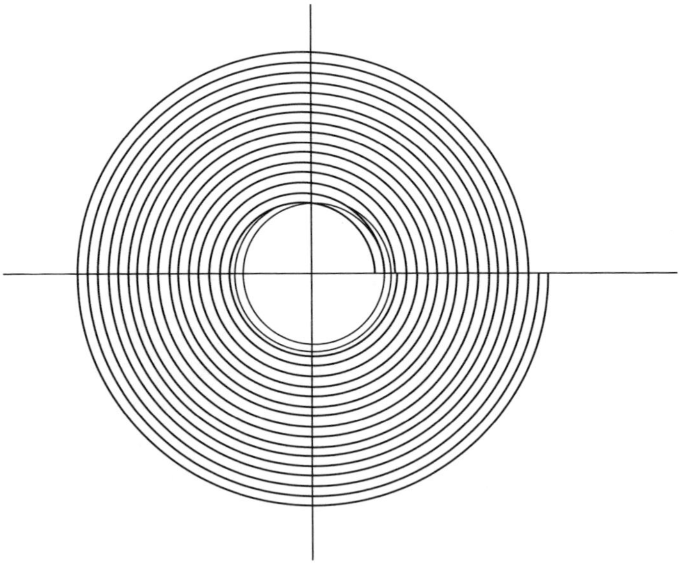

4.3 Double spiral generated from concentric arcs. The semi-circular arcs *above* the horizontal centre line are struck from an origin offset to the left from the vertical centre line. Those *below* are centred on the origin.

be machined separately. All semi-circular slots in the plate forming the lower part of the figure are centred at the origin. Slots in the upper half are offset relative to the vertical centre line. The upper plate is slid laterally to the extent of a single groove spacing relative to the lower, so that the respective sets of slots align to achieve the effect of a continuous spiral. Figure 3.2 of the previous chapter illustrated the bandage located in the grooves.

The innermost complete spiral of the combustion air passage communicates with the annular air passages in the main body (6) via holes drilled though the insulating plate.

4.3 Operation

The jet of gas induces primary air via the base of the venturi. The mixture burns on the surface of a flame-holder sitting on the venturi outlet. Secondary air reaches the flame via the annular space between the outside diameter of the burner tube (3) and the internal diameter of housing (2). The flame plays on the under-surface of the engine cylinder (the engine is inverted) before passing radially outwards via holes in the short cylindrical extension (10) to the cylinder. From there the hot exhaust travels upwards through a gap of 2mm radial width between cylinder and shroud (7). At the upper end of the shroud, a circumferential array of radial holes allows the exhaust to pass outwards into the innermost spiral of the pre-heater. The passage through the spiral involves several complete orbits, during which heat is exchanged with the incoming air travelling in the opposite direction through the adjacent channel in the spiral. The aim is for the exhaust to exit at a temperature close to ambient.

The innermost complete turn of the inlet air channel aligns vertically with annular gap (11) between the outside diameter of inner body (2) and the inside diameter of outer body (6). Through this the pre-heated air enters the combustion chamber to supply primary and secondary flows for combustion.

Ignition is by a spark between an electrode and the metal flame holder. The high-tension source is the piezo-electric igniter adapted from a domestic gas fire.

4.4 Materials

Inner and outer bodies are machined from Pyrophyllite – a naturally-occurring slate. In the natural state it can be turned, milled, filed and sawn, and it will take a thread. Working does, however, yield an abrasive dust. Upon heat treatment the material takes on certain of the properties of fired ceramics. It is apparently quarried in South Africa. In the UK supplies are available from Ceramic Substrates and Components (CSC, undated).

There are advantages to applying a post-machining heat treatment, a procedure calling for detailed attention to the supplier's instructions. The finished components shown in Fig. 4.2 have received heat treatment by the 'Ceramit 10' process (CSC, undated)

The stainless steel flame trap material specified for the bandage is manufactured and supplied by Amal Safety Systems UK Ltd., (IMI/Amal, undated). The corrugations are generated by rolling sheet material of 0.05 mm thickness (0.002 in.) between profiled rolls. A range of corrugation pitches and strip widths is available, but it is necessary to purchase a standard stock length of the chosen pitch – a generous 20 metres! The present recuperator is designed around a 38 mm width ($1\frac{1}{2}$ in.) and 0.6 mm (0.023 in.) corrugation pitch.

4.5 Preliminary operating experience

Trials were based initially on a flame-holder of perforated stainless steel similar to the cap of a pepper pot. Ignition was possible only with a layer of gauze beneath the pepper pot, and the flame was stable over a limited range of gas and air flow. Reduction of gas jet size to the minimum practicable by normal jobbing-shop drilling (0.25 mm, or 0.010 in.) brought improvement – but not a solution.

Stainless steel hypodermic tube is routinely manufactured with bore sizes down to one thousandth of an inch (CNW undated), and offers an alternative to specialist drilling. Specimen tube off-cuts in a range of bore sizes provided by the manufacturer were cemented into nozzles with a view to further trials.

Before these were carried out it was found that reliable ignition and flame retention occurred over a wide range of gas and air flows when the stainless-steel 'pepper-pot' was replaced by a porous flame holder consisting of a disc of Retimet™ metal foam. This evidently not only keeps local mixture velocity below flame propagation speed but provides effective pre-heat: The gas–air mixture burns on – or somewhat below – the upper surface of the foam, causing the latter to become incandescent. Heat conducted back through the metal matrix counter to the flow direction provides the pre-heat.

The majority of running has been with the engine replaced by a dummy cylinder having a transparent window in place of the cylinder cap. This allows direct visual observation of the flame. 'Pyrex' watch glasses of 50 mm diameter as used in the chemistry laboratory have been found suitable for the purpose, although they can sag during use. At favourable air/fuel ratios the visible part of the combustion reaction is a blue luminescence rather than a flame. In Plate 2 (between pages 134 and 135) the red halo is the perimeter of the Retimet™ flame holder. The small yellow flash is the electrode of the igniter.

Peak reaction temperature measured by a K-type sheathed thermocouple varies around 1100°C. Exhaust exits the Spirex pre-heater below 100°C accompanied by a spray of liquid water, suggesting that the HCV of the fuel is being exploited (but see Section 4.6.3).

The air pre-heats over a lower temperature difference than the exhaust cools, entering the combustion chamber at about 350°C. The measured asymmetry is consistent with the inherent asymmetry of theoretical temperature solutions (specimen at Fig. 3.3) – but not proportional to it. The discrepancy may owe something to inwards leakage of ambient air into both inlet and exhaust streams, resulting in dilution of both.

4.6 Second design iteration

Initial operating experience suggests that the heating capacity of the combustion chamber is limited by the ability of available 24-volt extractor fans to draw air into the scroll, and combustion products out. At peak air flow the smallest gas jet which could be made from the available hypodermic tubes gave a rich mixture, inconsistent with the role intended for the engine (power unit for Eco-Marathon vehicle). The immediate problem disappeared on discovering a supplier of gas burners who was disposing of a range of jets offering the benefit of an integral sintered-metal filter immediately upstream of the orifice. The smallest of these (marked #14, possibly indicating an orifice diameter of 0.14 mm (≈ 0.0055 in.)) gives consistent ignition and a flame lacking any yellow fringing, The feature suggests that the mixture, if not stoichiometric, is at least not rich.

The higher calorific value (HCV) of propane is 50,360 kJ/kg; that of butane 49,520 kJ/kg (Haywood, 1990). The arithmetic turns out to be a bit pedantic, but the apparent HVC of the propane/butane mixture is 0.35 *times* 50,360 + 0.65 *times* 49,520 = 49,814 kJ/kg. During a run timed at 11 minutes, 16.3g of fuel mixture were consumed. Potential heat release rate was thus 16.3×10^{-3} (kg) $\times$ 49,814 $\times$ 10^3(J/kg)/(11.0 $\times$ 60(s)) W, or 1.230 kW. Peak brake output of the engine measured previously under open-flame heating had been 35W (See Chapter 8), suggesting a brake thermal efficiency of 35/1230 = 2.84 percent.

It is worth comparing this interim achievement with the peak brake thermal efficiency of the Philips MP1002CA air engine. This boasted a sophisticated regenerator, extended heat transfer surfaces, charge pressure elevated to 14 atm. Moreover, it embodied the benefits of '*modern knowledge about heat transfer, flow resistance etc*' (de Brey *et al.* 1947) – and admittedly drove its own cooling fan and electric generator (on open-circuit). The peak brake thermal efficiency of 5.4 percent occurred at 1250 rpm (Organ 1962).

Combustion temperature and heat rate are independent, and the ultimate development aim includes matching both to the requirements of the engine.

Targets which need to be addressed simultaneously are recuperator exit temperature and, eventually, exhaust emissions. The proposed EcoMarathon application requires autonomous operation, meaning that power to auxiliaries, including the extractor fan, has to be a minimum. The air flow achieved at given fan power turns out to be a strong function of system temperatures (see below), providing an apt illustration of the inter-dependence of design priorities.

An indication in advance of likely thermal response to given air flow, fuel rate, fan pumping power etc. would maximize the chance of a successful outcome.

4.6.1 Stoichiometry

Atmospheric air comprises about 79 percent nitrogen (N_2) by volume to 21 percent oxygen (O_2). For reaction calculations this corresponds to 79 kmol of N_2 for each 21 kmol of O_2. The reaction balance with generalized hydrocarbon compound C_iH_j is:

$$C_iH_j + zO_2 + 79/21zN_2 \Rightarrow aCO_2 + bH_2O + 79/21zN_2$$

Each C atom of the fuel accounts for one O_2 molecule in the reaction to CO_2. On the other hand, *two* molecules of H_2O are formed for each atom of O_2 reacted, permitting numerical evaluation of a and b:

$$a = i$$

$$b = \tfrac{1}{2}j$$

$$z = i + \tfrac{1}{4}j$$

This allows the reaction equation to be re-expressed:

$$C_iH_j + (i + \tfrac{1}{4}j)O_2 + 79/21(i + \tfrac{1}{4}j)N_2 \Rightarrow iCO_2$$

$$+ \tfrac{1}{2}j\,H_2O + 79/21(i + \tfrac{1}{4}j)N_2 \qquad 4.1$$

According to the composition declared on the canister, the fuel used for experiments to date has been a mixture of 35 percent propane (C_3H_8) and 65 percent butane (C_4H_{10}). Equivalent i is 0.35 *times* 3 + 0.65 *times* 4 = 3.65, and equivalent j is 0.35 *times* 8 + 0.65 *times* 10 = 9.3. The stoichiometric calculation (only) may proceed *as though* the fuel mixture were $C_{3.65}H_{9.3}$.

Each kmol of fuel completely burned requires $(3.65 + \tfrac{1}{4} \times 9.3)$ kmol O_2, or 5.975 kmol O_2. Air being 21 percent O_2 by volume, the kmol of air required are $5.975 \times 100/21 = 28.45$, volumetric air/fuel ratio in this case is 28.45:1

In the case of liquid fuels, air/fuel ratio is more commonly expressed as *gravimetric ratio*, or mass ratio. Each kmol of fuel has mass $(12i + j)$ kg = 53.1 kg; each kmol of air has mass 28.98 kg. Gravimetric air/fuel ratio corresponding to the volume ratio calculated earlier is $(28.45 \times 28.98/53.1)$:1, or 15.52:1 (the parallel calculation on p. 424 of Organ 1997 is in error.)

The arithmetic serves as a start point for target air volume rate, which, however, has to be metered – and preferably without compounding the pressure loss through recuperator and combustion chamber.

4.6.2 Provisional air flow measurement

Figure 4.4 shows part of the orthographic drawing of an air flow sensor which, depending on point of view, imposes no pressure loss. The pre-heater requires an inlet port regardless, so adding an entry section flared and streamlined as indicated may in fact reduce inlet loss relative to an abrupt entry. A manometer connected to a tapping in the parallel section perpendicular to the flow direction detects static pressure p at that point. Bernoulli's equation relates total pressure (static p *plus* dynamic $\frac{1}{2}\rho u^2$) at the tapping to total pressure p_∞ of the stationary air at some horizontal distance upstream of the inlet. With subscript $_1$ to indicate conditions at the tapping:

$$p_\infty = p_1 + \tfrac{1}{2}\rho u_1^2$$

Manometer reading Δp corresponds to $\frac{1}{2}\rho u_1^2$, and may be read direct from a gauge reading in Pa, or in terms of head $\rho_f g h$ of a column of liquid of density ρ_f (kg/m^3):

$$\Delta p = p_\infty - p_1 = \tfrac{1}{2}\rho u_1^2$$

Inverting:

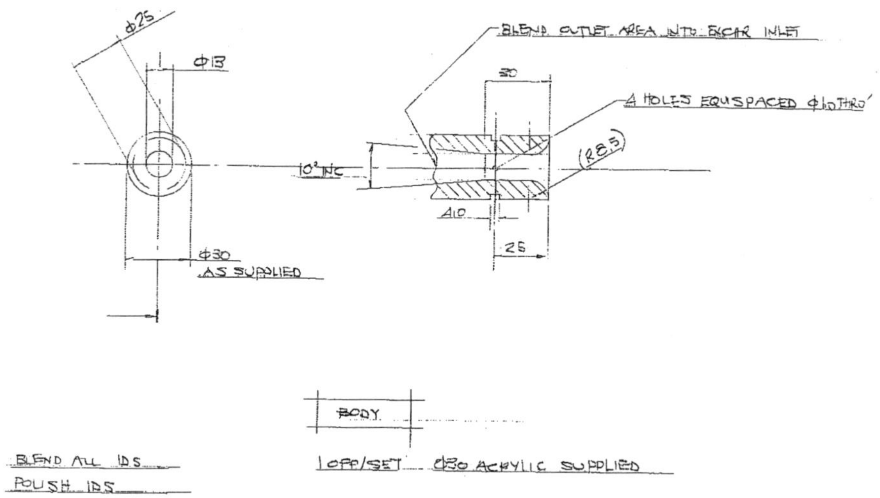

4.4 Venturi-meter to the author's original pencil drawing. Four radial holes in the parallel section connect to a manifold formed by an external annular groove sealed by a cylindrical sleeve. The latter is bonded into position using epoxy resin, and carries the connection to the manometer.

$$u_1 = \sqrt{(\Delta p / \tfrac{1}{2}\rho)} \qquad 4.2$$

With A_{ff1} to denote the free-flow area (m^2) at the measuring point, volume rate Q' is:

$$Q' = A_{ff1}\sqrt{(\Delta p / \tfrac{1}{2}\rho)} \ (m^3/s) \qquad 4.3$$

Mass rate m' is $\rho Q' = \rho u_1 A_{ff1}$ (kg/s). Air density ρ may be calculated using the ideal gas assumption, namely, $\rho = p_\infty / RT_\infty$, where T_∞ (K) is absolute temperature, e.g. 300K, and R is the specific gas constant for air, 287 J/kgK, leading to:

$$m' = A_{ff1}\sqrt{\{2p\Delta p / RT_\infty\}} \ (kg/s) \qquad 4.4$$

An instrument commonly employed for indicating small pressure differences (say, of a few hundred Pa) is the liquid manometer. Scale reading is in terms of 'inches of water column' (in. H_2O), where 1 in. $H_2O \approx 0.00278$ atm, or 277 Pa. The throat diameter of the venturi-meter of Figs 4.4 and 4.5 is 13.0 mm, and corresponding free-flow area 0.0001327 m^2. Denoting the manometer reading by h (in. H_2O) the calibration becomes:

$$m' = 0.00337\sqrt{h} \ (kg/s) \qquad 4.5$$

The finished meter is shown at Fig. 4.5, the graphical equivalent of the calibration at Fig. 4.6.

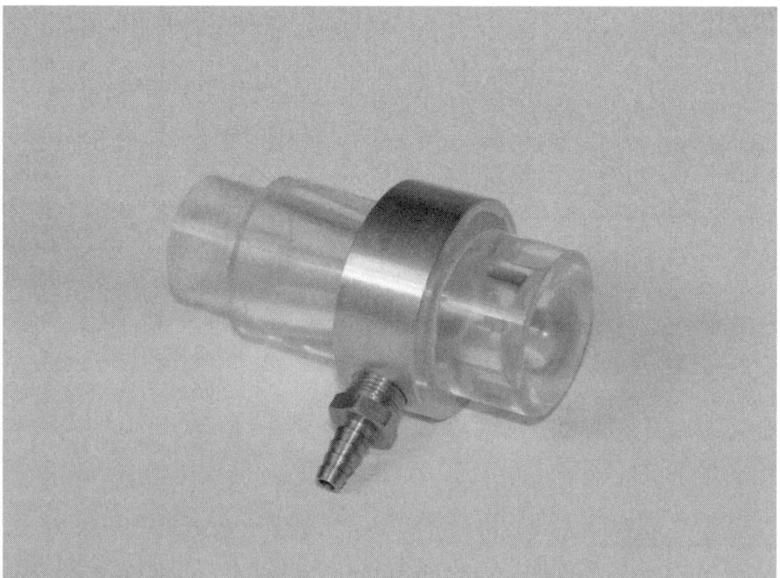

4.5 Siting the venturi-meter at the pre-heater inlet means it operates at ambient temperature, and can be made from a readily-machined material such as acrylic.

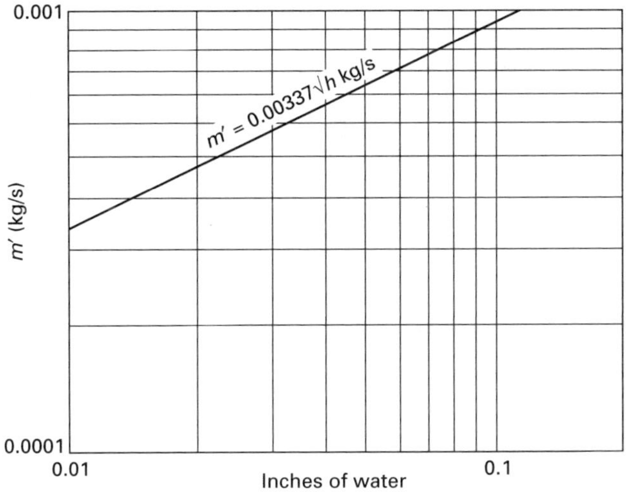

4.6 Venturi-meter calibration in terms of inches of water.

Manometric height *h* corresponding to Δp generated by the air requirement of 1 kW or so of heat release rate is a fraction of an inch of water. Accurate readout is facilitated by use of a manometer of the 'inclined' type. These are readily fabricated, but it is useful to know that such instruments are mass-produced, a common medium being plastic (Fig. 4.7). One routine commercial application is for the display of pressure drop across air filters used in spray-paint booths (a high pressure drop is the alarm signal that a contaminated filter requires replacement).

A reading of 0.042 in. H_2O on the manometer gives, from Eqn 4.5 or Fig. 4.6, $m' = 0.00069$ kg/s. At the estimated gravimetric air/fuel ratio, corresponding fuel rate should be 0.00069/15.52 or 0.04445 g/s. The measured rate of 16.3 g over 11 min. converts to 0.02424 g/s, or slightly greater than one half of the stoichiometric amount.

Widespread use of such gauges means that they are readily available. The Internet is a convenient source, and a few dollars covers the cost of instrument plus postage. Individual calibration and scientific precision are not features, but design and construction are not without sophistication. The fluid is commonly an oil rather than water. The specific gravity quoted in the instruction sheet for the oil of the instrument shown is 0.826, leading to the casual expectation that the vertical height of a scale division, measured with an accurate height gauge, would be a factor of 1/0.826 greater than the scale marking. An apparent discrepancy of some $8\frac{1}{2}$ percent in the case of the instrument shown is not an inaccuracy, but a factor designed in to compensate the variation in reservoir level which accompanies movement of the fluid within the display tube.

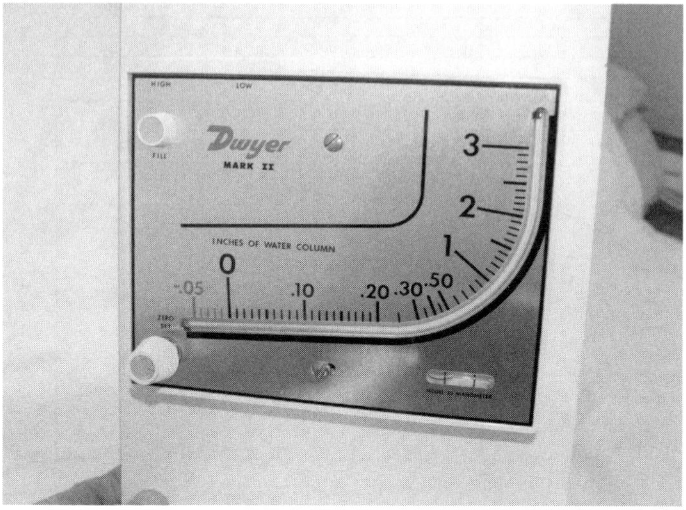

4.7 Inclined manometer, mass-produced in plastic. Relatively inexpensive as purchased via the Internet. Model shown is Dwyer # 25, Mk 2 on optional stand.

4.6.3 Combustion system simulation

The simulation described in Organ (1997) applies to the combustion system shown schematically in Fig. 4.8. This provided for variable rates of fuel m'_f, of primary air m'_{ap} and of secondary air m'_{as}. Combustion products were assumed to form at T_2 from fully mixed and vaporized reactants initially uniformly at T_1. After exchanging heat with the engine the products entered the pre-heater at temperature $T_{ex\ in}$, exiting at $T_{ex\ out}$ after exchange of heat with the air entering at T_∞. 'Generic' counter-flow exchange was provided for by arranging that temperature difference $T_{ex\ out} - T_{air\ in}$ related explicitly to outlet difference $T_{ex\ in} - T_{air\ out}$ in an expression linear in all the temperatures. With T_∞ for $T_{air\ in}$:

$$T_{ex\ out} - T_\infty = (T_{ex\ in} - T_{air\ out})$$

$$\exp\{-NTU[1 - \Sigma\, m'c_p\,|_{min}/\Sigma\, m'c_p\,|_{max}]\} \qquad 4.6$$

(and not as mis-printed in Organ (1997)). In Eqn 4.6 $\Sigma\, m'c_p\,|_{max}$ is the larger of the two enthalpy streams, $\Sigma\, m'c_p\,|_{min}$ the smaller (Kays and London 1964).

By contrast, the numerical solution for the Spirex described in Chapter 3 amounts to the function:

$$T_{ex\ out} - T_{air\ out} = S_x\{T_{ex\ in}, (T_\infty)\} \qquad 4.7$$

In Eqn 4.7 S_x represents the (non-linear) function *Spirex* which is solved numerically corresponding to input values of $T_{ex\ in}$ (and of environment

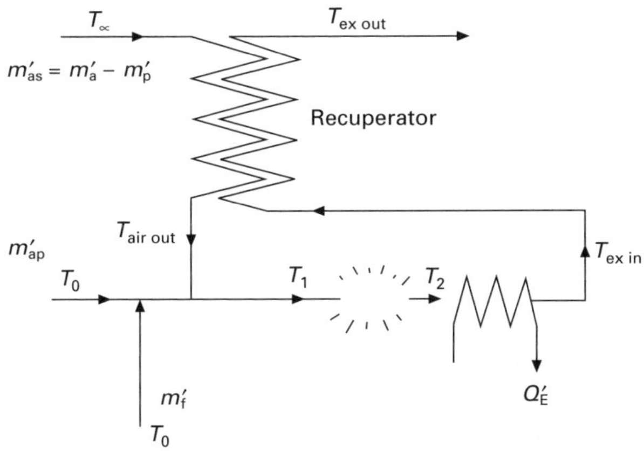

4.8 Notation for description of combustion chamber simulation.

temperature T_∞ – essentially invariant). Specific solutions of Eqn 4.7 are achieved by a process of matrix inversion internal to *Spirex*. Combustion simulation *combust* also inverts a matrix – a 6 × 6 array of the unknown temperatures. The two simulations invoke a common solution mechanism suggesting a choice of ways of combining them:

(a) expanding the 6 × 6 array of *combust* to include the $n \times n$ array elements for exhaust and air temperatures at the n nodes within the scroll channels of the Spirex, leading to $(6 + n) \times (6 + n)$ array of unknowns
(b) linearizing Eqn 4.7 about a likely value of $T_{ex\ in}$ and using the coefficients of the resulting (linearized) relationship between $T_{ex\ out}$, $T_{air\ in}$, $T_{air\ out}$ and $T_{ex\ in}$ as part of the 6 × 6 matrix of *combust*.

The latter method has been found to function in stable fashion. Being the simpler of the alternatives it has been implemented by expanding as follows:

$$T_{ex\ out} - T_{air\ out} \approx S_x(T_{ex\ in\ approx}) + \partial S_x(T_{ex\ in})/\partial T_{ex\ in}(T_{ex\ in} - T_{ex\ in\ approx}) \qquad 4.8$$

S_x and $\partial S_x/\partial T_{ex\ in}$ are evaluated numerically and treated as known coefficients of Eqn 4.8, which is thus linear in the three unknowns $T_{ex\ out}$, $T_{ex\ in}$, $T_{air\ out}$. The coefficients of these terms replace the corresponding expressions for the generic exchanger used in the earlier version of simulation *combust*.

Running the combined simulation with fuel and air flows of the bench tests results in peak temperatures 200–300°C higher than the peak measured value of 1200°C. At above 1500K (1227°C) dissociation (CO_2 to CO and O_2; H_2O to H_2 and O_2) is well-established (Rogers and Mayhew 1992). Consistent with the presence of CO and H_2, this endothermic process limits peak temperature, and prevents full release of the potential chemical energy in the fuel at that peak.

4.6.4 Further work under 'second iteration'

The advantage of the low conductivity of ceramics is offset somewhat by increased wall thickness relative to the metal counterpart, and correspondingly high thermal capacity. Use of these materials would almost certainly be counter-productive if short operating cycles are proposed, due to the losses involved in repeated heating and cooling.

For given electrical power to the extractor fan, the air-flow manometer reading h (in. H_2O) with the combustion chamber cold and unlit is about 0.09 in. H_2O – i.e., approximately twice the value with the system at full heat load. The effect of the combustion reaction and high temperatures is thus to attenuate the isothermal air flow by a factor of approximately $\sqrt{2}$. Mass rate at all locations between ambient air inlet and exhaust exit is roughly constant, so an explanation may lie with the temperature dependence of dynamic viscosity μ and flow velocity.

The Spirex simulation computes local Stanton number S_t in terms of Reynolds number R_e as a basis for the temperature solution. Local friction factor C_f follows from minor modification of the code, making it possible to endorse (or otherwise) the temperature/flow effect from first principles. An indication should follow as to how air flow – and thus thermal rating – may be increased.

It is likely that free-flow area needs to vary in some function of temperature. Any such geometry promises to be substantially more difficult to realize in practice than the uniform area achieved by offsetting concentric grooves (recall Fig. 4.3). This makes it desirable to have some confidence in the outcome before developing the fabrication techniques. Embodiment of the algebra of dissociation within the combustion simulation thus becomes an essential interim step.

Part II

Living with incompressible flow data

5
The regenerator problem brought down to size

5.1 Background

Regenerator theory has been slow coming to the aid of Stirling engine design. The arcane vocabulary of classic treatments ('reduced length', 'reduced period' etc.) is not helpful. Those who have sought guidance on hydraulic radius, r_h, or on thermal capacity ratio, N_{TCR} (= $\rho_w c_w T_{ref}/p_{ref}$) for design purposes will know that none exists.

On the other hand, solutions have recently become available (e.g. Organ (1997, 2000, 2005a) for transient thermal response at flush ratios typical of Stirling engine operation. These confirm that high values of *NTU* (number of transfer units = StL_{reg}/r_h) are achieved – in other words, that local difference in temperature ΔT between matrix and fluid may be held to a small value. Taking this to an extreme, i.e., supposing vanishingly small ΔT, leads to a particularly simple picture of regenerator operation which highlights the role of thermal capacity ratio N_{TCR}. The formulation yields temperature–time histories for any desired flush ratio N_{FL} and over any number of cycles between start-up and cyclic equilibrium. The algebra is sufficiently simple to permit extraction of a criterion on N_{TCR} in terms of target percentage matrix temperature swing ε percent.

5.2 Assumptions

Figure 5.1 represents a matrix of volume porosity $¶_v$ and free-flow area A_{ff} occupying a duct. The duct joins two spaces whose instantaneous volumes are controlled by movable pistons. It is assumed that:

- Flow friction does not materially affect the temperature solution; pressure at a given instant is uniform throughout the matrix.
- *NTU* are sufficiently high to justify setting temperature ΔT between fluid and matrix zero at the outset. It is crucial to note that the algebra does *not* amount to the earlier formulation with *NTU* set to infinity. The

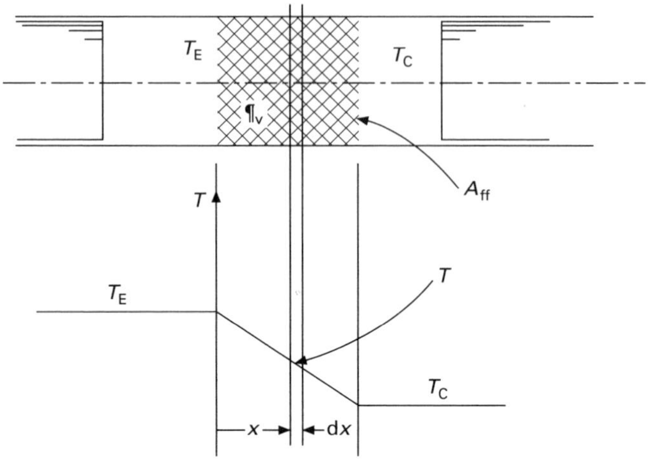

5.1 Physical model and notation for statement of energy conservation.

outcome should, of course, be identical. But there is no definitive discretization scheme which represents a differential equation with absolute accuracy, so an independent formulation of a special case ($NTU = \propto$) is a valuable check on the general integration scheme.

- The fluid behaves as a perfect gas, i.e., as $p/\rho = RT$.
- Fluid entering the matrix from the left does so at invariant temperature T_E. That entering from the right does so at invariant temperature T_C.
- Flow is 'slab', or 'one-dimensional': velocity at any instant is uniform over all cross-sections perpendicular to flow direction.
- Piston motion is some prescribed function of time, e.g., linear between switching, or simple-harmonic, provided only that the ratio of respective piston speeds, left-to-right, is in the ratio T_E/T_C so as to maintain pressure p constant. The analysis is readily modified for any variation of pressure with time.

To keep the important assumption of slab flow in mind, particle velocity is denoted by a symbol with underscore, namely, $\underline{u}$. In general, $\underline{u}$ at any instant will be different at different axial locations, reflecting the non-uniform density.

The remaining simplifications are made in the interests of presentation only. All are readily relaxed later in favour of assumptions better representing the real gas processes

- density ρ_w and specific heat c_w of the parent material of the matrix are invariant
- operation begins with a linear distribution of (common) temperature T between T_E and T_C

- flow passage geometry is that of a stack of rectangular-weave wire screens, where wire diameter is d_w and mesh number is m_w
- Mach number effects are negligible.

5.3 Defining equations

The element of matrix dx long and of volume porosity $¶_v$ forms a control volume having free-flow area A_{ff}. For a fluid flowing instantaneously at velocity $\underline{u}$, rate of energy exchange with the matrix is equal to rate of change of internal energy $\rho c_v A_{ff} dx \partial T/\partial t$, plus the difference between enthalpy rates out and in, namely, $\rho c_p \underline{u} A_{ff}(\partial T/\partial x)dx$. Under the overriding assumption ($\Delta T = 0$) this is equal to the rate of internal energy change of the matrix element. Envelope volume is $dxA_{ff}/¶_v$, so that the volume of matrix material is $dxA_{ff}(1 - ¶_v)/¶_v$, and rate of internal energy change (of the solid element) is $dxA_{ff}[(1 - ¶_v)/¶_v]\rho_w c_w \partial T/\partial t$.

Equating the difference in fluid enthalpy rate in and out to rate of increase of the internal energy of the control volume (fluid plus matrix):

$$\rho c_p \underline{u} A_{ff}\{T - [T + (\partial T/\partial x)dx]\}$$

$$= A_{ff} dx \partial T/\partial t\{[(1 - ¶_v)/¶_v]\rho_w c_w + A_{ff}\rho c_v\}$$

Simplifying throws up a term $\rho_w c_w/\rho c_p$. For the ideal gas and for the uniform-pressure case this is this is essentially $N_{TCR}T/T_{ref}(\gamma - 1)/\gamma$, where N_{TCR} is thermal capacity ratio $\rho_w c_w T_{ref}/p_{ref}$. Substituting and simplifying:

$$\partial T/\partial t + \gamma \underline{u} \partial T/\partial x = - N_{TCR}\{T/T_{ref}\ (1 - ¶_v)/¶_v\}\partial T/\partial t$$

Adding a term $\gamma \partial T/\partial t$ to both sides permits the left-hand side to be written as the substantial derivative $DT/dt = \partial T/\partial t + \underline{u}\partial T/\partial x$, an essential step if solution is to proceed along fluid particle paths:

$$DT/\partial t = \partial T/\partial t\{1 - (T/T_{ref})N_{TCR}(1 - ¶_v)/¶_v\}(\gamma - 1)/\gamma \qquad 5.1$$

The analysis is concerned with high-performance regenerators, in which fluid and wall temperatures are expected to differ little from the supposed linear distribution between T_E and T_C. On this basis the (unknown) dimensionless multiplier T/T_{ref} of the N_{TCR} term may be replaced by the (known) value of ideal temperature at that point. Eqn 5.1 states that the rate DT/dt of temperature change (common to fluid and matrix) with crank angle when following a fluid particle is equal to local rate of change with time $\partial T/\partial t$ multiplied by a term whose value is known as a function of location. The latter term is, as anticipated, a function of thermal capacity ratio N_{TCR}, specific heat ratio γ and volume porosity $¶_v$. If boundary conditions can be specified, Eqn 5.1 may be solved numerically to give temperature history for all crank angles and locations.

5.4 Boundary conditions

At the instant of flow reversal, working fluid enters the outermost gauze layer at nominal inlet temperature – T_E at the expansion end, T_C at the compression end. Thickness $2d_w$ of the single gauze layer (Fig. 5.2a) is small in relation to particle displacement over an integration step. This justifies replacing the $\rho c_p \underline{u} A_{ff}(\partial T/\partial x)dx$ of the earlier energy balance by the quasi steady-flow equivalent, namely, $\rho c_p \underline{u} A_{ff}(T_E - T)$. The approach has precedent in solution by the Method of Characteristics of unsteady flows involving abrupt changes in free-flow area (Shapiro 1954).

The control volume formed by the gauze layer instantaneously contains $2d_w \rho A_{ff}$ kg of fluid and $2d_w \rho_w A_{ff}(1 - \P_v)/\P_v$ kg of wire material. Considering the left-hand boundary, where entry temperature is T_E, and writing out precisely as for the earlier energy balance:

$$\rho c_p \underline{u} A_{ff}(T_E - T) = d_w A_{ff}\{\rho_w c_w (1 - \P_v)/\P_v + \rho c_v\}\partial T/\partial t$$

Noting that $\partial T/\partial t$ may be written $-\partial(T_E - T)/\partial t$ and rearranging leads to the differential relationship:

$$(T - T_E)^{-1}\partial(T - T_E)/\partial t = \text{RHS}$$

Integrating, deriving the constant of integration from the fact that $T = T_0$ when $t - t_0$, and applying to the integration interval of duration dt:

$$T - T_E = T_0 - T_E\, e^{-\text{RHS}\cdot dt} \qquad 5.2$$

With temperature ratio T_E/T_C denoted N_T:

$$\text{RHS} = \frac{|\underline{u}|/d_w}{\{[(1 - \P_v)/\P_v]N_{TCR}N_T + 1/(\gamma - 1)\}(\gamma - 1)/\gamma}$$

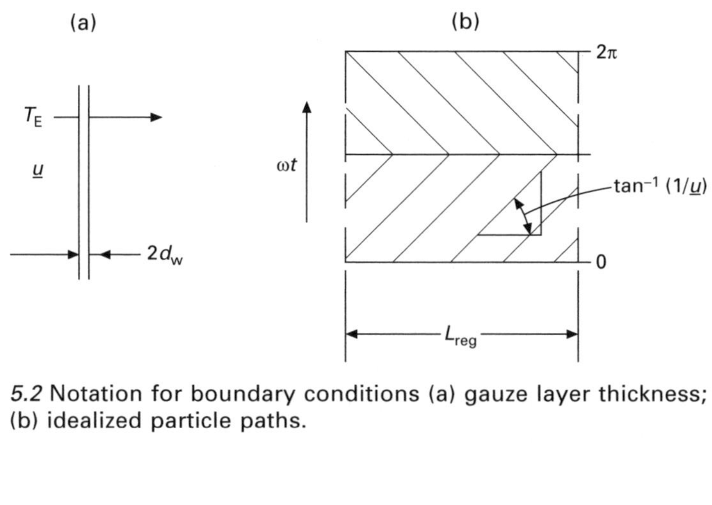

5.2 Notation for boundary conditions (a) gauze layer thickness; (b) idealized particle paths.

At the cold (compression) end, where inlet temperature is T_C, boundary condition and RHS are achieved by replacing temperature ratio N_T by unity.

5.5 Flush ratio

Equation 5.1 applies along the path taken by a fluid particle, and so may be applied along the particle tracks of a 'slab flow' model of the gas processes of the complete engine. Indeed, the treatment lends itself to modification (e.g. Organ 1997) to reflect cyclic pressure swing. On the other hand, the attraction of 'decoupling' the treatment is the opportunity of illustrating how the gap between traditional regenerator theory and the Stirling gas processes may be bridged – to the advantage of both. An essential link is the notion of flush ratio N_{FL}:

$$N_{FL} = \frac{\text{Mass throughput per blow}}{\text{Fluid mass content of matrix pores}}$$

For the simple picture of regenerator operation (Fig. 5.1) a numerical value of N_{FL} may be determined in terms of the maxima and minima of numbers of fluid particle tracks which do *not* interact with the matrix, compared with the number contained within the matrix. The notion extends to the regenerator *in situ* in the Stirling engine: Fig. 5.3 is a particle trajectory diagram produced by the methods of Organ (1997) for the gas circuit parameters of the Philips MP1002CA air engine. Successive trajectories subdivide total working fluid equally, so matrix mass content and cyclic mass throughput are estimated by counting as before. On the other hand the mass content of the matrix pores

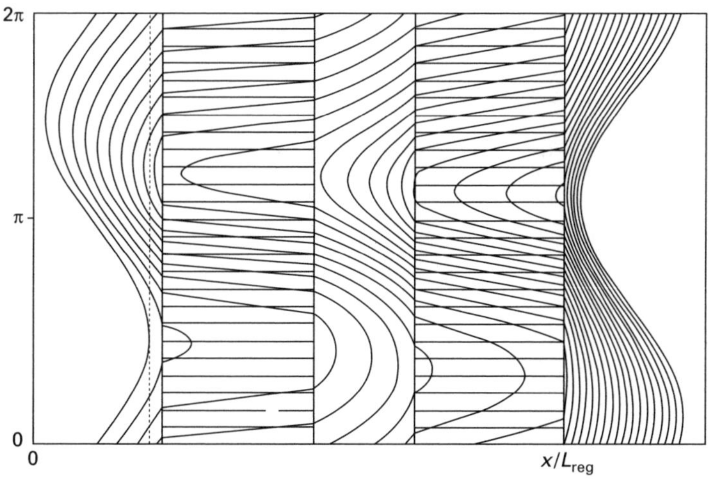

5.3 Ideal particle trajectory diagram for Philips MP1002CA air engine. Operation at rated conditions (Organ 1997).

varies somewhat over the cycle, so that N_{FL} must be a mean value. Alternatively, the value of N_{FL} may be bracketed by specifying an upper and a lower limit.

Table 5.1 lists median N_{FL} for a number of well-known Stirling engines for eventual use in establishing a criterion for regenerator design.

5.6 Integration algorithm

Achievement of meaningful temperature solutions calls for the numerical manipulation of Eqn 5.1 to reflect the underlying physical realities: Fig. 5.4 shows an element of particle path in the $t - x$ coordinates of the solution. The change DT of the left-hand side of Eqn 5.1 is that along the path line element over the increment dt of time (or crank angle). In discrete, numerical form it involves values of the 'known' T_j at current time, and the 'unknown' T_{j+1}^{i+1}. It might be thought, by analogy with schemes for dealing with the diffusion equation, that discretization of the right-hand term, $\partial T/\partial t$ would aim for a second-order average of the time rate of change by involving both unknowns T_{j+1}^{i+1} and T_j^{i+1}. That this would be incorrect is seen by noting that the triangular time–distance domain bounded by points (i, j), (i+1, j) and (i+1, j+1) relates to times, all of which, except the instant (i+1, j+1), lie *ahead in time* of the

Table 5.1 Thermal capacity ratio N_{TCR} of selected Stirling engines together with parameters determining nominal cyclic temperature swing. Note low N_{FL} of the USS P-40 having two regenerators per gas path: one half of regenerator nominal fluid mass content shuttles back and forth without exiting either end of the matrix

Designation	Fl. ratio (N_{FL})	Volume porosity ($\P_v$)	Specific heat ratio (γ)	Temperature ratio (N_T)	N_{TCR} $(1 - \P_v)/\P_v$
Philips					
400 hp/cyl	2.4	0.582	1.66	3.08	224
GPU-3	2.4	0.753	1.4	3.27	62
MP1002CA	2.7	0.8	1.4	2.92	219
USS P-40	0.5	0.685	1.4	3.07	40
USS V-160	2.0	0.69	1.66	2.72	47
Allison PD-46	1.24	0.69	1.66	2.67	58

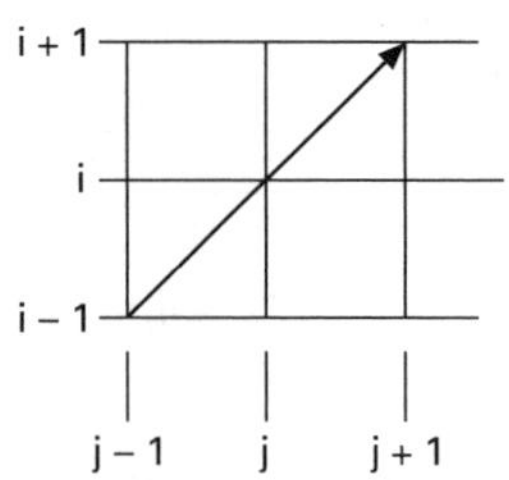

5.4 'Unit process' of the integration sequence.

processes followed by the particle in its path between (i, j) and (i+1, j+1), and cannot, therefore, influence the computation. Another way of looking at this is to note that the assumption of zero diffusion in the flow direction permits the particle travelling between (i, j) and (i+1, j+1) to be pursued at an infinitesimal distance by a temperature disturbance of arbitrary magnitude. A mid-interval difference relationship may be formulated corresponding to $\partial T/\partial t$, but the assumption of linearity between points i, j and i+1, j+1 means that this will be equivalent to the simple $(T^{i+1}_{j+1} - T^{i}_{j+1})/\Delta t$.

Although the formulation is sufficiently economical that a modern CPU will process more than 100 complete cyclic reversals per second, there is some satisfaction in minimizing the work of computation between start-up and attainment of cyclic equilibrium. A useful short-cut is achieved by starting the integration process at the $\frac{1}{2}$-blow ($\frac{1}{4}$-cycle) point: this is tantamount to assuming that the initial temperature distribution approximates the mid-blow distribution. If the expedient is not used, and if the initial distribution is subject to a complete blow, it is biased in the direction of that blow, and many cycles are required to re-establish mid-blow location.

5.7 Specimen temperature profiles

Simple-harmonic motion is a more appealing approximation to conditions in the Stirling engine than $\underline{u}$ = constant, and is such an elementary embellishment that it forms the basis for the example: Fig. 5.5(a) shows the particle paths for a flush ratio N_{FL} of unity, with integration mesh superimposed. The gradation in spacing left-to-right reflects the assumption of ideal gas behaviour with variation of density between expansion end conditions (T_E) and those at the compression end (T_C).

Figure 5.5(b) shows the computed envelope of fluid (and matrix) temperatures after attainment of cyclic equilibrium for volume porosity $¶_v$ of 0.8 and specific heat ratio γ of 1.4. It is necessary only to inspect Eqn 5.1 to be sure that different values of γ lead to different temperature swings.

Figure 5.6 is the corresponding temperature relief, with fluid particle paths ruled on the surface. This is seen to have all essential features in common with solutions (e.g. those of Organ (1997)) in which fluid and matrix temperatures are coupled by finite values of Stanton number *St*.

5.8 Design criterion for N_{TCR}

By re-expanding the substantial derivative $DT/\partial t$ to its constituent terms $\partial T/\partial t + \underline{u}\partial T/\partial x$, substituting into Eqn 5.1 and regrouping, it is possible to express local temperature variation with crank angle, $\partial T/\partial t$, in terms of gradient in the flow direction $\partial T/\partial x$. Assuming highly effective operation, nominal temperature gradient is $(T_C - T_E)/L_{reg} = T_C(1 - N_T)/L_{reg}$ and substantially

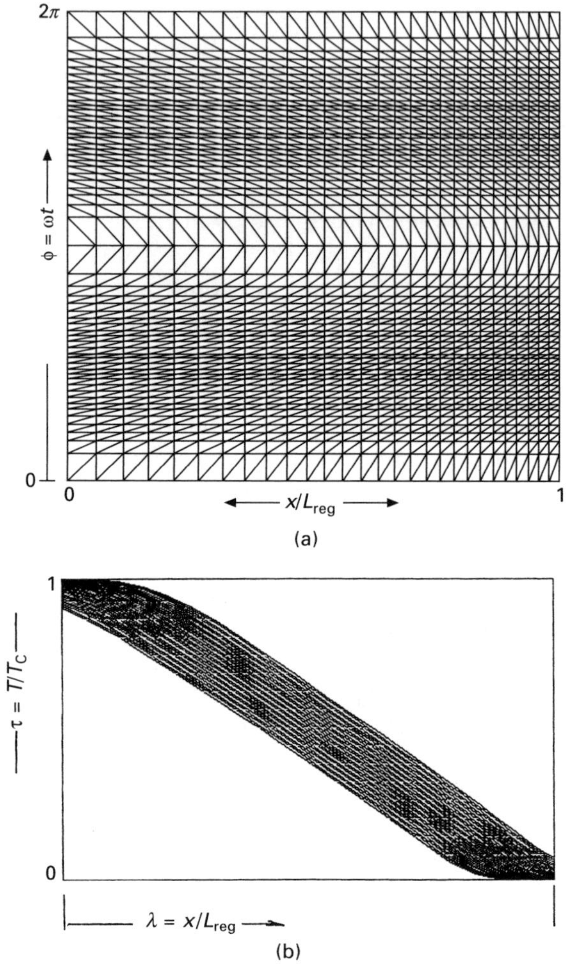

5.5 Specimen fluid particle trajectory map and temperature solution at cyclic equilibrium. Thermal capacity ratio N_{TCR} = 50, $¶_v$ = 0.8, N_{FL} = unity and γ = 1.4. (a) particle trajectories corresponding to simple-harmonic flow of ideal gas at flush ratio N_{FL} of unity (integration mesh superimposed); (b) envelope of common temperature T.

invariant, allowing a criterion to be extracted for N_{TCR} as a function of operating conditions. The criterion is in terms of allowable fractional temperature swing ε, estimated as rate of swing per radian $\partial T/\partial t$ multiplied by radians per blow. For symmetrical operation the latter is self-evidently π radians. Inversion to make N_{TCR} explicit calls for an expression for $\underline{u}$. The classic picture of regenerator operation has $\underline{u}$ uniform between switchings. In Fig. 5.3 absolute particle amplitude during a blow is $L_{reg}N_{FL}$. With t_{blow} (s) for time per symmetrical blow, the modulus of mean velocity $\underline{u}$ is $L_{reg} \times N_{FL}/t_{blow}$. Noting that $\omega t_{blow} = \pi$:

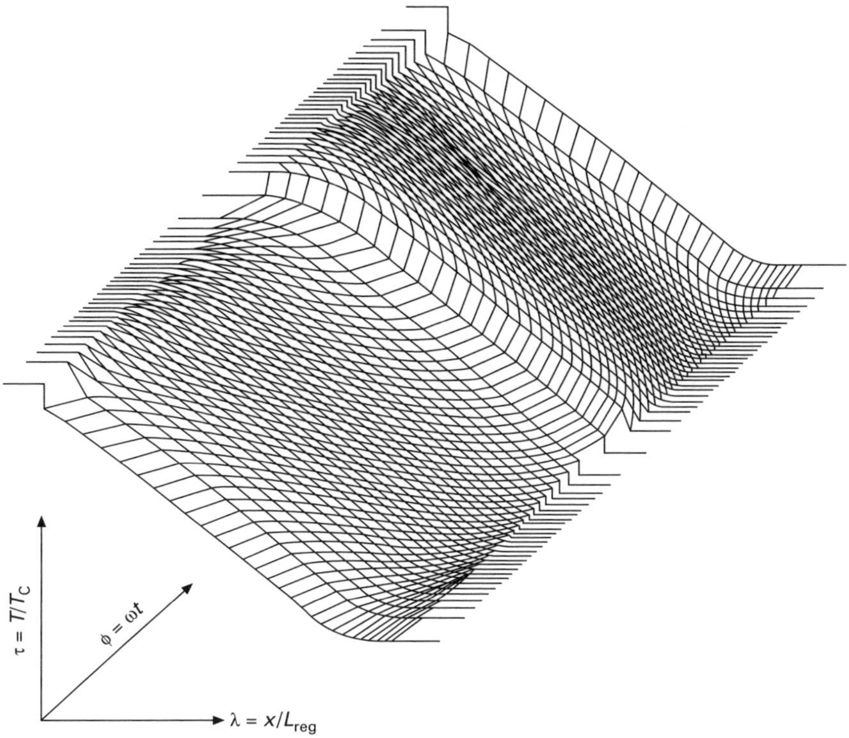

5.6 Relief of common temperature *T* corresponding to Fig. 5.5.

$$\underline{u} = L_{reg}\omega N_{FL}/\pi \tag{5.3}$$

Replacing $\partial T/\partial t$ of Eqn 5.1 by $\Delta T/\Delta t$, inverting to make N_{TCR} explicit and substituting Eqn 5.3:

$$N_{TCR}(1 - ¶_v)/¶_v \geq \frac{1 + (-1 + N_{FL}/\varepsilon)}{\frac{1}{2}(N_T + 1)} \gamma/(\gamma - 1) \tag{5.4}$$

(N_{TCR} is evaluated as the *modulus* of the right-hand side.)

Temperature swing ε depends on location, but a value at the mid-point of the matrix is representative. This has been expressed in Eqn 5.4 by using the nominal value of common temperature at location $\frac{1}{2}L_{reg}$, namely, $T/T_C = \frac{1}{2}(N_T + 1)$.

Equation 5.4 suggests that, for the values of N_T, N_{FL} and N_T which determine operation, the value of the compound parameter $N_{TCR}(1 - ¶_v)/¶_v$ required to hold regenerator fractional temperature swing below a specified value is that given by evaluating the right-hand side (absolute value).

Figure 5.7a assumes temperature ratio $N_T = 3.0$ and specific heat ratio $\gamma = 1.4$ as for diatomic gases, and plots minimum $N_{TCR}(1 - ¶_v)/¶_v$ from Eqn 5.4 as a function of mid-regenerator temperature swing ε. The parameter is flush ratio N_{FL}. The value of the term $\gamma/(\gamma - 1)$ for diatomic gases is 3.5, that for monatomic gases is 2.51, supporting the observation (Organ 1992a) that, all other things being equal, diatomic gases such as air impose the higher thermal loading on the regenerator. Figure 5.7b repeats for the monatomic case ($\gamma = 1.66$), and confirms that generally smaller values are required to limit temperature swing to a given percentage.

The numerical value of $N_{TCR}(1 - ¶_v)/¶_v$ for the Philips MP1002CA plotted into Fig. 5.7(a) suggests a temperature swing comparable to that for the USS P-40 – but it is not difficult to find instances of quite different predicted swings – cf the point for the GPU-3 in Fig. 5.7(b). The large swing which might otherwise be anticipated due to the uniquely low N_{TCR} of the USS P-40 is largely offset by the (uniquely low) flush ratio of that machine.

Inspection of Eqn 5.4 reveals that, for realistic temperature excursion ε, the dominant term is that involving N_{FL}. Neglecting the other terms, and supposing $N_T = 3.0$ and $¶_v = 0.75$ there emerges a tentative design rule for N_{TCR} in terms of allowable percentage ε:

$$N_{TCR} \geq CN_{FL}/\varepsilon \text{ percent} \qquad 5.5$$

In Eqn 5.5 the constant C takes the value 545 for diatomic gases, such as air, and 375 for monatomic gases. For example, a limit of 4 percent on temperature swing for the MP1002CA ($\gamma = 1.4$) suggests the need for $N_{TCR} \geq 708$. This compares with a value of 876 from the engine specification (Table 5.1).

5.9 Alternative formulation in corroboration

Equation 5.1 may be recast in terms of $\partial T/\partial x$ by substituting for the $\partial T/\partial t$ of the substantial derivative. The result is an expression having in the denominator the term in curly braces from Eqn 5.1, in which the group involving N_{TCR} dominates for all values of practical interest. Exploiting this feature to expand by the binomial theorem leads to the approximation:

$$\mathrm{D}T/\partial t \approx \partial T/\partial x\{1 + [(1 - N_{TCR}(T/T_C)$$
$$\times (1 - ¶_v)/¶_v\}(\gamma - 1)/\gamma]^{-1}\}L_{reg}\omega N_{FL}/\pi \qquad 5.6$$

For large N_{TCR} temperature profiles calculated by discretizing and solving Eqn 5.6 are essentially the same as those already obtained, so are not replotted. There is, however, reassuring additional insight: the right-hand side is proportional to temperature gradient $\partial T/\partial x$. Close to entry $\partial T/\partial x$ = zero (cf Fig. 5.5), and Eqn 5.6 ensures that this remains the case as the blow develops. In the mid-region and towards exit, $\partial T/\partial x$ is more or less uniform, and $\mathrm{D}T/\mathrm{d}t$ correspondingly little changed as the particle is tracked through the matrix. Large values of N_{TCR} in Eqn 5.6 (N.B. N_{TCR} appears in the denominator) give:

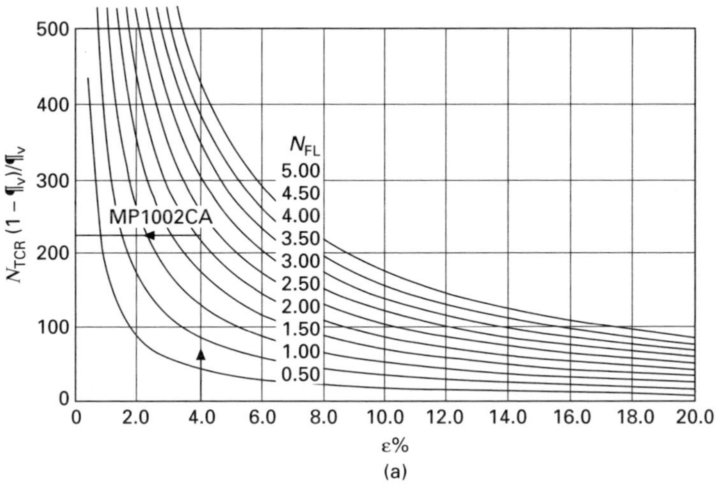

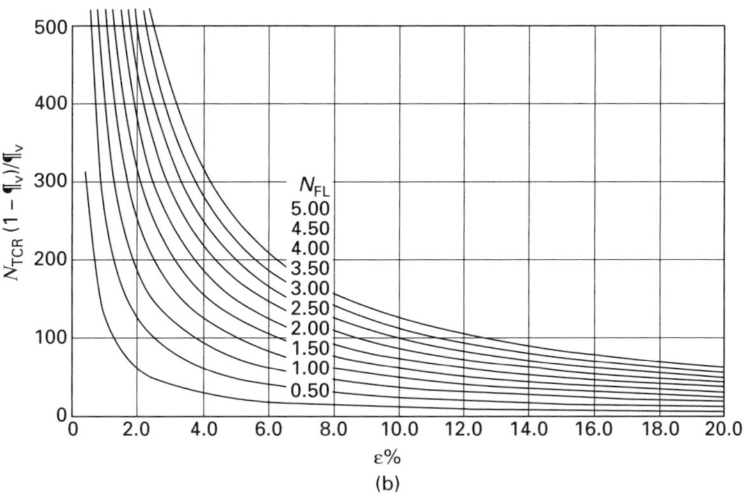

5.7 Solutions of Eqn 5.4 with flush ratio N_{FL} as parameter. The curves give values of the compound parameter $N_{TCR}(1 - ¶_v)/ ¶_v$ required to limit temperature excursion ε to a given percentage (a) specific heat ratio $\gamma = 1.4$ for diatomic fluid, e.g. air; (b) specific heat ratio $\gamma = 1.66$ for monatomic fluid, e.g. helium.

$$DT/\partial t = \partial T/\partial x(L_{reg}\omega N_{FL}/\pi) \qquad 5.6(a)$$

The right-hand side is equal by definition to $\underline{u}\partial T/\partial t$. The only way this identity can be consistent with the definition of the substantial derivative is for $\partial T/\partial t$ to be zero, i.e., for there to be no variation of local temperature with time. This precisely accords with expectation for infinite thermal capacity ratio.

The writer knows of no precedent for a formulation which so completely expresses the anticipated shape of the temperature solutions.

5.10 Conclusions

- Analysis of the thermal regenerator which pre-supposes unrestricted convective heat transfer focuses attention on the factors which determine matrix temperature swing – in particular, volume porosity $¶_v$ specific heat ratio γ and flush ratio N_{FL}.
- Temperature solutions have all essential features in common with earlier treatments which assumed limited heat transfer, except that solutions for fluid and matrix now coincide, consistent with the starting assumption of high *NTU*.
- For regenerators operating more or less in conformity with the assumptions of the analysis, a rule-of-thumb emerges to relate thermal capacity ratio N_{TCR} with flush ratio N_{FL} and allowable temperature swing ε percent.
- The differential equation defining the change of common temperature (fluid and matrix) along a particle path may be cast in a way which explicitly predicts the general shape of the temperature solutions.

6

The regenerative annulus and shuttle heat transfer

6.1 Introduction

The simple 'hot-air' engine is noteworthy not for its specific power or for its thermal efficiency, but for the fact that it has been with us, little changed since 1816 (Sier 1995), and will probably always be with us.

Power density when operating at atmospheric pressure is about $\frac{1}{500}$ th of that of the two-stroke internal combustion engine of comparable displacement (Organ 1995). The only scope for improvement, short of forfeiting the delightful mechanical simplicity – lies in (a) raising compression ratio or (b) pressurizing or (c) both.

Doubling charge pressure doubles mass rate m' and, all other things being equal, halves *number of transfer units* ($NTU = (hA_{ff}/m'c_p)(L/r_h)$). None of the options (*a*), (*b*) or (*c*) above will be effective without compensating for the shortfall. The best chance of doing this lies in an improved understanding of the unique heat exchange system – the 'regenerative annulus' between displacer and cylinder shell.

If theoretical study suggests scope for improved conversion of heat to work, then it may become worthwhile to address the lamentable thermal efficiency which accompanies open-flame heating. If there is a market for 25W (say) of quiet mechanical output, then 125–150W of heat input should be sufficient to achieve it. The associated thermal load of exhaust recuperation is within the scope of a Spirex exchanger, and natural convection might suffice to induce the combustion gases to flow. If so, then the hot-air engine could become part of twenty-first century technology with hardly an increase in mechanical complexity.

Chapter 7 will take this enquiry further by showing how *NTU* can be increased, even at elevated charge pressure, by superimposing a spinning component on displacer motion.

6.2 Background

The heat transfer phenomenon at issue has been called 'shuttle heat transfer'. It arises to some extent in almost all machines operating on the Stirling cycle, but is of particular significance in the 'hot-air' engine, since it is also the regenerative process.

An inevitable outcome of achieving the regenerator function in this fashion is that the gas path is between two opposed heat transfer surfaces. Both surfaces sustain severe temperature gradients in the flow direction, and are in reciprocating motion relative to each other. The relative motion leads to cyclic variation of wall temperature difference at all axial locations, with a positive and a negative peak at each cycle. This makes reversible heat exchange unachievable, and it can be predicted without formal analysis that peak attainable thermal recovery ratio will lie below that of the corresponding stationary regenerator.

In an early treatment, Zimmerman and Longsworth (1970) neglect fluid motion and radiation (terms in T^4 are absent) and assume invariant, linear distributions of temperature in both moving and stationary solid elements. A closed-form, algebraic expression results for heat shuttled per cycle, q_s. In view of the assumptions, the result is likely to be more relevant to the clearance gap around the displacer of high-power-density Stirling engines (the 'appendix gap') than to the regenerative annulus.

Harness and Newmann (1972) undertake to extend the Zimmerman and Longsworth approach. The treatment starts from an energy balance in differential form for an element of gas instantaneously in contact with both cylinder and displacer. The formulation is two-dimensional to the extent of taking account of thermal conduction in the axial direction in the cylinder ('though not, for some reason, in the displacer). Radiation is not considered. Evaluation of their achievement is difficult: not only does the paper lack a list of notation, but details of the integration scheme are withheld, and there are no temperature solutions.

In an approach which is more ambitious analytically, Rios (1971) specifically addresses the appendix gap problem (flow into and out of the gap, but not through it). The development accounts for fluid motion, although with the assumption that 'the gas may be considered to have an average temperature equal to the mean of the displacer and cylinder surface temperatures on any plane perpendicular to the axis.)' Radiation is not considered. For his specimen results the effect of gas motion is ignored altogether. It is difficult to be happy about a heavily mathematical development where the opportunity is not taken of displaying temperature solutions as a check on plausibility.

Urieli and Berchowitz (1984) report the most comprehensive formulation of the problem to this point. Analytical solution is claimed for fluid temperature difference $T - T_s$ as a function of time over a cycle, where T_s is 'local, mean temperature'. By hypothesis the axial distribution of T_s (their Assumption 5) and the radial distribution of $T - T_s$ (their Assumption 6) are both linear. Temperature solutions are not displayed, but the assumption of linearity makes the treatment more appropriate to the appendix gap than to the regenerative annulus.

Without appraising the precedents, Martini (1978) distils the final results of Zimmerman and Longsworth and of Rios into empirical, closed-form algebraic equations for shuttle heat transfer per cycle.

The temperature solutions for the stationary regenerator (e.g. Organ (1997)) do not apply without modification.

6.3 Reformulation

Figure 6.1 represents the cylinder and displacer. To provide an initial condition, temperature varies linearly between T_E and T_C in both members, although the formulation does not rely on linearity. The axial distances over which the variations take place are not generally the same, so even if variation is linear, *gradients* will differ, as indicated.

To set the starting condition it is assumed that the temperature profiles superimpose when the displacer is at mid-stroke (Fig. 6.1(a)). When this member moves to inner dead-centre (left-most position – Fig. 6.1(b)) the temperature profile carried with it will lie *below* that of the cylinder, suggesting a flow of heat from the latter to the former. When it moves to the opposite dead-centre position (Fig. 6.1(c)) the temperature difference is reversed, suggesting reversal of the heat flow direction – i.e., from displacer to cylinder. Cycling the displacer back and forth gives rise to the 'shuttle' process. This adds to the axial flow of heat driven by thermal conduction, and both losses compound the performance penalty of reduced recovery ratio anticipated in the previous section.

Acquiring the fluid temperature history in the regenerative annulus calls for formulation from first principles.

6.4 Assumptions

For the initial stages of the development, the simplifying assumptions of standard regenerator theory are adopted (see, e.g. Jakob (1957)). Once the formulation is in place many of these provisional assumptions may be relaxed. For the moment, pressure, density and specific heats (of fluid and matrix) are uniform and constant, and mass rate is uniform and constant between switching. When flow enters the annulus from the hot end it does so at uniform temperature T_E. When it enters from the cold end it does so at T_C. Matrix thermal diffusivity is infinite in the direction perpendicular to flow, but finite parallel to that direction. Thermal diffusivity in the fluid is finite perpendicular to the flow direction but zero parallel to it.

Further assumptions are made, largely to keep presentation manageable:

- consideration of the effects of thermal radiation is postponed
- coefficient of convective heat transfer, h, is the same for displacer as for cylinder shell
- flow is one-dimensional (i.e. is 'slab' flow)
- there is circumferential symmetry: fluid flow, heat flow, together with fluid and solid surface temperatures at a given circumferential point in the annulus characterize those at all circumferential points which lie at a common distance from the point of fluid entry/exit
- (initial condition) the temperature gradient $\partial T_{wc}/\partial x$ in the cylinder shell (Fig. 6.1(a)) is initially $(T_C - T_E)/(L_r + S_d)$, where S_d is the stroke of the displacer. That in the displacer shell $\partial T_{wd}/\partial x$ is initially $(T_C - T_E)/L_r$.

6.5 Analysis

It can be helpful for the formulation of a new problem to have as much as possible in common with established treatments. With this in mind, instantaneous velocity u_d of the displacer is subtracted from that of the fluid, $\underline{u}$, and from the (zero) velocity u_c of the cylinder. In the new coordinate system the displacer is stationary. The local, instantaneous velocity $\underline{u} - u_d$ is that which would apply in the traditional regenerator model where the regenerator is stationary, and where fluid motion is generated by the movement of pistons left and right. In the transformed system all points on the enclosing cylinder have the common velocity (minus) u_d.

The goal of the analysis is an energy balance in *Lagrange* coordinates, i.e., appropriate to fluid particle of fixed identity. The earlier, purely algebraic formulation (Organ 1997) applies. On the other hand, an intuitive approach has since been shown to yield identical algebra (Finkelstein and Organ 2001).

This involves firstly considering what happens to the fluid particle when heat transfer coefficient, h, is zero. Figure 6.2(a) indicates an element of fluid of axial thickness dx in its relationship to the (now stationary) temperature profile of the displacer. A particle entering from the left with initial temperature T_E proceeds through the gap with T_E unaltered. Temperature *difference*, ΔT_c, between fluid element and displacer shell at the flow front is $T_g - T_{wd}$.

With $\underline{u}$ and u_d both reckoned positive from left to right, particle *relative velocity* is $\underline{u} - u_d$. Coefficient of heat transfer h is still zero for the moment, so that temperature difference, ΔT_d, between fluid element and displacer shell, $T_g - T_{wd}$, increases at rate $-(\underline{u} - u_d)\partial T_d/\partial x$, i.e., at $-(\underline{u} - u_d)(T_C - T_E)/L_r$, where $\underline{u}$ is the velocity of the fluid element.

$$D\Delta T_d/dt = -(\underline{u} - u_d)\partial T_{wd}/\partial x \qquad 6.1$$

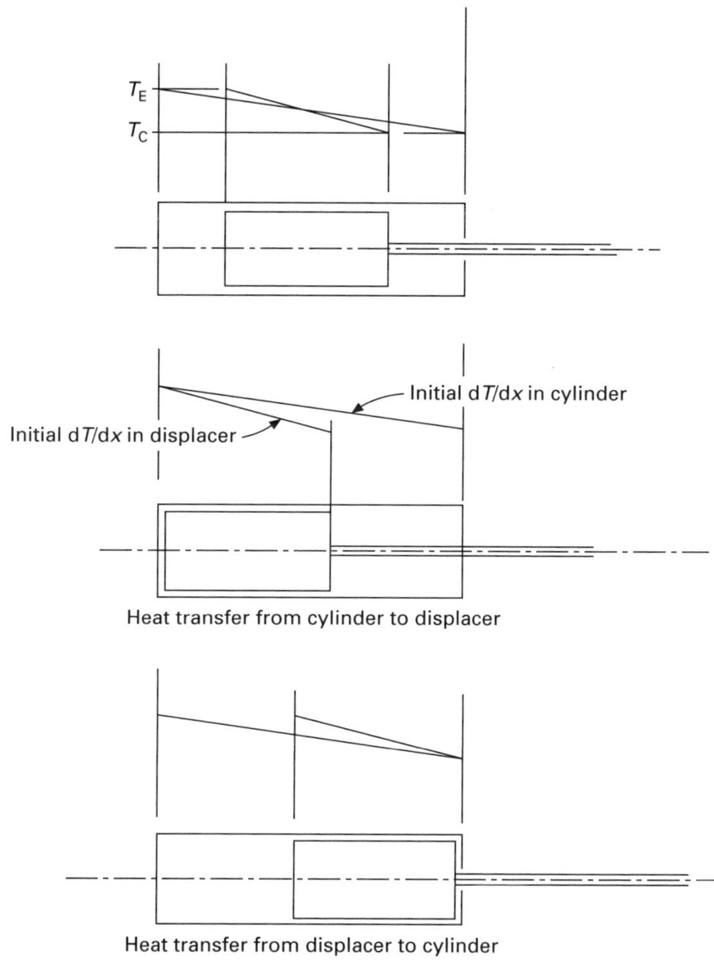

6.1 Shuttle phenomenon in terms of initial conditions – temperature distributions linear with x in both cylinder and displacer. At eventual cyclic equilibrium neither distribution is necessarily linear.

Use of the substantial derivative $D/dt \equiv \partial/\partial t + (\underline{u} - u_d)\partial/\partial x$ expresses the fact that fluid element under examination is in (relative) motion at velocity $\underline{u} - u_d$.

Heat transfer coefficient h is now allowed to take a realistic (finite) value. The rate of increase in ΔT_c now reduces:

- because the fluid is adjusting downwards towards the displacer temperature profile and
- because at the same time, the temperature distribution of the displacer is adjusting.

The rate of adaptation of T_g is proportional to the net rate of heat exchange with both displacer and cylinder. With reference to Fig. 6.2(b):

$$\rho\pi D \mathrm{d}r \mathrm{d}x c_p \mathrm{D}T_g/\mathrm{d}t = h\pi D\mathrm{d}x\Delta T_c + h\pi D\mathrm{d}x\Delta T_d$$

Extracting $\mathrm{D}T_g/\mathrm{d}t$ and noting that hydraulic radius r_h is equal to half of the radial gap dr:

$$\frac{\mathrm{D}T_g}{\mathrm{d}t} = \frac{h}{\rho|\underline{u}|c_p}\frac{|\underline{u}|}{2r_h}\{\Delta T_c + \Delta T_d\}$$

$$= \tfrac{1}{2} St \frac{|\underline{u}|}{r_h}\{\Delta T_c + \Delta T_d\} \qquad 6.2$$

Equation 6.2 reflects the slab-flow view of convective heat transfer. When formulation is complete it will be seen that a more sophisticated description can be substituted.

In the case of the cylinder shell, the ring of cylinder shell material wetted by the fluid element adjusts upwards in temperature according to the heat balance:

$$\pi D t_c \mathrm{d}x \rho_c c_c \, \mathrm{d}T_{wc}/\mathrm{d}t = h\pi D\mathrm{d}x\Delta T_c$$

Extracting $\mathrm{d}T_{wc}/\mathrm{d}t$ and introducing the abbreviation $\rho_c c_c/\rho c_p = N_{TCRc}$

$$\frac{\mathrm{D}T_{wc}}{\mathrm{d}t} = \tfrac{1}{2} St \frac{|\underline{u}|}{r_h}\frac{2r_h/t_c}{N_{TCRc}}\Delta T_c \qquad 6.3$$

As in the case of the fluid element, a more sophisticated balance may eventually substitute Eqn 6.3. In this way, unsteady radial conduction and/or longitudinal conduction can be considered.

One of the earlier simplifying assumptions allowed local, instantaneous heat transfer coefficient h to be common to both components of heat exchange – that between fluid and displacer and that between fluid and cylinder. Eventual calculation will be in terms not of h *per se*, but of corresponding dimensionless Stanton number, $St = h/(\rho|\underline{u}|c_p)$. A common, local value of St calls for use

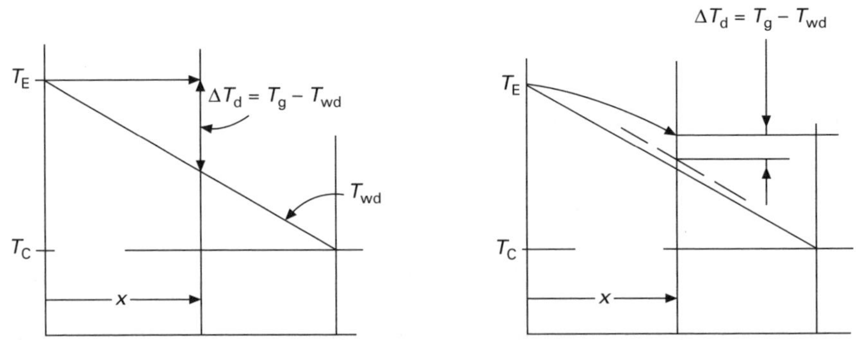

6.2 Stages in picture of regenerator operation leading to Eqns 6.1 and 6.2 (a) heat transfer zero; (b) finite heat transfer.

of a representative value of $\underline{u}$. Given the small free-flow area of the annulus relative to displacer cross-section it may be expected that, except for u close to zero, $|\underline{u}| >> |\underline{u}_d|$ and will be used to evaluate St for both elements of the heat exchange. The heat balance for the annulus of displacer material in contact with the fluid element may then be written down by analogy with Eqn 6.3

$$\frac{dT_{wd}}{dt} = \tfrac{1}{2} St \frac{|u|}{r_h} \frac{2r_h/t_d}{N_{TCRd}} \Delta T_d \qquad 6.4$$

The earlier (adiabatic) rate of change of ΔT_d (Eqn 6.1a) is reduced by the local rates of change in T_g *and* in the T_w due to exchange of heat.

Net rate of change of change of ΔT_d is obtained by addition:

$$D\Delta T_d/dt = -(\underline{u} - u_d)\partial T_{wd}/\partial x - \tfrac{1}{2} St(|\underline{u}|/r_h)\{\Delta T_c + \Delta T_d\}$$
$$-\tfrac{1}{2} St(|\underline{u}|/r_h)\Delta T_d\, 2r_h/t_d N_{TCRd} \qquad 6.5$$

The same argument applies to the development of the temperature difference between the fluid element and the cylinder shell, leading to:

$$D\Delta T_c/dt = -\underline{u}\partial T_{wc}/\partial x - \tfrac{1}{2} St(|\underline{u}|/r_h)\{\Delta T_c + \Delta T_d\}$$
$$-\tfrac{1}{2} St(|\underline{u}|/r_h)\Delta T_c\, 2r_h/t_c N_{TCRc} \qquad 6.6$$

But for the terms $t_d N_{TCRd}$ and $t_c N_{TCRc}$ Eqns 6.5 and 6.6 are identical in the coefficients of the sum of the temperature differences $\Delta T_c + \Delta T_d$. The prospect of a temperature solution makes it attractive to arrange for numerical values to be the same, namely, $t_d N_{TCRd} = t_c N_{TCRc}$, in which case it becomes possible to add the equations:

$$D(\Delta T_c + \Delta T_d)/dt = -\underline{u}\{\partial T_{wc}/\partial x + (1 - u_d/\underline{u})\partial T_{wd}/\partial x\}$$
$$- St|\underline{u}|/r_h\{\Delta T_c + \Delta T_d\}\{1 + r_h/t_d N_{TCRd}\} \qquad 6.7$$

This is a first-order total differential equation for the sum $\Delta T_c + \Delta T_d$:

$$D(\Delta T_c + \Delta T_d)/dt + P(\Delta T_c + \Delta T_d) = Q \qquad 6.8$$

In Eqn 6.8:

$$P = -St|\underline{u}|/r_h\{1 + r_h/t_d N_{TCRd}\}$$
$$Q = -\underline{u}\{\partial T_{wc}/\partial x + (1 - u_d/\underline{u})\partial T_{wd}/\partial x\}$$

Thus the equation has precisely the *form* (although not the detail) of that for the stationary regenerator (Organ 1997). It may be solved for the sum of the unknowns, $\Delta T_c + \Delta T_d$, over a time interval Δt in terms of the value start of the interval:

$$(\Delta T_c + \Delta T_d)_{t+dt} = Q/P + \{(\Delta T_c + \Delta T_d)_t - Q/P\}e^{-Pdt} \qquad 6.9$$

The result required is the variation of T_g for all x and t. This follows from Eqn 6.9 without further manipulation because $\Delta T_c + \Delta T_d = T_g - T_{wd} + T_g - T_{wc} = 2T_g - T_{wd} - T_{wc}$, and because both T_{wd} and T_{wc} are known on specifying the location of the particle at T_c.

If in Eqn 6.9 $|\underline{u}|dt$ is x considered a fraction $|\underline{u}|dt/L_r = x/L_r$ of matrix length, L_r, the exponent becomes $-NTU(x/L_r)$.

There is a choice between two solution strategies for Eqn 6.9. One is general, and applicable whether or not shell temperatures are treated as variable, and involves evaluating Eqn 6.9 on the mesh points of a combined Euler–Lagrange integration grid. The requirement stems from the fact that the fluid particle is in motion (Lagrange grid); the changes in metal temperatures occur at points fixed in relation to rigid surfaces (Euler grid), and the solution points must coincide. The solution technique has been comprehensively developed in the context of the conventional regenerator problem.

Where displacer and cylinder temperature distributions are assumed fixed (e.g. linear), the solution process follows selected fluid particles, solving Eqn 6.9 for $\Delta T_c + \Delta T_d$ at locations where T_{wd} and T_{wc} are known as solutions earlier in the sequence.

Figure 6.3a shows the fluid particle trajectory map superimposed on the cylinder temperature profile as 'seen' by the gas content of the regenerative annulus. Unsurprisingly, there is wholesale up and down motion of the profile once per cycle.

The corresponding temperature history for the gas (Fig. 6.3(b)) shows similarities with solutions for the conventional (stationary) regenerator. This is because the present solution process does not distinguish the independent exchanges of heat with the two different temperatures of displacer and cylinder. The temperature disturbances generated at flow reversal propagate with the incoming gas as for earlier regenerator solutions.

6.6 Cyclic shuttle loss

Over any given integration interval the displacer shell experiences a net gain of heat dq_s evaluated by summing local values of $h\pi D dx(T_g - T_{wd})dt$ over the axial length L_r of the shell. Multiplying dq_s by the algebraic increment $u_d dt$ in displacer motion yields a quantity with units of Joule-metre (Jm). Summing over one forward and one reverse blow and dividing by twice displacer stroke S_d yields a value for heat shifted per cycle by the shuttle mechanism over net distance S_d.

It has been deduced (e.g. Finkelstein and Organ 2001) that values of *NTU* (number of transfer units) for stationary regenerators consisting of woven wire screens can be as high as 120. Values arising in the context of the regenerative annulus appear not to be documented. On the other hand, hydraulic radius r_h is readily shown to exceed that of the typical wire screen by a factor

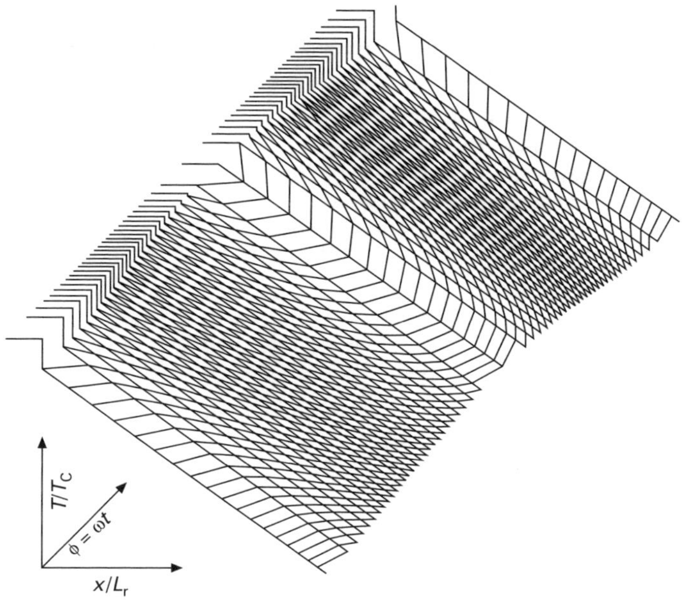

(a)

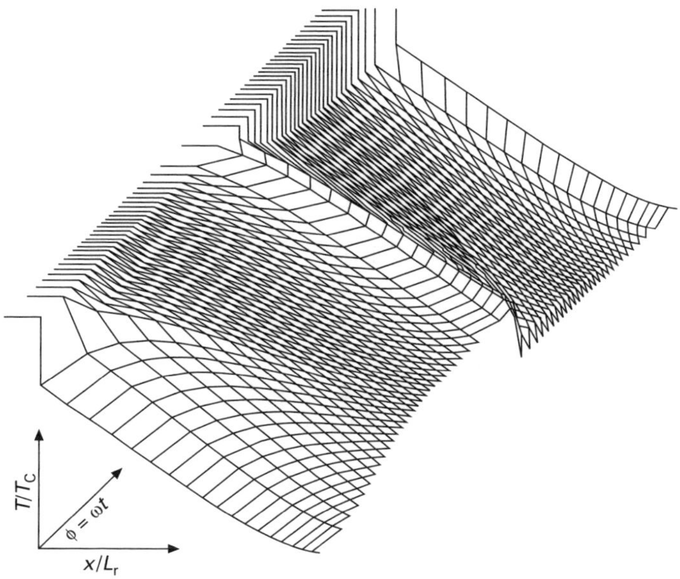

(b)

6.3 Specimen solutions of Eqn 6.9 plotted for parameter values of Table 6.1 (a) fluid particle trajectory map superimposed on cylinder temperature profile as 'seen' by the gas. Not surprisingly there is wholesale up and down motion of the profile once per cycle; (b) temperature solution for the gas, with superimposed particle trajectory plot. At the low *NTU* achieved the 'flush' phenomenon is in evidence.

Table 6.1 Data used to generate temperature solutions

N_T	temperature ratio T_E/T_C	3.0
N_{FL}	flush ratio	2.0
N_{TCR}	thermal capacity ratio	(variable – see text)
NTU	number of transfer units	(variable – see text)
γ	specific heat ratio	1.4
t_c/r_h	radial thickness ratio, cylinder	1.0
t_d/r_h	radial thickness ratio, cylinder	1.0

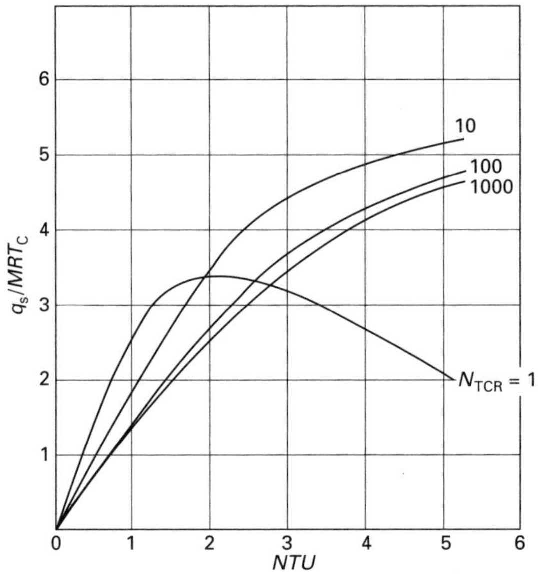

6.4 Specific shuttle loss as function of *NTU*. Parameter is thermal capacity ratio N_{TCR}.

of about 20. *NTU* being defined as StL_r/r_h, a first guess at peak *NTU* might be 120/20 = 6. On this basis Fig. 6.4 plots dimensionless shuttle loss as a function of *NTU*. The parameter is thermal capacity ratio, N_{TCR}.

Reassuringly, shuttle loss is zero for *NTU* zero. The figure suggests that increasing N_{TCR} brings about a modest decrease in q_s, although there appears to be little benefit in exceeding $N_{TCR} > 1000$. Except for the lowest value of N_{TCR}, at which the temperature swing of the solid material clearly influences temperature recovery, increasing *NTU* appears to lead to continuously increasing specific shuttle loss. In practice there is an upper limit to q_s, since it cannot possibly exceed cycle heat input.

In the design of a conventional regenerator the aim is to achieve the highest possible *NTU* consistent with acceptable pumping loss, since this promises the highest recovery ratio. Design of the optimum regenerative annulus evidently calls for a more subtle balance.

7

The rotating-displacer air angine

7.1 Résumé

The designation 'hot-air' engine describes a Stirling-type engine in which the regenerative effect is achieved by heat exchange in the gap between displacer and cylinder. The type does not make use of extended heat transfer surfaces. Except in rare cases where crank mechanism is enclosed, mean cycle pressure is atmospheric.

There is probably a greater number of hot-air engines in existence than true Stirling engines. Those of recent manufacture are generally toys, but some are relics from the late nineteenth and early twentieth centuries, originally marketed for performing light duties such as driving a ventilation fan.

Although the type is mechanically simple, the flows of heat and fluid in the crucial gas path element – the 'regenerative annulus' – is strongly two-dimensional. This factor removes the possibility of making use of the 'slab flow' simplification, and ironically leads to greater analytical difficulties than those posed by the true Stirling engine. (A look ahead to the chapter on the pulse-tube or 'thermal lag' engine will reinforce the view that the simpler the mechanical arrangement the more inscrutable are the gas processes by which heat is converted to work.)

Unswept volume in the hot-air engine tends to be small – especially in the coaxial or 'beta' configuration, suggesting:

- compression ratio high by Stirling standards. All other things being equal, large pressure swing implies a large adiabatic component of temperature excursion and, in turn, large swings in temperature difference between gas and enclosure. Such enhancement may be expected to afford partial compensation for the meagre surface area of the type. Indeed, it may provide part of the explanation as to why this relatively unpromising variant functions at all.
- fluid particles from one variable-volume space traverses the entire length of the regenerative annulus to enter the variable-volume space at the

opposite end. This contrasts with conditions in the high-pressure Stirling engine, where much of the working fluid never reaches – let alone traverses – the regenerator.

Figure 7.1 is a combined particle trajectory map and ideal 'isothermal' pressure trace for a representative hot-air engine of coaxial configuration, and illustrates both of the above points.

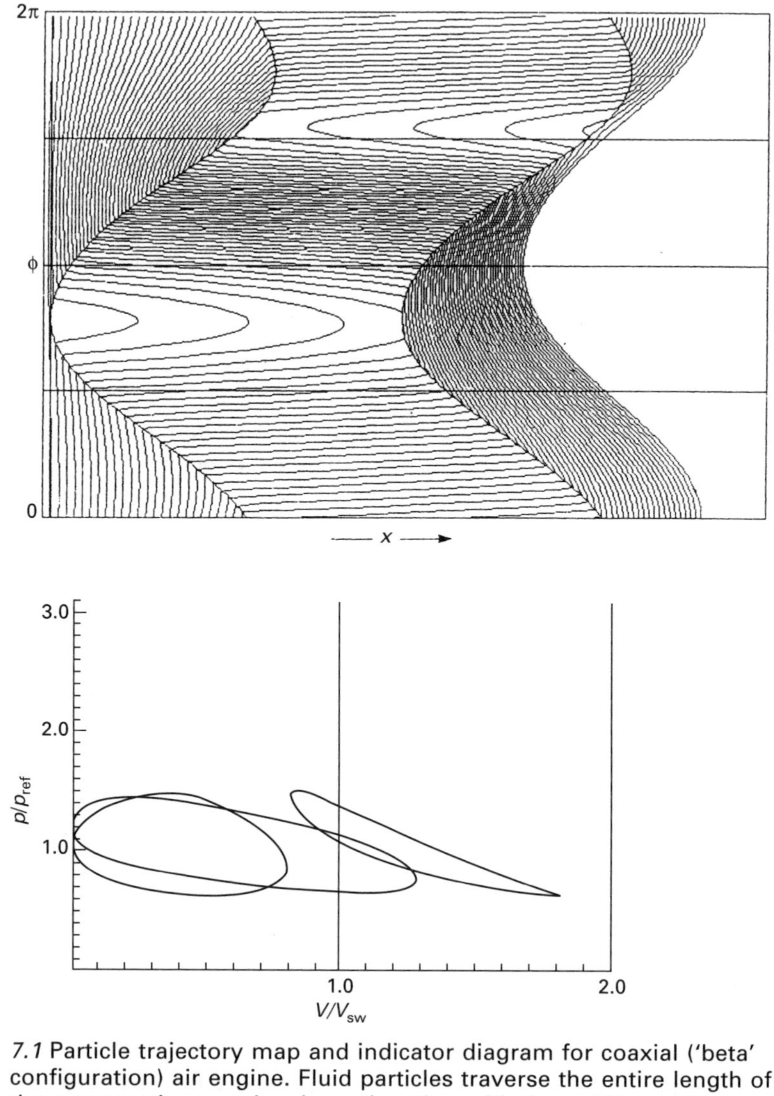

7.1 Particle trajectory map and indicator diagram for coaxial ('beta' configuration) air engine. Fluid particles traverse the entire length of the regenerative annulus. In conjunction with the *p–V* loops the map confirms the potential of the type for high compression ratio.

A difficulty in making sense of hot-air engine performance is lack of data relating measured output to internal specification. For 'real' Stirling engines, gas path parameters, suitably grouped into dimensionless parameters have been shown to correlate well with operating conditions similarly grouped. Comparable correlations have yet to emerge for the air engine. Fortunately, the *NTU* concept has certain inevitability wherever there is convective heat exchange having a steady-flow component. This is the justification for tentatively plotting data from two hot-air engines of proprietary origin into correlations for 'proper' Stirling engines, as at Fig. 7.2. The Kyko drove a 16-inch diameter fan at something like 250 rpm, which is taken as its

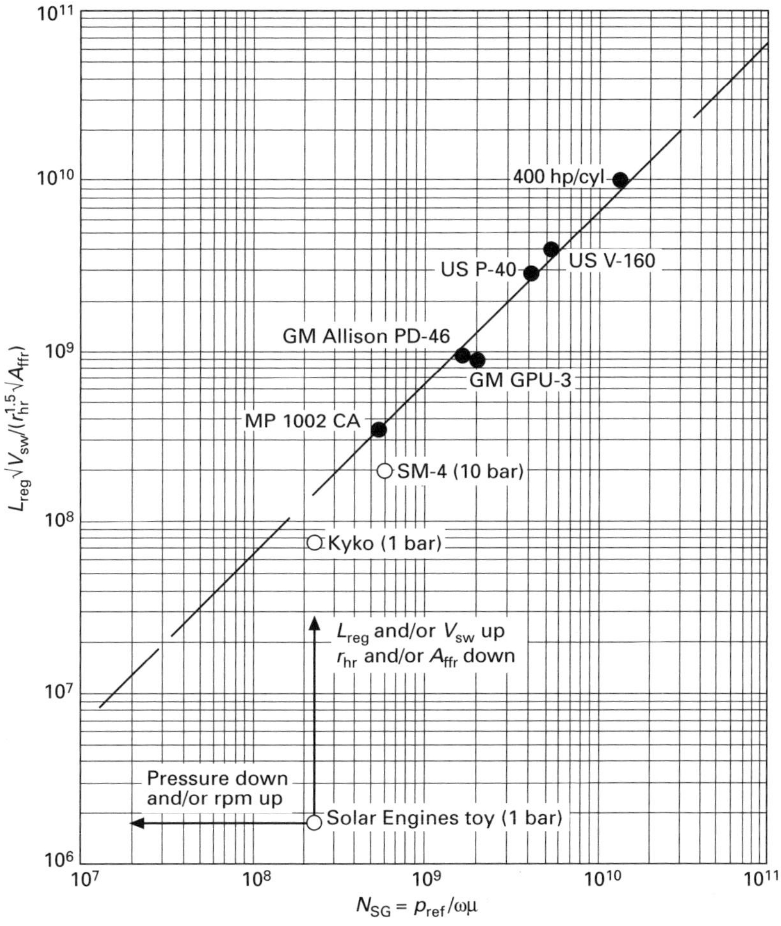

7.2 Performance points for Kyko and toy air engine plotted into correlation for pressurized Stirling engines. Increasing pressure without compensating flow passage geometry pushes the performance point to the right and to a region of higher N_{SG} and increasingly deficient *NTU*.

'performance point'. Originally manufactured in quantity by the Model Engineering Company of London (1974) it is something of a benchmark, and is illustrated to scale in the two orthographic views of Fig. 7.3. The archetypal toy counterpart is currently marketed by the Cambell Tools Co. (2005). Having been in production in one proprietary form or another since an incarnation with Davies Charlton in the 1950s it, too, is a point of reference for the genre. The version marketed by Solar Engines in 1977 is shown to scale in Fig. 7.4. Unloaded it runs at 600–800 rev/min, which reduces to 200 rev/min under a token load at the flywheel rim. A 'rated' performance point is evidently somewhat arbitrary.

Correlations of this type separate geometric parameters (y-axis) from operating parameter $N_{SG} = p_{ref}/\omega\mu_{ref}$ (x-axis). Thus it is seen that any increase in nominal charge pressure p_{ref} promises to push the operating point further to the right and away from the correlation. Moreover, since between functionally-similar gas paths NTU in the regenerator are proportional to $1/\sqrt{N_{SG}}$, migration of the operating point to the right is consistent with a reduction in NTU.

The only way to re-establish the operating point on the correlation by conventional means is to change regenerator passage geometry so as to increase the value of the combined geometric parameter plotted on the y-axis. An increase in flow passage length L or a reduction in hydraulic radius r_h (or both) would, in principle, achieve this. But as in the case of most medicines, there are side effects: all other things being equal, pressure drop is proportional to L/r_h, and an increase in pumping loss is threatened. Even if this were not a problem, scope for a ten-fold change in L/r_h necessary to compensate the drop NTU following a ten-fold increase in p_{ref} is just not there, and other means must be sought.

7.2 An alternative

Adding a rotational component to the reciprocation motion of the displacer suggests the following effects:

- the rotary component causes the gas particles to take a helical path between expansion and compression spaces, increasing effective flow path length without change to displacer or cylinder surface geometry
- some control is gained over effective Reynolds number R_e, independently of crankshaft speed
- it is an established phenomenon that rotation of the inner member of an annulus relative to the stationary outer has a destabilizing effect on the entrained flow, leading to vortices which circulate between inner and outer surfaces. Onset of instability is determined in terms of a 'Taylor parameter' N_{Ta} (see Section 7.3). The instability point is not a counterpart of the laminar/turbulent transition of pipe flow, although transition to turbulence does occur at elevated values of N_{Ta}.

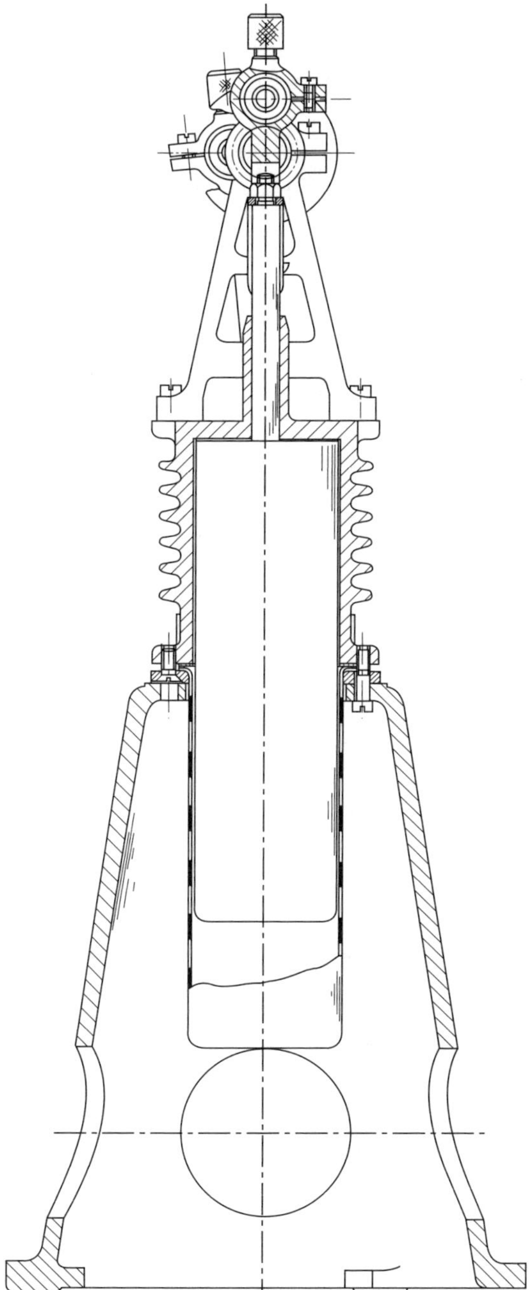

7.3 Section through 'Kyko' engine. The drawing shows the iron casting which at some stage has replaced the original strip- and sheet-metal stand. Fan and guard are not shown.

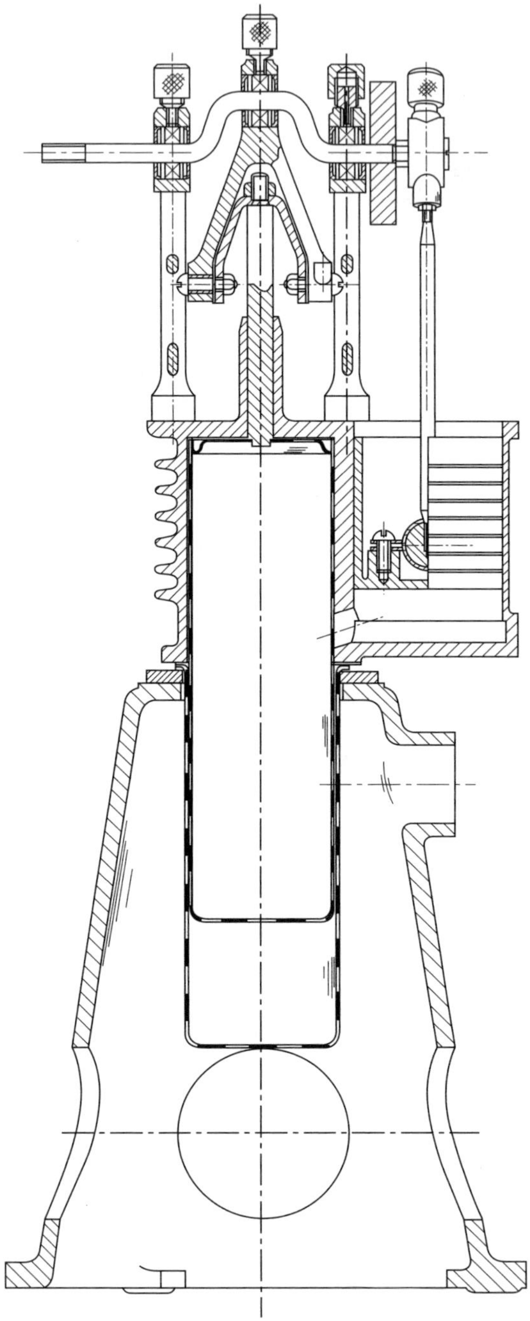

7.3 Continued

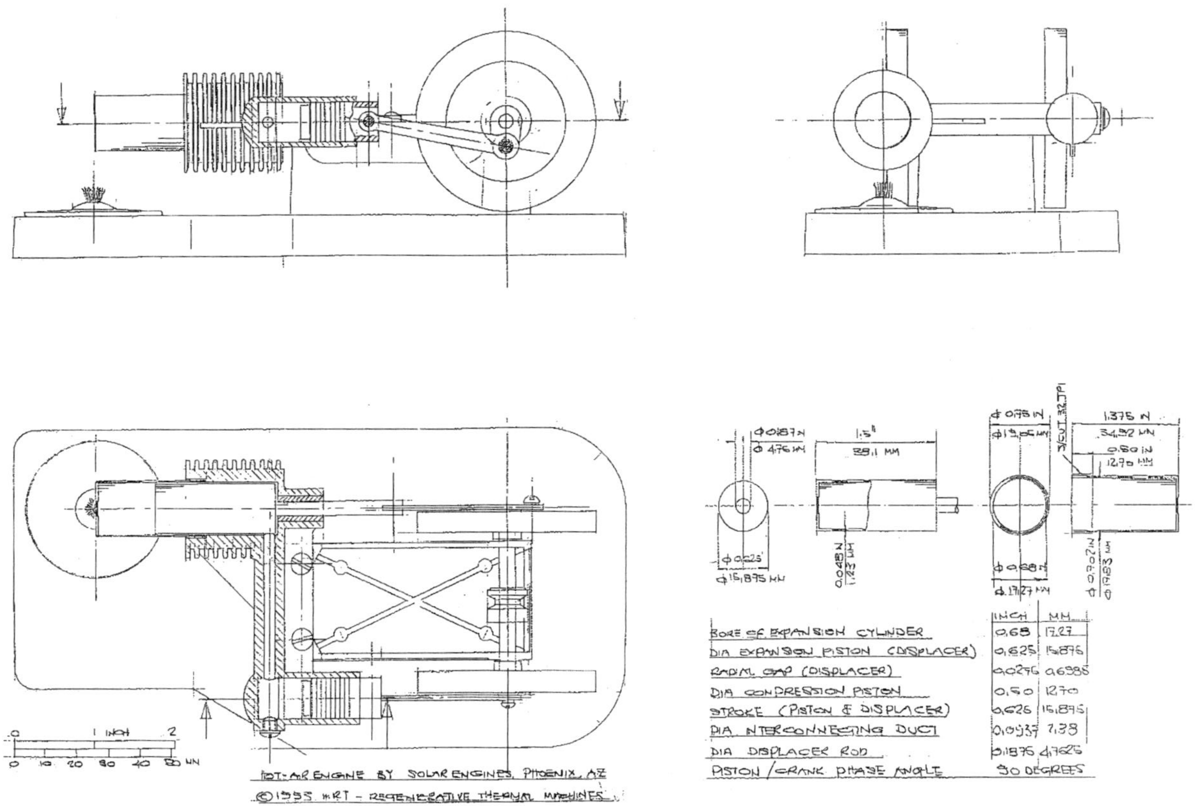

7.4 Toy hot-air engine in production in the 1950s and still commercially available. This scale drawing was made by dismantling and measuring a specimen manufactured in 1977 by Solar Engines Inc.

7.3 Taylor parameter

Flow in the annulus between rotating concentric cylinders has been much studied. Taylor points out that, in this context, measurement of velocity distribution alone suffices to allow experimental discrimination between momentum transport and vorticity transport in turbulent flow, whereas when flow is between parallel planes or through ducts, simultaneous measurement of temperature and velocity distribution is called for. Two substantial papers to the Royal Society (Taylor 1935, 1936) cover only part of his wide-ranging experimental and theoretical investigation. White's 1974 account of the topic he calls 'centrifugal instability' is characteristically authoritative and readable.

Taylor measured the torque required to hold the outer cylinder stationary while the inner was rotated at ω_i rad/s; likewise that required to hold the inner cylinder stationary while the outer was rotated at ω_o. For a given value of radial gap c torque normalized by the product ρ(rev/sec)^2 and plotted against ρ(rev/sec)/μ is evidently the counterpart of the friction factor $C_f - R_e$ correlation for pipe flow. Figure 7.5 is an example for a radial gap of 1.375 percent of nominal cylinder diameter. The upper curve is for the inner cylinder rotation, the lower is for outer. The destabilizing effect of the former case is clearly in evidence, as is the reverse (stabilizing) effect of the latter.

Taylor developed an analytical picture of particle velocity in the annulus. If the annular gap c is small compared to the diameter d_i of the inner cylinder, stability is dependent on a single parameter, the Taylor number N_{Ta}:

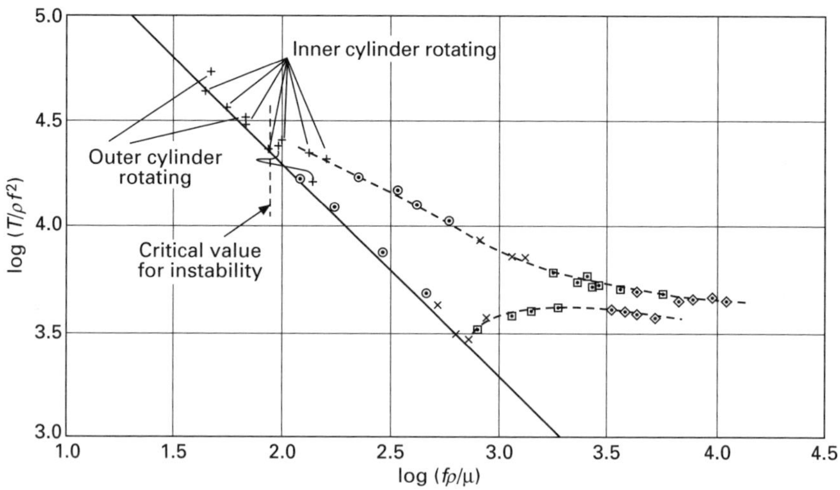

7.5 Specific friction torque ($T/\rho f^2$) vs flow parameter $\rho f/\mu$ (by analogy with C_f vs *Re*). Radial gap c is 1.375 percent of nominal diameter. After Taylor (1936) with permission of the Royal Society of London.

$$N_{Ta} = \tfrac{1}{2}d_i c^3 \rho^2 (\omega_i^2 - \omega_o^2)/\mu^2 = c/d_i Re_{rot}^2 \quad 7.1$$

The value of N_{Ta} at which flow becomes unstable is $N_{Ta_{crit}} = 1708$ – for which a working value is 1700.

According to White, the phenomenon had attracted the interest of Rayleigh in 1916, and had been explained in physical terms by von Kármán in 1934: if the square of the circulation of a flow field increases outwards, a radially-displaced particle encounters insufficient pressure gradient to exert a restoring effect. Figure 7.6 is White's elegant depiction of the Taylor vortices.

Taylor's picture allows no flow component in the axial direction. A treatment of the inviscid case by Ludwig (1964) takes it into account, but there is no pictorial counterpart of Fig. 7.6. But it is not too difficult to imagine adding an axial velocity to the inner cylinder, a conceivable result of which is particle paths in the 'coiled coil' form of the electric bulb filament.

In the context of the air engine, the vortex motion may be expected to enhance the heat transferred by convection between displacer and cylinder. At any given instant and location the temperatures of the two members are different. No first-principles guidance is available as to whether enhancing heat transfer coefficient in these circumstances offers a net benefit. But there is always experiment …

7.4 A rotating-displacer air engine

Research contract GR/F25204 granted to the author in July 1900 by the UK Science and Engineering Research Council (SERC – since renamed) allowed construction of the engine shown in Fig. 7.7 together with its combustion chamber. This was a hot-air engine in all essential features, but with provision for pressurization via the crank case to 10 bar gauge. Axial actuation of

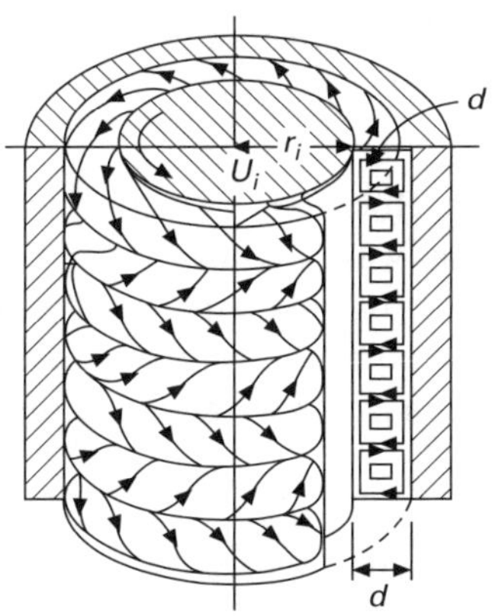

7.6 Depiction by White (1974) of Taylor vortices in Couette flow between rotating cylinders. With permission of McGraw Hill Book Co.

7.7 Engine purpose-built to examine the effect of displacer rotation at charge pressures up to 10 bar. Design by D. H. Rix.

piston and displacer was by rhombic-drive. Rotation of the displacer to 6000 rev/min was by electric motor. Power output was measured by a Plint electric dynamometer of 2 kW rating. The engine could be motored by the dynamometer when indicated work was insufficient to overcome mechanical friction, allowing exploration of the effect of displacer rotation over a range of charge pressure and crankshaft speed.

The design was highly conservative, as may be inferred from the diameter of the cylinder in relation to the proportions of the crank-case. The proportions might be associated with working pressures up to ten times the modest 10 bar for which the crucial hot components were stressed. Without the provision for heat transfer enhancement (displacer rotation), mechanical and thermodynamic features would have been palpably out-of-balance.

Design, construction, development and preliminary test were completed within a budget pruned to £18$\frac{1}{2}$K from a requirement of £25K. Time scale was a tight two years. An application for funding to complete the test program was declined.

It is not surprising that, with zero rate of displacer rotation, net shaft power remained below zero for all pressures and rev/min. The continuous curves of Fig. 7.8 are representative. Raising displacer rotational speed to 5200 rpm caused an increase in indicated work over the entire range of crankshaft rpm. Over a limited range, brake power was positive, the peak occurring close to the point of maximum angular speed ratio, displacer to crankshaft.

The modest benefit depicted does not promise to revolutionize the fortunes of the air engine – particularly when it is considered that the power to rotate the displacer must be deducted from brake output. On the other hand the measurements depicted in Fig. 7.8 were obtained whilst the sealing provision between piston and displacer drive shaft was under development. The unusual requirement is for this to function under combined sliding and rotation – and to do so without excessive friction and heating. As the development effort

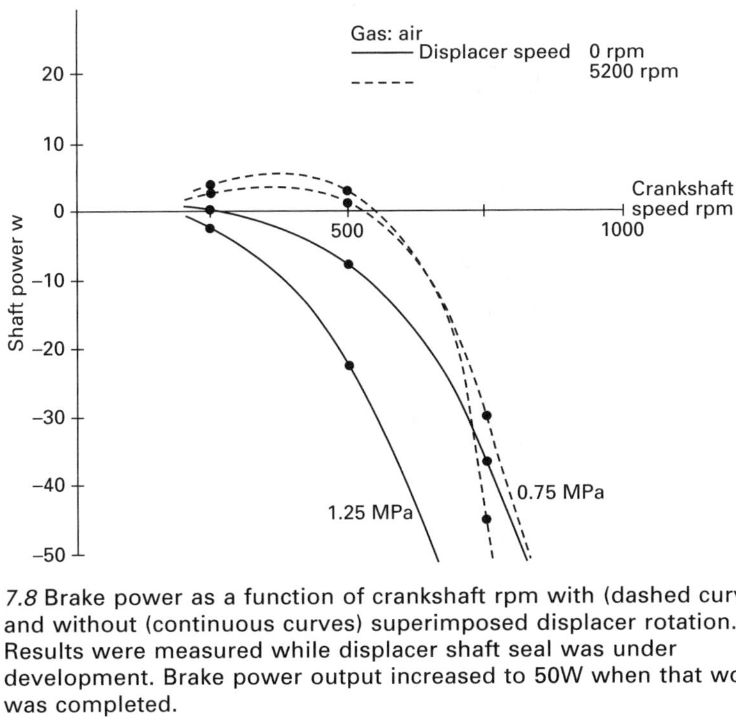

7.8 Brake power as a function of crankshaft rpm with (dashed curves) and without (continuous curves) superimposed displacer rotation. Results were measured while displacer shaft seal was under development. Brake power output increased to 50W when that work was completed.

progressed, and friction and leakage were reduced, brake output of 50W was measured at 750 crankshaft rpm and maximum displacer rpm.

With the benefit of hindsight it would have been advisable to carry out exploratory tests on an existing engine of more modest construction. The antique shown in Fig. 7.9 survives in the thermodynamics laboratory of the Engineering Department at Cambridge University. Hot cylinder diameter is only slightly less than that of the more massive test engine. It would have been awkward to pressurize, but several ways can now be envisaged of spinning the displacer via the exposed end of the drive shaft.

7.9 Proprietary hot-air engine held by the University Engineering Department, Cambridge. The generously-finned chassis appears to be die-cast. Fabrication connotes wider range of production resources than would be necessary for hot-air engine manufacture alone.

The engine is an interesting specimen. It is one of few of the *genre* to be equipped with appropriate – and apparently adequate – provision for heat dissipation: fins of substantial surface area are cast integral with the non-ferrous-alloy chassis and exposed to air disturbed by the fan. The main journal bearings locate relative to the ornate, cast fan shroud by slender radial spokes. All is heavily crackle-enamelled – except for the connecting rods, which are sheet pressings accurately riveted around rolling-element bearings. There is nothing to identify the manufacturer, but construction is distinctive. Many features are reminiscent of sewing machine technology. (In the early twentieth Century, the Singer company diversified into film projectors – of unusual 28 mm gauge). Construction of the engine of Fig. 7.9 is consistent with diversification by a well-equipped manufacturing company.

A viable prime mover needs to drive its own displacer. Any enhancement of pumping loss due to the rotation is then a penalty on the apparent gain in shaft power. An opportunity to explore further was provided by the teaching syllabus of the author's university engineering department. This provided for a range of guided design projects covering the major sub-disciplines – mechanical, thermal, structural, control etc. Over the years the author generated a number of projects in the mechanical category – miniature two-stroke compression-ignition aero engine, 'low delta-T' air engine, implantable heart pump and a vehicle for competing the Shell Mileage Marathon (since renamed Eco-Marathon).

A desirable feature of a project is that it should provide practical exercise in as much as possible of the 'chalk-and-talk' syllabus – three-dimensional graphical statics, kinematics, stress analysis, engineering drawing to BS 308, materials selection, ISO tolerances and fits, etc. The instructor who takes on a new project completes a specimen design, and the instrument shop makes it. Commercially competitive performance of the finished artefact is of less consequence than that it should function sufficiently well to demonstrate the working principle and the effect (adverse or positive) of design decisions taken.

A small, unpressurized air engine with rotating displacer promises to satisfy all the criteria. The author accordingly worked through a design in which displacer rotation was provided by gearing from the crankshaft. Provision was incorporated for operation with or without the gearing engaged.

7.5 Academic design exercise

The concept owes much to an engine designed and built by Ross (1989), who had model aircraft propulsion in mind. Figure 7.10 shows his 'gamma-configuration' engine (two parallel cylinders), with displacer and power piston actuated from a wobble plate. Angular constraint of the wobble plate was achieved by running one of the arms against a guide in the crank-case. The

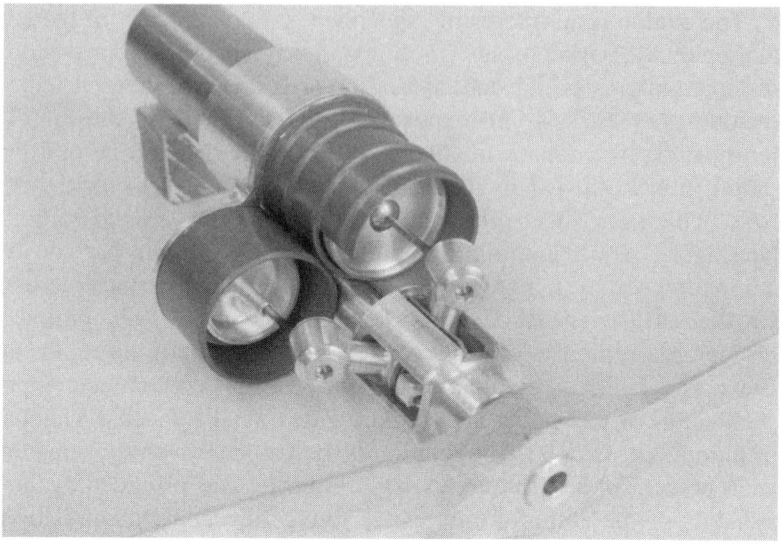

7.10 Engine designed with model aircraft propulsion in mind. Reproduced with permission of M. Andrew Ross.

guide is aligned with the engine axis. The crank drive alone offers a comprehensive academic exercise in graphical projection and three-dimensional statics.

The draft assembly drawing of Fig. 7.11 shows modifications to cause rotation of the displacer. The displacer drive rod is guided both where it pierces the working space and at the gear end. This suggests sufficient lateral stiffness to restrain one of the wobble plate arms against rotational movement, i.e., to take over the role of the ptfe-lined guides of the Ross engine (a friction penalty is likely). A gear wheel mounted directly on the crankshaft main journal drives a pinion mounted on a single ball race. This arrangement, too, relies on the displacer rod to oppose the out-of-plane forces – those generated at the gear teeth.

The wobble plate runs on a single deep-groove rolling-element bearing which is thus subject to a component of axial load applied to the periphery of the outer race. This is not a standard operating condition, and the proposal is speculative. To reduce the possibility of premature failure, the bearing is selected for a higher combined radial/axial load-carrying capacity than would be appropriate if the same load components were applied conventionally.

Availability of a numerically-controlled wire profiling machine made possible a one-piece crank-case with integral cooling fins. The latter are aligned axially in anticipation of receiving a flow of air from a propeller driven by the crankshaft. The arrangement allows power take-off at displacer shaft speed.

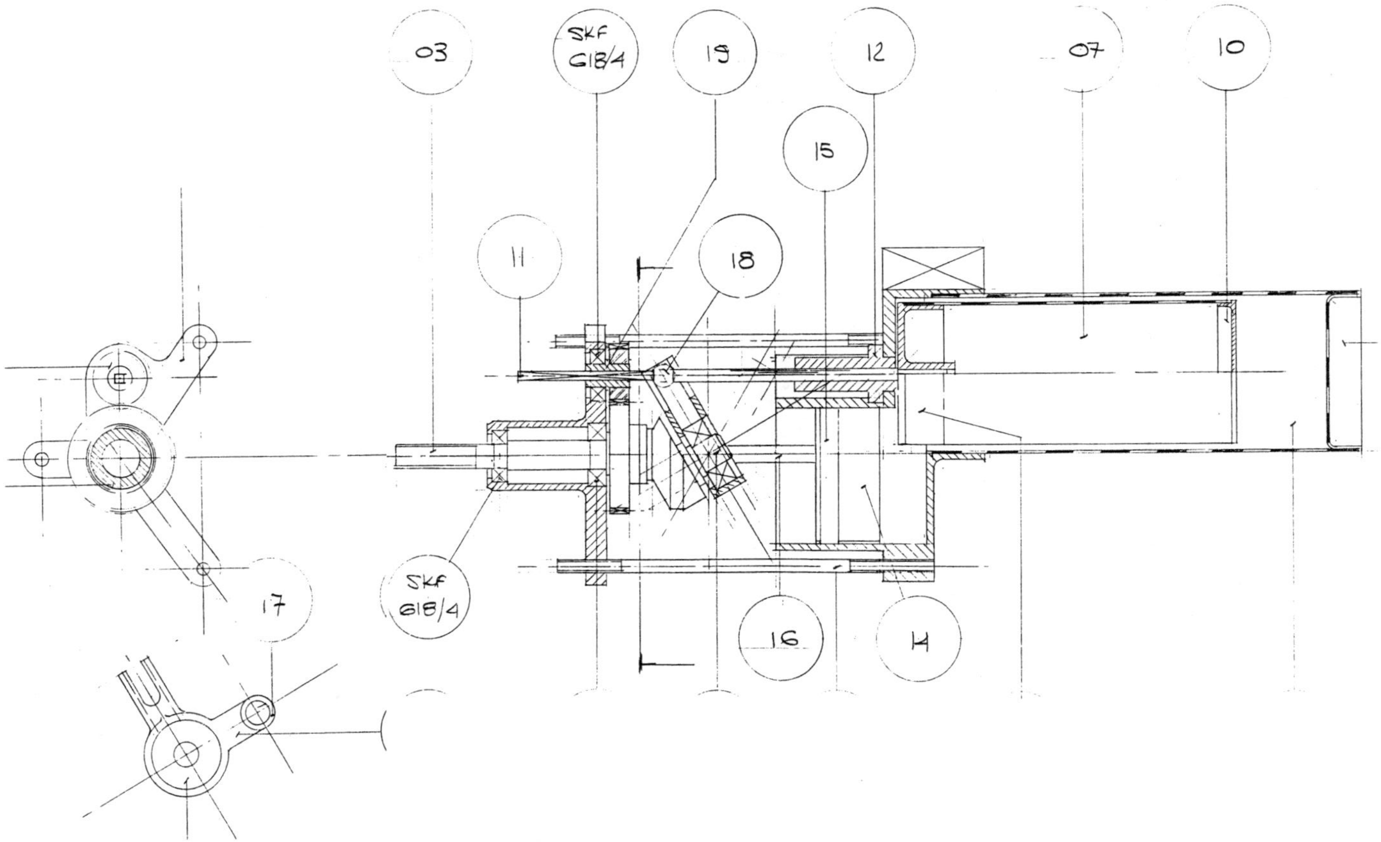

7.11 Draft arrangement drawing of academic exercise in design of self-contained drive for displacer rotation.

In the provisional embodiment illustrated, nominal charge pressure is atmospheric. Full evaluation of the concept calls for pressurization, but the additional design and machining work associated with crank case and seals design threatens to exceed the academic timetable allowance. The net benefit (if any) due to displacer rotation will therefore show up as increased shaft power and/or rpm at rated pressure rather than as an ability take advantage of increased working fluid mass.

The obvious material for cylinder and displacer shells is stainless steel. At the low pressures envisaged, wall thickness based on estimated stress is so small (Organ 1995) that manufacture and handling promise to be major problems. Seamless extruded stainless steel tube is commercially available in attractive wall thicknesses. (Watch out for tubing marketed as seamless but fabricated by winding flat strip and welding in barber's pole fashion. If remachining of a diameter is called for, do not use this material.) Precision seamless extruded hard brass tubing ('telescope tubing') is available ex-stock (Rollet 2005) in a range of diameters permitting choice of annular gap. Cost per unit length of both brass and stainless steel is modest, but most suppliers insist on minimum-order quantities, which can make the outlay prohibitive. Wellington Engineering (2006) appears to be an exception, offering small-order quantities of precision tubing in stainless-steel (AISI 304, 316

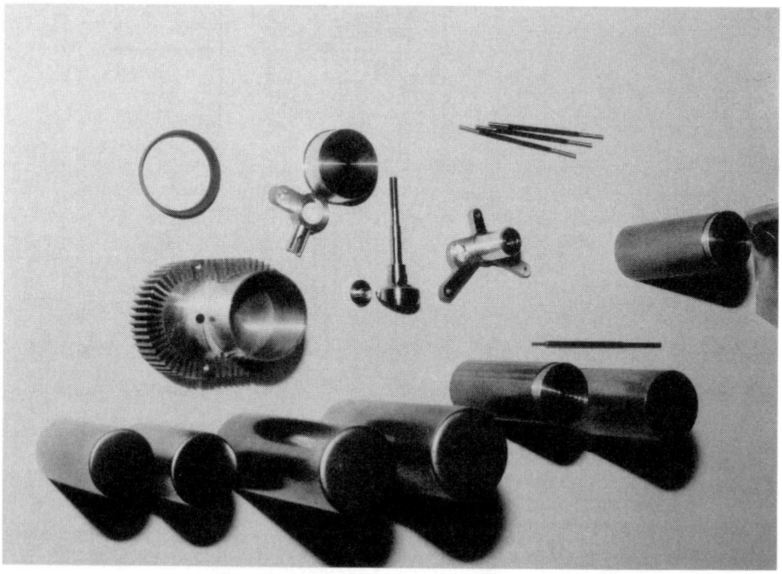

7.12 Selected components of small rotating-displacer engine. The concept was under exploration for possible addition to a portfolio of teaching projects under the 'guided mechanical design' heading. It was eventually judged too demanding for the timetable hours allocated.

and 321), cupro-nickel and aluminium alloys. According to the web-based stock list, stocks are mainly in UK imperial (inch) sizes.

One of the angular speed step-up ratios provided for in the design study (Fig. 7.11) is $2\frac{1}{2}$:1. At a crankshaft speed of 1000 rev/min this gives ω_i = 2500, or approximately 250 rad/s. Using μ = 0.017E-03 Pas for air, and d_i = 25.0E-03 m (approximately one inch) these conditions call for $c \geq 0.75$ mm for the required instability. One of the combinations of brass tubing from the supplier already identified gives $c = 0.8$ mm. This provides a margin, so that instability may still be achieved if rpm falls slightly short of target, or if the axial component of flow tends to suppress instability.

Figure 7.12 shows selected components – but is unfortunately unable to do justice to the exemplary standard of machining achieved by the staff of the instrument shop while working to a diminutive scale.

8

The strange case of the self-regulating air engine

8.1 Background

The simple, unpressurized hot-air engine is nothing if not a paradox.

Countless professionals and amateurs owe their involvement in Stirling engine technology to an encounter with a toy air engine. But a less likely ambassador for the 'serious' power unit would be difficult to imagine. Claims for quiet operation are valid, but for wrong reasons: no prime mover – gas turbine or diesel – has any business making a noise whilst producing a few watts of mechanical output. Power-to-weight ratio is a factor of 500 below that of two-stroke internal combustion engine of comparable swept volume. The alleged ability to operate from a wide range of heat sources is eclipsed by the unrivalled capacity of the typical unit to waste input energy.

When commitment to a concept prevails in the face of damning reality there has to be a reason. In the case of the hot-air engine this appears to be a faith in as-yet unexploited *potential*. It happens that the author shares that faith: an unpressurized hot-air engine of 60cc swept volume is under development for participation in the Shell Eco-Marathon, in which lightweight, streamlined vehicles compete to improve upon a fuel economy record currently standing at some 10,000 miles/gallon. This equates to a mean power consumption of about 20W under the conditions of the event. The modest rating is attributable partly to the 'burn-and-coast' driving technique, whereby the power plant is cycled on and off so as to maintain mean speed of 15 miles/hour.

The power level is within the scope of an unsophisticated air engine. Availability of a complete vehicle (Fig. 8.1), originally designed around an internal combustion engine, was an invitation to put the proposition to the test. Towing the vehicle with a spring balance confirmed a steady-speed power requirement in the region of 25W.

---

*In collaboration with Ian Larque, who manufactured and tested the engine.

8.1 Shell 'Eco-marathon' vehicle powered by high-compression air engine.

The principal design decision lies between pressurized and unpressurized cycles. The former offers compactness and a choice of working fluids, but introduces potential problems of sealing, compounded by concerns for driver safety. The latter promises simplicity, but at the cost of bulk and the necessity of designing around the thermodynamic properties of air at atmospheric pressure. On the other hand, there is scope for increasing compression ratio as an alternative to pressurization.

The vehicle has already run on the competition track (Rockingham) under hot-air power, endorsing the decision to take this latter course.

8.2 Some realities

The mechanical output of a heat engine is proportional, all other things being equal, to the mass of working fluid which it can process per second. The fact explains the prodigious propulsive power of the gas turbine – in both full-size and model form. It also explains why a reciprocating engine limited to a compression ratio of about 2:1 cannot deliver competitive specific power *unless* it can be persuaded to produce torque *at very high rpm.*

The only regenerative reciprocating machine with potential for elevated cycle speed is the pressure-wave engine (see Chapter 13). The hope for the unpressurized reciprocating air engine lies in *increasing effective compression ratio.*

Conventional drive linkages do not permit compression ratio to be set at will. The role of compression ratio has been studied before – but in the abstract and on the basis of 'ideal' cycle processes. The argument goes that compression and expansion spaces are *adiabatic* (no heat transfer between the air and the variable-volume spaces). There are no separate heat exchangers, so by contrast with the isothermal ideal, the compression phase causes a rise in temperature throughout. The higher the degree of compression – so the argument continues – the closer does post-compression temperature approach the temperature at which *expansion* is supposed to begin. When compression ratio is such that the two temperatures coincide, the indicator ($p - V$) diagram is of zero area, and work per cycle is also zero. There is only one thing wrong with this argument: an adiabatic engine is an impossibility – regardless of compression ratio – for the simple reason that 'no heat in' means 'no work out'.

The starting point was an engine described by Organ *et al.* (2001). The design had been intended to combine unprecedentedly high compression ratio with an element of scavenging of the working fluid charge to assist the heat rejection phase. The combination turned out to be less promising in practical implementation than in computer simulation: the many irreversibilities had been foreseen individually, but their cumulative interaction under-estimated. However, many of the components were suitable for reuse in a more conservative embodiment.

The goal of maximum compression ratio excludes all but the coaxial or 'beta' configuration. Inverting the entire engine allows operation from simple heating systems pending development of a definitive combustion chamber.

8.3 Constructional details

Figure 8.2 is part-schematic to emphasize the detail of the crank linkage. The patent application reproduced as Appendix I describes its characteristics as a planar mechanism. Foremost among these is considerable kinematic flexibility. Achievable compression ratio appears higher than that afforded by separate cranks, or by a conventional bell-crank mechanism. There are the additional benefits of near straight-line actuation of both piston and displacer, and the accompanying reaction of the majority of side-thrust by rolling-element bearings. Appendix II is the algebra of the kinematics, which may be of use to those interested in coding the volume variations as input to computer simulation.

The piston drives twin connecting rods. These pivot on twin opposed gudgeon pins, each integral with, and cantilevered out from, plane arms secured by button-head socket screws against flats recessed into the sides of the piston skirt. Both ends of the connecting rod carry phosphor bronze bushes. Piston drive is via twin L-shaped bell cranks, while the displacer drive is from the single straight arm. The crank pin drives the twin bell-cranks through a drawn-cup needle roller race. The bell-cranks are pivoted

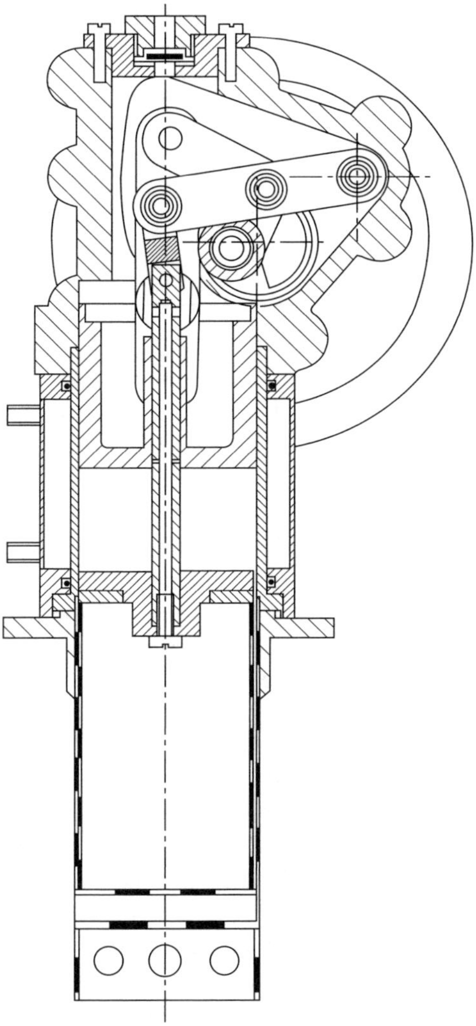

8.2 Crank and drive linkage. The small diameter of the crank web limits scope for dynamic balance, but size considerations prevailed for consistency with the aim of minimizing crank case volume. Outline specification in Table 8.1.

to the drive arm linkage at a deep groove ball race. The beam link pivots on a deep groove ball race at a post located in a crank-case side plate.

The crank-case is of parallel, ruled shape inside and out, allowing manufacture by numerically controlled wire erosion. The internal shape is sculpted around the extremes of travel of the mechanism. With a view to keeping under-the-piston volume to a minimum, the internal cut-out is reused by dividing it into two identical halves and bolting one half to each side

Table 8.1 Outline specification of hot-air engine for Shell 'Eco-Marathon' vehicle

Swept volume 65.5 cc
Bore (piston diameter) 55 mm
Crank shaft offset 7.88 mm
Stoke of piston 27.6 mm
Displacer diameter 54 mm
Regenerative gap 0.6 mm
Bore/stroke ratio 1.99
Unloaded rotational speed 900 rpm (approx.)
Expansion-end temperature 873 K (approx.)

closure plate as boss-cum-spigot. Each boss is machined back so that, when the side plates are in position, a small lateral clearance remains either side of the mechanism.

The piston is fitted with a one-way flap valve of small flow capacity; the crank-case with a light-weight plastic disc valve. In conjunction, the two one-way valves ensure that minimum cycle pressure and peak crank case pressure coincide. The elevated mean cycle pressure calls for a rotary seal on the crankshaft. Peak pressure differential is not large, and the installed lip seal of a standard 2RS rolling-element bearing has served so far.

Displacer and expansion cylinder cap are fabricated from 0.5 mm thick drawn stainless steel tubing to AISI 316L specification. End plates are electron-beam welded into position. The annular gap around the displacer which serves as the regenerator began as a nominal 0.6 mm. The ideal value is that affording the optimum balance between the benefits of thermal regeneration and the penalties of dead space and flow resistance. Incrementally machining small amounts from the outside diameter of the displacer has already confirmed the sensitivity of brake output to gap size.

The ringless piston is of aluminium alloy HE15-TF. In the early, water-cooled arrangement shown in Fig. 8.3 it runs in a honed, cast-iron liner. The displacer rod is guided within a cylindrical boss integral with the piston crown. The displacer rod is attached at a spigoted flange to the base of the stainless steel displacer. The opposite end of the displacer rod attaches to the drive arm link via a clevis which accommodates the departure from straight-line motion. The angular excursion can be reduced ad lib by increasing the centres of the clevis.

The thicker of the crank-case side plates carries the main bearing housing with radial location by a cylindrical spigot. The housing is of aluminium alloy HE15-TF, and has two deep-groove ball races supporting the fabricated, cantilever crankshaft. The latter is of BS En-16 steel (range T) with case-hardened crank pin press-fitted with an interference. The outer ball race (only) is of 2RS (two rubbing-seal) type and acts as pressure seal.

8.3 Engine as originally assembled with water jacket as compression-end cooling provision.

8.4 Exploratory power and torque measurement

The flywheel is fitted with a polished cylindrical extension of 12 mm diameter around which a simple friction brake can be clamped. Load adjustment is by tightening clamp bolts via springs which maintain constant clamping force independently of slight thermal expansion. Torque is indicated by a 'Salter' spring balance attached at 100 mm radius and graduated in increments of 5 g. An inductance pick-up senses one pulse per revolution from a magnet attached to the flywheel rim, allowing rpm to be indicated by a digital cycle speedometer.

The engine *per se* was capable of operation before the matter of heating provision had been pursued any further than the choice of fuel. The combustion chamber and pre-heater described in Chapters 3 and 4 had not been conceived of, and the design sequence: engine first; heating provision second – would not be pursued again, and for two reasons:

1. A combustion chamber with pre-heater – even with the relatively compact Spirex – adds height and bulk to the bare engine: a unit which promised to fit readily within space vacated by the original petrol engine of the Eco-Marathon vehicle now barely does so.

2. The combustion products must flow over (or through) the expansion exchanger with some degree of symmetry if hot-spots are to be avoided. Admission and outlet channels immediately either side of the exchanger have to be of a geometry which achieves this *and* must simultaneously form suitable transition passages with the air pre-heater. If it sounds straightforward, try it!

With the benefit of hindsight, the author suggests either of the following:

- design burner, expansion exchanger and air pre-heater as a system. Check compatibility with the engine and installation envelope, iterating as necessary
- Look into rating, cost and size of proprietary combustion systems. The Internet carries many entries for the 'flox' (FLameless OXidation) type. Uniformity of temperature and low emissions are benefits claimed for the high levels of exhaust gas recirculation.

In the circumstances, exploratory testing went ahead with the aid of an open propane torch bolted below the expansion cylinder cap. Temperature control was rudimentary, and took the form of manual control of the gas supply in relation to the colour of the heated expansion end.

With crank-case open to atmosphere (i.e. without the benefit of under-the-piston pre-compression) brake power and torque varied with rpm according to the curves through the solid points of Figs 8.4(a) and (b), peaking at 17 watts at 0.37 Nm respectively.

In an intermediate configuration, the crank-case was sealed, but the inserts for minimizing volume not used. The non-return valves were not functional, but crank-case and cylinder pressures were allowed to equalize at the point of minimum cycle pressure. This was achieved via small cross-drillings in the hollow displacer rod. The piston exposed these at its point of maximum separation from the displacer. Maximum power was 24 watts, and peak torque 0.45 Nm.

In a further configuration, the crank-case was packed to minimum dead volume, the main shaft seal was in place and the one-way plate valve installed. A pressure tapping in the gas path had not been allowed for at the design stage, so the pressure ratio achieved in the two-stage compression is not yet known. On the other hand, turning the flywheel past compression by hand with the engine cold now required the full effort of a single hand, and resulted in repeated 'bounces' on release. Power and torque varied with rpm according to the curves through the open points of Figs 8.4(a) and (b), power peaking at 32 W at torque at 0.59 Nm.

The process of acquiring further data was interrupted by increasing awareness that, at a given value of rev/min the engine could be induced to produce a range of torques. The phenomenon suggests a capacity for a degree of 'self-regulation'.

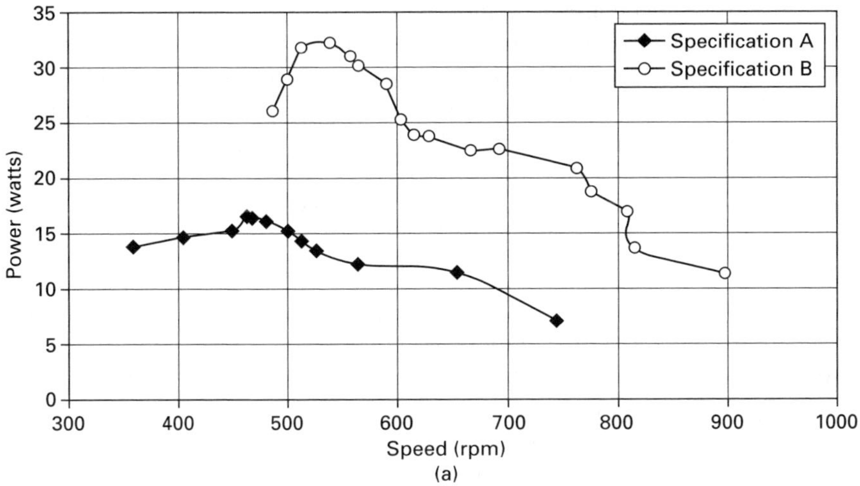

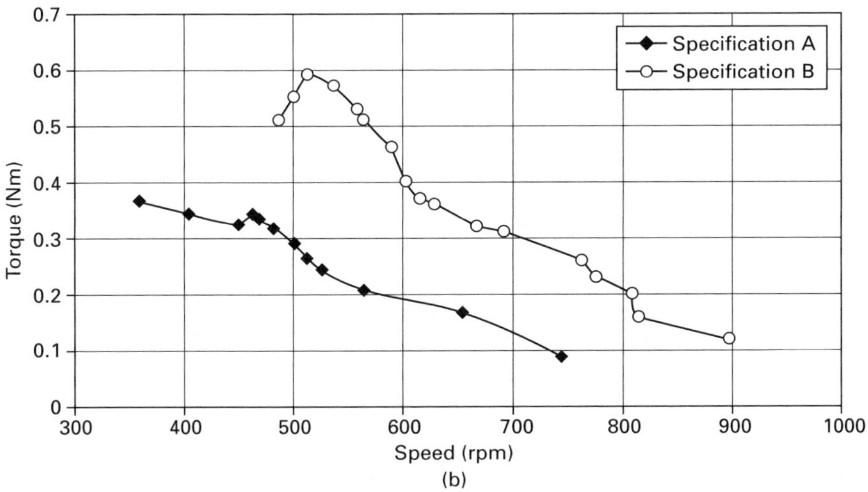

8.4 Exploratory brake power and torque measurements with open-flame heating. (a) brake power (W) vs rev/min; (b) torque (Nm) vs rev/min. Solid points: no crank-case pre-compression; open points: minimum cycle pressure boosted to peak crank-case pressure. After Larque (2002) with permission.

8.5 'Self-regulation'

Application of increasing brake load causes a continuous reduction in rpm in accordance with expectation. After a short period at peak imposed load, however, rpm gradually increase in the direction of the original no-load value. The increase occurs with no intentional augmentation of heat provision. The engine may now be progressively loaded again, starting at the previous peak value. Speed decreases with increasing load, but if loading is interrupted, speed again tends to return to original, no-load value.

Further increase in brake load brings about a repetition of the phenomenon until a 'plateau' is reached, at which the engine responds in the anticipated manner – i.e., by a reduction in rpm for each increase in braking force up to the point of stall.

If the brake load is now suddenly removed, speed surges, surpassing by a wide margin the initial no-load value. Then, however, it gradually settles back to that initial value.

The burner in use to this point was unsophisticated, there being no means for sensitive adjustment of heat input to engine load. At face value, therefore, the phenomenon appears to reflect an ability of the internal gas processes to adapt, within limits, to the external load.

Figure 8.5 plots a representative series of tests. The self-compensating feature means that there are essentially three curves of rpm vs brake load, namely, *a–b*, *c–d* and *e–f*, apparently independent of each other. On the

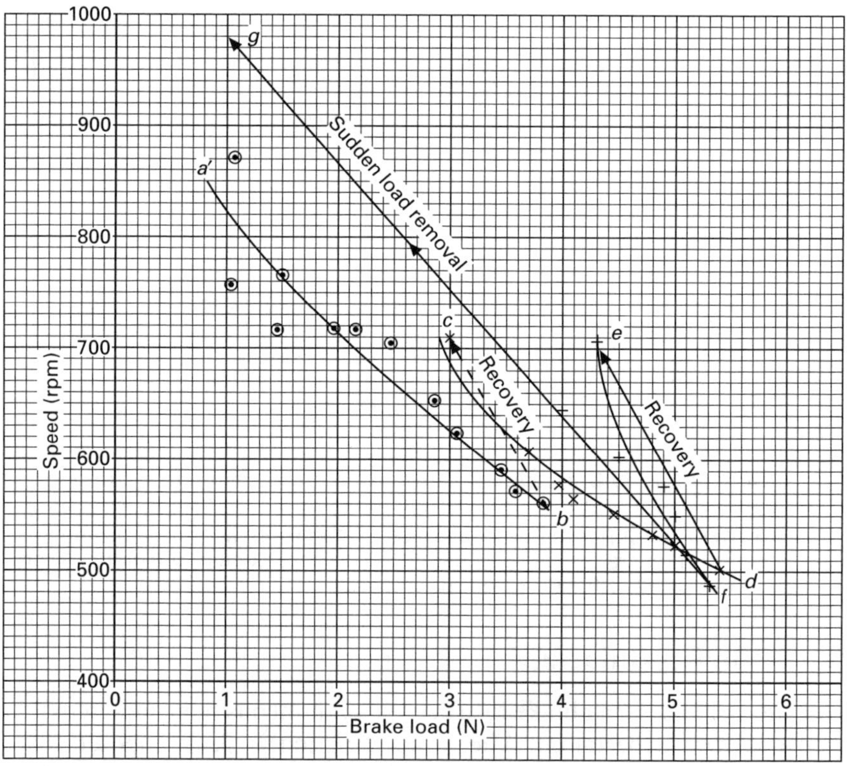

8.5 Independent curves *a–b*, *c–d*, and *e–f* of rpm vs external torque with no apparent difference in heating or cooling between the curves.

graph, dashed lines represent the 'recovery' phase between the minimum speed under one set of loading and the higher speed with which the loading recommences. A different view of the phenomenon might be expected if torque were increased at much reduced rate, possibly provoking the recovery phase to take place during loading.

8.6 Tentative explanation

The phenomenon is reported to be a familiar feature of the operation of a three-cylinder, unpressurized air engine for boat propulsion (Bomford 2004). Similar behaviour has been noted independently (Feulner 2003). As far as the writer is aware, it is observed only in engines of the 'hot-air' type, and not in those having multi-tube gas paths and fixed regenerators.

Any explanation should therefore be in terms of some critical difference between the gas processes of the two categories. Several come to mind, but the most substantial concerns the regenerator provision.

The only known symbolic statement of gas particle temperature in the annular gap is the coupled pair of differential equations developed in Chapter 6. The formulation is in terms of ΔT_d, (difference between the local, instantaneous temperature of the gas particle and that of the adjacent displacer) and of ΔT_c, (corresponding difference referred to the adjacent cylinder surface). Both differences are functions of particle velocity $\underline{u}$, displacer speed, u_d, axial temperature gradient in the displacer shell, $\partial T_{wd}/\partial x$, axial temperature gradient in the cylinder wall $\partial T_{wc}/\partial x$, Stanton number, St, hydraulic radius of the annulus, r_h, radial thicknesses, t_d, t_c, of displacer and cylinder shells respectively, and of thermal capacity ratios N_{TCRd}, N_{TCRc} of displacer shell material and cylinder material. In reality, the solutions are strong functions also of instantaneous rate of pressure variation, $\partial p/\partial t$:

$$\Delta T_d, \Delta T_c = f\,\{\underline{u}, u_d, \partial T_{wd}/\partial x, \partial T_{wc}/\partial x, St, r_h, t_d, t_c,$$
$$N_{TCRd}, N_{TCRc}, \partial p/\partial t\} \qquad 8.1$$

The hydraulic radius of the annular gap r_h is an order of magnitude larger than that of the representative wire screen (0.25 mm vs say 0.03 mm), so that St (Stanton number) values are always relatively low. This suggests that cyclic temperature swing is substantial, and determined as much by $\partial p/\partial t$ and by particle and displacer motion as by heat transfer.

Over and above the gas exchange processes in the annular gap, little else goes on in these simple machines, so an explanation is sought in terms of the regenerative processes.

Imagine the engine running unloaded at a steady 800 rpm, as at point *a* on Fig. 8.5. A torque is now applied at the output shaft. Speed drops, but *work per cycle, W*, increases, because finite brake power is now being delivered.

W can change only if the area of the indicator (pressure-volume, *p*–*V*) diagram increases. The increase may be due to increased pressure swing, or to a changed relationship between the cyclic variations in *p* and *V* – or to both effects. Any such change will affect local heat transfer coefficient via a changed *St*, but also implies a change in the phase between *p*, $\underline{u}$ and u_d.

To investigate further, note that the cyclic processes are subject to the First Law of Thermodynamics:

$$W = Q_E - Q_C \tag{8.2}$$

Increased *W* calls for an increase in the difference $Q_E - Q_C$. The only ways this can be achieved are by an increase in Q_E, a decrease in Q_C – or by a combination of both. Q_E can increase only if the temperature gradient through the metal enclosure increases – essentially in direct proportion to the increased rate of heat flow. If, as assumed, the temperature of the external surface remains constant, increased Q_E means a *reduction* in the temperature of the *inside* surface. To this extent the increase in *W* is in contradiction to elementary cycle analysis – and, indeed to intuition. At first sight there seems no reason to expect the alternative – a decrease in Q_C. Indeed, a Carnot view would insist on the opposite – an *increase* in heat rejection per cycle.

But now consider the regenerator. It seems safe to assume that the annular gap functions rather poorly as a regenerator, i.e., regenerator *recovery ratio* or *effectiveness* η_r has a low numerical value. The main penalty of low recovery ratio is high thermal loading on the compression-end exchanger (high Q_C). An improvement, for whatever reason, in η_r increases the energy 'recycled' by the regenerator and, all other things being equal, *decreases* Q_C. Could the increased brake loading lead to increased η_r?

A definitive answer is not possible, but the following considerations arise:

- Even the fixed, high-performance wire-mesh regenerator does not function according to classic regenerator theory. The cyclic pattern of storage and release of heat is influenced by the cyclic swing of pressure and its phase in relation to the alternating flow of the traditional picture of operation.
- The regenerative processes of the annular gap are almost certainly more sensitive to the effect of pressure swing than those of the fixed, matrix regenerator. The complex interactions are further compounded by the 'shuttle' phenomenon.

Retarding shaft speed by increasing torque may be expected to alter the phase between internal pressure and temperature events. Regenerator recovery ratio almost certainly changes in sympathy. If the change is favourable i.e. if there is improved utilization of Q_E and a corresponding reduction in Q_C, then work/cycle *W*, may be expected to increase. If the increase exceeds the work/cycle demanded by the brake, then the excess goes into accelerating the flywheel.

The response to further step increases in load is explained in the same way. But what of the speed surge on removal of the load? The only way regenerator effectiveness can change at given rpm is by a redistribution of temperature in the solid components – in the displacer shell, or in the cylinder shell, or in both. If changed regenerator recovery ratio is an explanation for the alleged self-compensation, then the temperature distribution(s) at the instant of load removal is/are different from the no-load values. Thermal capacity ratio N_{TCR} (see Notation) in an unpressurized engine has a high value, meaning that a large number of cycles takes place before the temperature profile stabilizes to a new regime. It is possible that, while the profile(s) retain their peak load form, no-load operation benefits from the high recovery ratio, and responds with temporarily increased no-load speed.

8.7 Conclusions

The simple air engine adapted to a high compression ratio appears capable of useful power output. It does, however, exhibit a characteristic which most other reciprocating prime movers do not – an ability to develop increased

8.6 Air cooling has replaced water cooling as the engine has been prepared for reinstallation in the 'Eco-Marathon' vehicle. Finned compression cylinder shown separately.

power in response to increasing external load with no apparent demand for increased temperature of heat supply. There are applications in which this feature might advantageously reduce problems of governing.

For the simple air engine to have a future, specific power must be improved by a factor of at least 10. A factor of 2 has been demonstrated merely by exploiting the peak pressure arising in an enclosed crank-case. Scope remaining includes:

- achievement of a deeper insight into the internal gas processes. This is commonly sought through thermodynamic analysis and computer simulation. There is now a stringent test for the utility of such analytical and simulation work: *it must adequately predict the self-regulation phenomenon.*
- further increase in compression ratio consistent with decreasing the fraction of air mass in the expansion space during compression (and vice-versa). Exploration of the kinematic flexibility of the present crank mechanism continues with the aid of computer simulation. The investigation to date suggests that increasing compression ratio follows a law of *increasing* – rather than diminishing – marginal returns.

Evolutionary development continues in parallel. An efficient, recuperative combustion chamber supersedes the *ad hoc* heating arrangements. A heavily-finned, air-cooled cylinder has replaced the water-cooled original, with considerable saving in overall installed volume and weight (Fig. 8.6).

A technique has been developed of machining displacer and cylinder shells to almost any desired wall thickness. Performance being sensitive to the radial thickness of the regenerative gap, the optimum value of the latter can now be established. There is the bonus of considerable reduction in displacer mass below that dictated by the wall thickness of proprietary tubing.

Power measurements are now made on a dynamometer on which the engine itself is free to swing about the crankshaft axis (by contrast with the more common arrangement whereby the brake is trunnion-mounted).

9 Some light on the inner workings of the 'thermal lag' engine

9.1 The concept

Devices operating on the pulse-tube principle are attracting increasing interest, possibly on account of their mechanical simplicity. The prime mover variant has become known as the 'thermal lag' engine, but exploits the same principle as the reversed-cycle counterpart – the pulse-tube cooler. Both dispense with the displacer, leaving only one* moving component per cylinder – the piston. Perversely from the point of view of the would-be designer, the gain in mechanical simplicity is offset by increased inscrutability of the gas processes.

It is tempting to address the conceptual challenge by taking the Stirling cycle as the start point. This would be misleading: the displacer-less devices rely for their operation on phenomena ('second-order' effects) which the practical Stirling device would perform better without. Conversely, an analysis sensitive to the gas processes of the pulse-tube offers a new 'angle' on the orthodox Stirling cycle.

The principal second-order effects are temperature transients and flow friction. It is possible that thermal buoyancy has an influence in some configurations. Such phenomena are problematic to probe experimentally, and not particularly straightforward to isolate by analysis. The best chance of achieving the insight required for design is through parallel exercises in experiment and simulation, and by carrying out the experimental side on a variant whose performance is highly sensitive to gas path design and input conditions.

In the case of the cryo-cooler the forcing function is operating speed (or frequency). Output response is heat rate at expansion exchanger temperature. This cannot be resolved into parasitic and 'legitimate' components. The

* Some pulse-tube coolers feature twin opposed pistons with a view to minimizing mechanical vibration – but the gas in the main working spaces responds to the *net* of the volume variations in combination.

unloaded pulse-tube prime mover, by contrast, reaches a variable equilibrium speed highly sensitive to the parameters of operation and design. Drawing up an accurate heat balance is not easy, but at least the heating provision can be kept constant. Accordingly, the present study focuses on the prime-mover embodiment.

9.2 'Thermal lag' engine

In 1995 a US patent was granted to Tailer for a 'thermal lag' engine. The device comprises a cold chamber sealed by a piston and connected to a hot chamber (Tailer, 1995). Inwards motion of the piston displaces air from the cold chamber to the hot, simultaneously compressing the entire air mass. If the process is rapid, the air experiences an element of adiabatic temperature rise. However, a temperature difference persists between the displaced cold gas and the surface with which it has newly come into contact. The 'dwell' period about the inner dead centre position of the piston allows heat transfer from the heated walls and causes a rise in pressure. Rapid return of the piston to outer dead-centre causes heated air to return more or less adiabatically to the cold chamber, delivering greater work to the piston than was absorbed by the earlier compression. During the dwell around outer dead centre the expanded air cools and pressure falls. (If the delay at this point were sufficient, pressure would return to the start value.)

Grant of the Tailer patent appears inconsistent with comprehensive prior disclosure of the principle by West (1993) and by Wicks and Caminero (1994), and with the fact that the internal functioning is essentially that of the Gifford pulse-tube cooler (Gifford and Longsworth, 1964).

West and Wicks both undertook to predict the gas processes. Neither treatment achieved its objective. Many accounts of 'table-top' thermal lag engines have since appeared, none focusing on the underlying thermodynamic principles. On this basis, internal design would not be critical. Indeed, it has been claimed (Warbrooke, 2005) that 'almost any sort of engine will run'. Conversely, this author has found both output power and no-load speed to be highly sensitive to the respective hydraulic radii r_h of regenerator and pulse-tube sections: increasing heat transfer capacity of the pulse-tube by reducing r_h results in failure to function.

The criterion of efficiency and power-producing potential in conventional engines – Otto, Brayton, Stirling etc. is the respective ideal, air standard cycle. In each case this is defined in terms of equilibrium processes – isochors, isothermals, etc. If such ideal cycles may be said to 'function', they do so independently of time and thus independently of speed. The 'thermal lag' engine, by contrast, functions by virtue of *lack* of equilibrium. The gas processes are *time-dependent*. This precludes construction of a cycle of equilibrium processes. Such cycles have already been dismissed as arbitrary (Chapter 2). Abandoning a perverse tradition can only mean progress.

9.3 Ideal gas process sequence

Figure 9.1 shows the gas path 'straightened out' into the sequence (right-to-left):

1. variable-volume compression-space
2. cooler
3. pulse-tube
4. heater
5. regenerator.

It is assumed:

- the temperature gradients in the regenerator matrix and in the walls of the pulse-tube are prescribed – e.g. linear, as in Fig. 9.1
- in the regenerator (only) *NTU* (number of transfer units) is sufficiently high that gas temperature is equal to matrix temperature *for all time*
- the *NTU* of pulse-tube is limited, allowing a difference in temperature between gas and adjacent solid surface to persist temporarily.

At start-up, gas temperature T_g is equal at all points to local enclosure temperature T_w

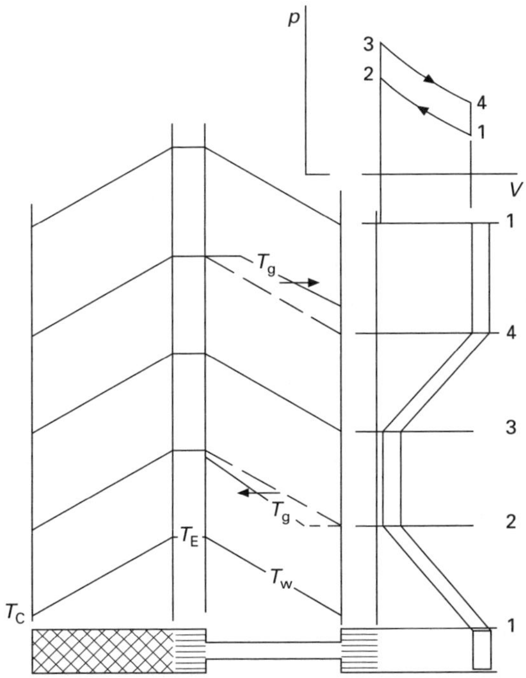

9.1 Stylized picture of gas events in the 'thermal lag' engine. The inset figure is an *indicator* (*p-V*) diagram – *not* a sequence of process paths.

1–2 **Compression**: The gas is shifted bodily to left. The temperature profile in the pulse-tube section is transported leftward, creating a difference in temperature, with T_g generally below T_w. During the transfer and reduction in volume, pressure increases. Due to an element of adiabatic temperature rise, the magnitude of difference $T_g - T_w$ is less than it would be due to translation alone.

(Gas in the regenerator redistributes, but there is no effect on gas temperature profile *within* this element because of the high *NTU*.)

2–3 **Relaxation:** T_g in the pulse-tube adjusts over time towards T_w by heat addition from tube wall to gas. A slight redistribution of gas causes an overall rise in pressure p.

3–4 **Expansion**: The gas temperature profile in the pulse-tube shifts bodily rightwards, causing T_g generally to exceed T_w. The element of adiabatic cooling accompanying expansion again causes temperature difference $T_g - T_w$ to be less than that due to translation alone. Mean pressure during expansion exceeds that which prevailed during compression, so there is an excess of work.

(The temperature profile in the regenerator gas is unaffected due to the high *NTU*.)

4–1 **Relaxation**: The gas temperature profile readjusts *over time* to that of pulse-tube wall by convection of heat to the latter. Pressure falls in the direction of the initial value (but achieves that value only if the relaxation period is long enough).

It is interesting to note:

- the gas processes of thermal lag and Stirling cycles doubtless *overlap*
- the practical Stirling cycle functions best with minimum restriction on heat transfer. The thermal lag engine *requires* heat transfer to be restricted, and *would not function* if the restriction were removed.

The 'relaxation' phases are defined to this point only in terms of their end states. To explore further it is necessary to quantify the exchange processes.

9.4 A detailed model of the thermal processes

9.4.1 Assumptions

- The working fluid is an ideal gas with $p/\rho = RT$.
- Flow is one-dimensional ('slab' flow): a fluid particle initially lying in a plane perpendicular to the direction of motion (x-direction) remains in that plane, and characterizes all particles in the same plane. (Buoyancy effects may upset this assumption, given the large unobstructed volume, relatively modest cycle rate and the temperature extremes across the

pulse-tube section. With engine axis vertical and heated end uppermost, such effects seem inevitable.)

- The distribution of enclosure temperature T_w does not vary with time.
- Total working fluid mass M remains constant (there is no leakage).
- Kays' and London's heat transfer and flow friction correlations for wire mesh screens are applicable. (Earlier sections of this book have recommended *not* making this assumption in the context of Stirling engines.) The same concern almost certainly applies to the thermal lag types. It is not implemented here because substituting appropriate correlations, when available, is straightforward.

The analysis will be developed in stages, simplifying assumptions being discarded one-by-one.

Figure 9.2 shows a typical 'thermal lag' engine schematically. A representative distribution of T_w with axial location x is suggested.

The formulation will use the Lagrange reference frame. Lagrange coordinates are the loci in the $t - x$ plane of selected particles of working fluid as they move in response to piston motion and to corresponding changes in temperature and pressure.

9.4.2 Symbol representation of flow model

The gas is subdivided into $n - 1$ parcels, where n is the number of bounding faces, including those of piston and closed end. Following a fluid particle amounts to following a pair of bounding faces. Subdivision may be into equal sub-volumes, equal sub-masses $\Delta m = M/(n - 1)$, or into arbitrary subdivisions of M. For convenience of illustration, initial subdivision will be

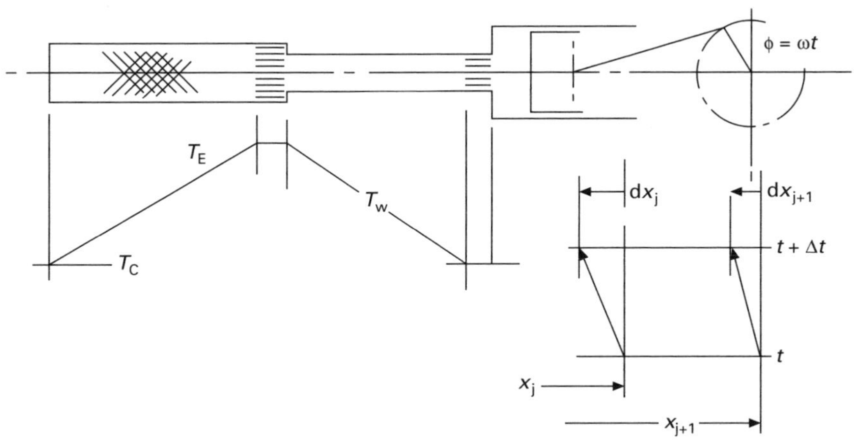

9.2 Schematic representation of 'thermal lag' engine, showing enclosure temperature and initial subdivision of working fluid.

into equal increments of axial length. Later it will be found advantageous to subdivide *cumulative volume*, v (rather than length).

An increment of motion of the piston inwards from its initial position induces an element of leftward motion in each of the bounding faces of the gas elements. After an increment of time Δt the typical dividing face indicated in the inset to Fig. 9.2 has moved from x_j to $x_j + dx_j$, while pressure p has increased by Dp. In reality, pressure p varies not only with time t (and thus with crank angle $\phi = \omega t$), but also with location x at given t. For the machine to function $\partial p/\partial x$ must be small for all t, so it is assumed *provisionally* that $\partial p/\partial x = 0$.

Mass Δm contained between faces at j and $j + 1$ is changeless, so that $D(\Delta m) = 0$. Operator 'D' denotes the 'total' change in a variable along a particle path in the $t-x$ plane, D being the total (or 'substantial' or 'substantive') derivative $D = \partial/\partial t + u\partial/\partial x$. For volume V of ideal gas at temperature T:

$$D(pV/T) \approx Dp/p + DV/V - DT/T \approx 0$$

The symbol for 'approximately equal' ≈ is used to emphasize that the algebraic expansion becomes accurate only as Dp, etc. tend to zero.

In the notation of Fig. 9.2, which represents a parcel of ideal gas contained in the elemental volume of free-flow area A_{ff} between plane faces a distance $x_{j+1} - x_j$ apart:

$$D\{pA_{ff}(x_{j+1} - x_j)/\tfrac{1}{2}R(T_{j+1} + T_j)\} = 0$$

Differentiating by the chain rule, and cancelling constant multipliers:

$$(x_{j+1} - x_j)Dp + p\{(dx_{j+1} - dx_j)$$
$$- (x_{j+1} - x_j)(T_{j+1} + T_j)^{-1}(DT_{j+1} + DT_j)\} = 0 \qquad 9.1$$

There are three dependent variables – dx, Dp and DT. These may be reduced to two. The reduction mechanism is most readily appreciated by provisionally discarding the (essential) assumption of restricted heat transfer, and by provisionally allowing T to adapt instantaneously to local wall temperature T_w. The idealizations are discarded later.

$$DT_j = dT_w/dx|_j dx_j \qquad 9.2$$

Substituting Eqn 9.2 into Eqn 9.1 gives a relationship between the unknowns dx_j, dx_{j+1} and Dp. At this stage in the development the latter is independent of location, and may be written dp.

First, however, it is noted that, where the volume element shown in the inset to Fig. 9.2 straddles a discontinuity in free-flow area A_{ff}, then Eqn 9.1 is inappropriate. The problem is handled by using *cumulative volume*, v, as independent variable instead of x. The gas path of Fig. 9.1 may be laid out in terms of v reckoned from an origin of $v = 0$ at the left-hand end, and subdivided into $n - 1$ initially equal *volumes*. Eqn 9.1 becomes:

$$(v_{j+1} - v_j)\mathrm{d}p + p\{(\mathrm{d}v_{j+1} - \mathrm{d}v_j)$$
$$- (v_{j+1} - v_j)(T_{j+1} + T_j)^{-1}(\mathrm{D}T_{j+1} + \mathrm{D}T_j)\} = 0 \qquad 9.1(a)$$

Eqn 9.2 becomes:

$$\mathrm{D}T_j = \mathrm{d}T_w/\mathrm{d}v\,|_j \mathrm{d}v_j \qquad 9.2(a)$$

Substituting and using $\underline{T}$ to represent the sum $T_{w\,j} + T_{w\,j+1}$

$$\mathrm{d}v_{j+1} - \mathrm{d}v_j - (v_{j+1} - v_j)\{\mathrm{d}T_w/\mathrm{d}v\,|_{j+1}\mathrm{d}v_{j+1} + \mathrm{d}T_w/\mathrm{d}v\,|_j\mathrm{d}v_j\}\underline{T}^{-1}$$
$$= - \mathrm{d}p/p(v_{j+1} - v_j) \qquad 9.3$$

Collecting coefficients of $\mathrm{d}v_j$, $\mathrm{d}v_{j+1}$:

$$\mathrm{d}v_{j+1}\{1 - \underline{T}^{-1}(v_{j+1} - v_j)\mathrm{d}T_w/\mathrm{d}v\,|_{j+1}\}$$
$$- \mathrm{d}v_j\{1 + \underline{T}^{-1}(v_{j+1} - v_j)\mathrm{d}T_w/\mathrm{d}v\,|_j\}$$
$$+ \mathrm{d}p/p(v_{j+1} - v_j) = 0 \qquad 9.4$$

Equation 9.4 furnishes the illustration of the Lagrange formulation, and provides a test of correct implementation. It embodies the proposition that the temperature of a particle adjusts instantaneously to that of the immediately adjacent solid enclosure. This should guarantee that compression and expansion curves superimpose, and that no heat will be converted to work. When the integration scheme has been shown to yield this outcome, formulation can be taken a stage further to reflect limited heat transfer – and eventually to cope simultaneously with flow losses.

9.4.3 Choice of integration scheme

Achievement of best engine performance is going to depend on striking the right balance between the two components of temperature swing as seen at a fixed point – the adiabatic element and that due to wholesale translation of the working fluid. On the other hand, nothing whatsoever has emerged to this point to suggest what the optimum balance might be; nor, indeed, how any such balance – if it could be defined – might be struck in terms of choice of gas path proportions.

A program of parallel experimentation and computer simulation is in progress under the author's direction, but explicit design guidelines have so far failed to emerge. Recourse to mechanized optimization appears inevitable. Optimization demands the highest degree of realism in the symbolic model – and a massive number of computational iterations. The combination calls for an efficient numerical algorithm and rapid numerical convergence.

If the demand on CPU resources is not an issue, formulating and coding an integration scheme is relatively straightforward. Coding for computational efficiency tends to be more demanding. Embodiment of checks of correct functioning calls for correspondingly increased subtlety. The reader proposing to pursue the simulation approach might welcome an introduction to integration schemes in both categories.

All schemes to be discussed will deal with the volume sub-elements of Fig. 9.1 one at a time, starting at the left-hand end and proceeding to the piston face at the right. Each such sweep will work with values of the dependent variables p, $v_{j+1} - v_j$ known at time t, and at the end of the sweep will advance to time $t + dt$ having calculated the dp, dv_{j+1} and dv_j corresponding to dt.

Equation 9.4 is simplified by the introduction of two abbreviations. These are computed from quantities having known values at the start of an integration sweep:

$$f_{Rj} = 1 + 2\underline{T}^{-1}(v_{j+1} - v_j)dT_w/dv\,|_j \qquad 9.5(a)$$

$$f_{Lj+1} = 1 - 2\underline{T}^{-1}(v_{j+1} - v_j)dT_w/dv\,|_{j+1} \qquad 9.5(b)$$

In Eqns 9.5 $\underline{T}$ = mean temperature $\frac{1}{2}(T_{w\,j} + T_{w\,j+1})$

Implicit scheme

All terms in Eqns 9.5(a) and (b) are 'old' values – and therefore 'known'. Eqn 9.4 becomes:

$$dv_{j+1} f_{Lj+1} - dv_j f_{Rj} + dp/p(v_{j+1} - v_j) = 0 \qquad 9.4(a)$$

A gas path divided into n nodes has $n - 2$ unknown dv_j. At the extreme left ($j = 1$) $dv_1 = 0$. At the extreme right ($j = n$) dv_n is equal to the change in cylinder volume, and is prescribed. A total of $n - 1$ equations of the form of Eqn 9.4a are available for solution of the $n - 2$ unknown dv_j and for the (single) value of dp/p. The set of $n - 1$ equations forms an array. A representative row – row i – for an internal point ($2 < j < n - 2$) is:

$$a_{i,j-1}dv_j + a_{i,j}dv_{j+1} + \ldots a_{i,n-1}dp/p = b \qquad 9.6$$

Because the left-most unknown dv is not dv_1 but dv_2 the subscript index of coefficient a lags 1 step behind that of dv.

$$a_{i,j-1} = -f_{Rj}$$

$$a_{i,j} = f_{Lj+1}$$

$$a_{i,n-1} = v_{j+1} - v_j$$

$$b = 0$$

First and final rows ($i = 1$, $i = n - 1$) have finite values of b because the (known) dv_1 and dv_n transfer to the right-hand side together with their coefficients, changing sign in the process.

Five nodes ($n = 5$) would be impracticably few, but this number permits illustration of the relationship between the unknowns and the array of coefficients $a_{i,j}$:

dv_2	dv_3	dv_4	dp/p		RHS	
$a_{1,1}$	0	0	$a_{1,4}$	=	0	$(= -f_{R1}dv_1 = -0.f_{R1} = 0)$
$a_{2,1}$	$a_{2,2}$	0	$a_{2,4}$	=	0	
0	$a_{3,2}$	$a_{3,3}$	$a_{3,4}$	=	0	
0	0	$a_{4,3}$	$a_{4,4}$	=	$-f_{Ln}dv_n$	

Values of v_j, v_{j+1} and p following the integration increment are:

$$v_j^{new} = v_j + dv_j,\ v_{j+1}^{new} = v_{j+1} + dv_{j+1},\ p^{new} = p(1 + dp/p)$$

In the computer-coded equivalent, volumes are normalized by swept volume V_{sw} and temperatures by reference (ambient) temperature T_C.

The implicit approach, whether in the above specimen formulation or with provision for limited heat transfer and flow loss, requires a numerical routine to deal with the matrix. These implicit schemes of this account are illustrated with solutions acquired by use of SIMQX (Anon., undated, b) – an unsophisticated algorithm based on pivotal condensation. If the complete gas process model is to be used in optimization studies, it becomes essential to substitute an algorithm offering more favourable computational efficiency. The Nag™ routines F01BRF (decomposition) in combination with F04AXF store and manipulate only the finite coefficients, and have been found to be a great improvement over SIMQX.

A single matrix inversion acquires all the dv_j for $2 \le j \le n - 1$ and the dp. Values of p and v_j at the new time t (or crank angle) are formed by addition: $p_{t+dt} = p_t + dp$, $v_{j,t+dt} = v_{j,t} + dv_j$.

Quasi-explicit scheme

The assumption of uniform pressure, p, is retained for the moment. Eqn 9.4(a) is rewritten in anticipation that a prior estimate of dp will be available at each increment in crankshaft rotation:

$$dv_{j+1} f_{Lj+1} = dv_j f_{Rj} - dp/p(v_{j+1} - v_j) \qquad 9.7$$

At the left-hand (closed) end of the gas path $j = 1$ and $dv_1 = 0$. For this unique case:

$$dv_2 f_{Lj+1} = - dp/p(v_{j+1} - v_j)$$

Applying Eqn 9.4(a) leads to an explicit expression for the dv_2 corresponding to the assumed dp/p. The remaining $n - 3$ values of dv_j follow through application of Eqn 9.7 as a recurrence relationship based on repeated use of the same dp/p.

Starting the integration cycle at the outer (or inner) dead-centre location of the piston ensures that the initial dp is negligible, allowing use of the approximation dp = 0 for the first computational sweep. The resulting solution is adjusted by noting that use of the correct dp would lead to volume increment dv_n at the piston face equal to the (prescribed and thus exact) increment in cylinder volume dV.

The initial approximation to dp is denoted by dp', and initial gas path volume by V_0. Then $(p_0 + \mathrm{d}p')(V_0 + \mathrm{d}v_n) \approx (p_0 + \mathrm{d}p)(V_0 + \mathrm{d}V)$. Making d$p'$ explicit and discarding products of small differences:

$$\mathrm{d}p \approx \mathrm{d}p' + p_0(\mathrm{d}v_n - \mathrm{d}V)/V_0 \qquad 9.8$$

The adjustment (Eqn 9.8) and the recurrence (Eqn 9.7) are applied successively as many times as are required to achieve convergence.

Explicit scheme

This will be illustrated by assuming that the cycle begins with gas temperature distribution as before, but with heat exchange entirely suppressed. The resulting 'adiabatic' cycle is of interest because it introduces and illustrates the interaction of the two phenomena by which the engine functions – temperature profile translation and simultaneous temperature rise or fall in sympathy with pressure change.

With mechanized optimization in prospect it becomes crucial to anticipate that each cycle simulated is going to have to be tested for closure (convergence). Cyclic equilibrium of the physical processes does not *necessarily* coincide with numerical closure, but achievement of the latter in a benchmark case - such as the adiabatic cycle – increases the confidence with which the computation is tested for achievement of the former.

With this in mind second-order terms resulting from expansion of the adiabatic relationship pV^γ = constant are retained:

$$pV^\gamma = (p + dp)(V + \mathrm{d}V)^\gamma$$

$$= p(1 + \mathrm{d}p/p)[V(1 + \mathrm{d}V/V)]^\gamma$$

$$= p(1 + \mathrm{d}p/p)V^\gamma(1+ \mathrm{d}V/V)^\gamma$$

$$(1 + \mathrm{d}p/p)(1 + \mathrm{d}V/V)^\gamma = \text{unity}$$

The term $(1 + \mathrm{d}V/V)^\gamma$ is expanded by the binomial theorem (or in a Taylor series) as far as terms of second order. The result is multiplied out, and the dV for the volume elements expressed in terms of cumulative volume v of the earlier stages:

$$\mathrm{d}v_{j+1} = \mathrm{d}v_j - \mathrm{d}p/\gamma p \frac{v_{j+1} - v_j + \gamma(\mathrm{d}v_{j+1} - \mathrm{d}v_j)/(v_{j+1} - v_j)}{1 + \frac{1}{2}(\gamma - 1)(\mathrm{d}v_{j+1} - \mathrm{d}v_j)/(v_{j+1} - v_j)} \qquad 9.9$$

Equation 9.9 is not strictly explicit in the unknown dv_{j+1} and dv_j, but use of respective values from the previous time step for occurrences on the right-hand side makes it so. Iteration deals with any discrepancy.

Successive applications of the recurrence between dv_{j+1} and dv_j amount to a summing of the right-hand side of Equation 9.9 – but with dp/p as a multiplier of each term in the sum – which may therefore be accumulated independently of dp/p.

Applied to the right-hand piston face dv_{j+1} is dv_n, and has a pre-set value. Equation 9.9 can thus be inverted to yield dp/p. All the dv_j are then available by back-substitution. Local temperatures T_j follow by application of the differential form of the isenstropic relationship $T/p^{(\gamma-1)/\gamma}$ = constant, namely, by application of $dT/T = (\gamma - 1)/\gamma \, dp/p$.

Table 9.1 is a numerical description of a small 'thermal lag' engine and a set of assumed operating conditions. Figure 9.3 is a temperature relief obtained by solution of Eqn 9.9 for the data of the table. The original temperature profile is seen being displaced leftward over the compression phase, and returning rightward corresponding to expansion. The right-most particle locus indicates the extent of overall temperature rise and fall associated with the rise and fall of pressure.

When integration proceeds at crank angle increments of 1 degree (360 sweeps/cycle) pressure after a single cycle returns to the start value within an accuracy better than 0.001 percent. Specific cycle work – ideally zero – is returned as (minus) 0.000076 (–).

The temperature reliefs which illustrate this account are computed at lower resolution – 4 degree intervals, or 90 steps per cycle, and with the gas path sub-divided by 40 values of cumulative volume v. CPU time per cycle to compute the coordinates for Fig. 9.3 was 15.6 ms. Time to convert to the screen plot was 31.2 ms.

The explicit formulation may be taken a step towards the ultimate objective – a solution reflecting limited heat transfer – by formulating the 'temperature-determined' case. For a gas parcel between two bounding faces (particle paths) this requires pV/T = constant. The integration algorithm is acquired via the equality:

$$(p + \mathrm{D}p)(V + \mathrm{D}V)(T + \mathrm{D}T)^{-1} = pV/T$$

The left-hand side may be expanded via the approximation $(T + dT)^{-1} = T(1 + dT/T)^{-1} \approx T^{-1}(1 - dT/T)^{-1}$. On the other hand, van Rijn (2006) has found that multiplying up before expanding results in the more convenient form:

$$(p + \mathrm{D}p)(V + \mathrm{D}V) = (T + \mathrm{D}T)pV/T$$

Carrying out the multiplication and retaining terms of second order:

$$\mathrm{D}V - V\mathrm{D}T/T = -\mathrm{D}p/p(V + \mathrm{D}V) \qquad 9.10$$

Table 9.1 Geometric and operating parameters used in simulation of 'thermal lag' engine of Figs 9.12 and 9.13

V_{sw}	swept volume		(cm^3)	19.9
p_{ref}	charge pressure		(Pa)	10^5
rpm	revolutions per minute		(1/min)	1000
Operating temperatures (K)				
	T_E	expansion		900
	T_C	compression		300
Free-flow areas			(mm^2)	
	A_{ffr}	regenerator (housing X-sec. area × volume porosity $¶_v$)		175.0
	A_{ffxe}	expansion exchanger		43.0
	A_{ffpt}	pulse tube		78.6
	A_{ffxc}	compression exchanger		43.0
	A_{ffc}	compression cylinder		831.0
Flow passage lengths			(mm)	
	L_r	regenerator		120.0
	L_{xe}	expansion exchanger		8.0
	L_{pt}	pulse tube		130.0
	L_{xc}	compression exchanger		8.0
Hydraulic radii			(mm)	
	r_{hr}	regenerator		0.04
	r_{hxe}	expansion exchanger		0.425
	r_{hpt}	pulse tube		2.5
	r_{hxc}	compression exchanger		0.425
	r_{hc}	compression cylinder		8.0
working fluid			(–)	air
γ	specific heat ratio		(–)	1.4
R	gas constant (J/kgK)		287	
μ_{300K}	coeff. of dyn. visc. at amb. Temp.		(Pas)	0.017×10^{-3}
$¶_v$	volume porosity of regenerator mesh		(–)	0.8
T_{su}	Sutherland temperature		(K)	112
'Benchmark' integration subdivisions:				
n	no. gas path nodes		(–)	40
n_ϕ	no. crank angle steps/cycle		(–)	90

Scalar volume V is once again expressed in terms of *cumulative volume*, v, reckoned from an origin at the left-hand extremity of the gas path. Change in temperature DT_j of a particle following track j is again $dT_w/dv\,|_j dv_j$, as per Eqn 9.2(a). Re-using the abbreviations f_{Rj} and f_{Lj+1} (Eqns 9.5(a) and (b) Eqn 9.10 becomes, for $p = p(t)$ only:

$$dv_{j+1} = dv_j\, f_{Rj}/f_{Lj+1} - dp/p(v_{j+1} - v_j + dv_{j+1} - dv_j)/f_{Lj+1} \qquad 9.11$$

This is essentially the same as Eqn 9.4(a) for the implicit case, but with $dv_{j+1} - dv_j$ retained on the right-hand side to achieve 'second-order' precision. Values from the previous time step serve for these 'quasi-knowns' on a first iteration.

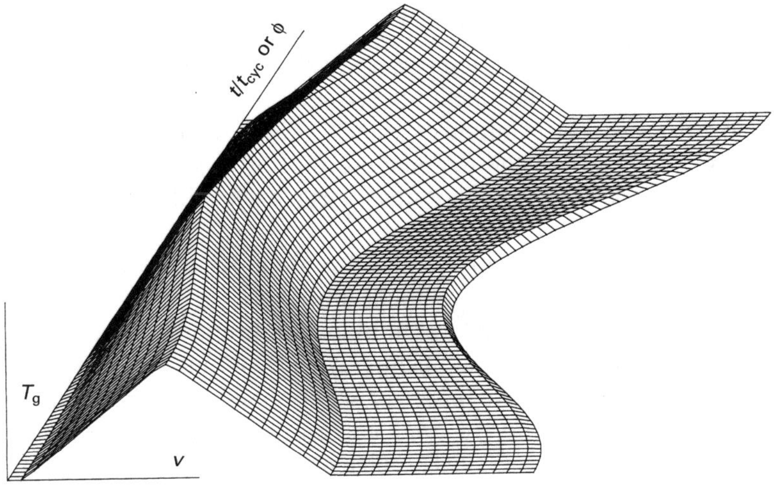

9.3 Explicit solution of Eqn 9.8 (adiabatic case) for parameters for data of Table 9.1. Initial temperature distribution is shifted back and forth. Overall rise and fall in temperature accompanying compression and expansion are indicated by the right-most particle path.

The extreme left-hand end of the system is closed and stationary, so that f_{R1} = zero.

$$dv_2 = -\, dp/p(v_2 - v_1 + dv_2 - dv_1)/f_{L2} \qquad 9.12$$

Applying the recurrence again:

$$dv_3 = dv_2\, f_{R2}/f_{L3} - dp/p(v_3 - v_2 + dv_3 - dv_2)/f_{L3}$$

dv_2 is available from Eqn 9.12:

$$dv_3 = \{-dp/p(v_2 - v_1 + dv_2 - dv_1)/f_{L2}\}\, f_{R2}/f_{L3}$$

$$-\, dp/p(v_3 - v_2 + dv_3 - dv_2)/f_{L3}$$

$$= -dp/p\{(v_2 - v_1 + dv_2 - dv_1)\, f_{R2}/f_{L2} + (v_3 - v_2 + dv_3 - dv_2)\}/f_{L3}$$

Both v_1 and dv_1 are zero – although the terms representing them have been allowed to remain to illustrate the algorithm in general form:

$$dv_n = -dp/p\{(v_n - v_{n-1} + dv_n - dv_{n-1})$$

$$+ \Sigma(v_j - v_{j-1} + dv_j - dv_{j-1})\, f_{Rj-1}/f_{Lj-1}\}/f_{Ln} \qquad 9.13$$

In Eqn 9.13 the summation Σ runs from 1 to $n - 1$.

The system is closed at the extreme right (particle path $j = n$) by the piston face, whose location and motion are known in terms of $v_n + dv_n$, meaning that dp/p is available from Eqn 9.13:

$$dp/p = -f_n dv_n / \{\Sigma(v_{j+1} - v_j + dv_{j+1} - dv_j)\} \qquad 9.14$$

With dp/p known, the various dv_j are available by back-substituting into Eqn 9.11, starting at the left-hand (closed) end of the system.

Figure 9.4 is a solution of Eqn 9.14 based on the data of Table 9.1. Whereas in Fig. 9.3 the temperature profile was transported wholesale at particle speed, the particles now wash back and forth through a static temperature profile.

CPU time per cycle differs little from that for the earlier, adiabatic case, and closure is to similar numerical limits.

Equation 9.14 is a counterpart, in Lagrange coordinates, of the celebrated 'isothermal' analysis of the Stirling cycle (Schmidt 1871), which is adapted to the thermal lag engine in a later section. Schmidt's is a closed, algebraic solution for p in function of crank angle ϕ (= ωt). It may be used to verify the differential analysis at its current stage of development. The Lagrange solution based on 40 volume subdivisions or more yields a cyclic variation of p coincident with that predicted by the Schmidt solution.

9.5 Limited heat transfer

With the formulation duly verified to this point, the dependent variable DT may be reintroduced as an unknown – i.e. the arbitrary T may be replaced by that computed from an instantaneous heat balance.

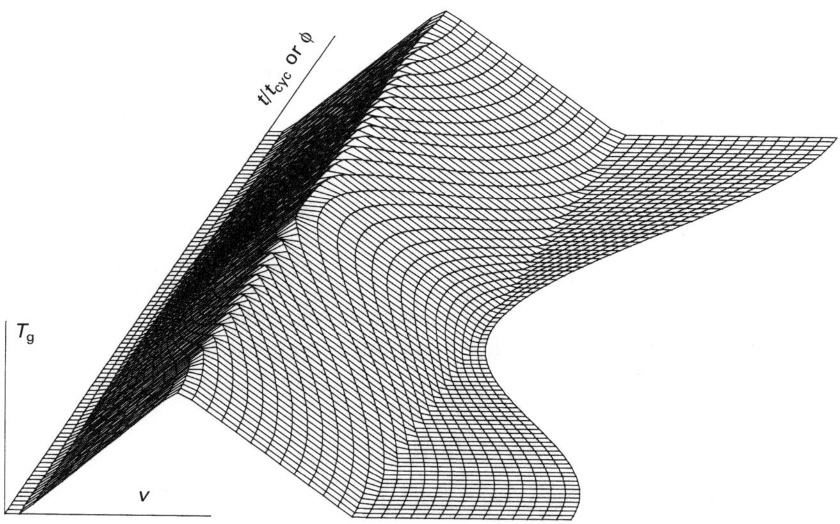

9.4 Solution of Eqns 9.13 and 9.14. Temperature is determined by location regardless of motion history – opposite extreme from adiabatic case of Fig. 9.3.

The problem of heat exchange between a duct with surface temperature gradient dT_w/dx in the flow direction and an adjacent fluid particle in motion at instantaneous velocity $\underline{u}$ and temperature T_g has been addressed by the author in the context of the regenerator. The original solution (e.g. Organ 1997, p. 247) is in terms of local, instantaneous temperature difference $\Delta T = T_g - T_w$, but may be re-expanded:

$$(T_g - T_w)_{t+\Delta t} = Q/P + \{(T_g - T_w)_t - Q/P\}e^{-P\Delta t} \qquad 9.15$$

$$Q = -\underline{u}\,dT_w/dx$$

$$P = |\,u\,|NTU/L$$

Equation 9.15 is rearranged to make explicit the increment in T_g over time interval dt:

$$T_g|_{t+\Delta t} - T_g\,|_t = DT$$

$$= (dT_w/dx)\underline{u}dt + (1 - e^{-P\Delta t})\{Q/P - (T_g - T_w)_t\} \qquad 9.16$$

Along a particle path $(dT_w/dx)\underline{u}dt = DT_w$. Velocity $\underline{u}$ converts to the present coordinate system (cumulative volume) via $\underline{u}dt = A_{ff}^{-1}dv$. Occurrences of D$T$ in Eqns 9.5(a) and (b) may be replaced by corresponding right-hand sides of Eqn 9.16. For pressure p a function of t only, Eqn 9.16 is compatible with all three formulations so far discussed – implicit, quasi-explicit and explicit.

The result in each case is a 'two conservation law' or CL-2 model of the gas processes, reflecting conservation of mass and energy. A convenient shorthand designation is CL-2ME (Organ 1992a). The five graphical illustrations which follow were generated by the quasi-explicit method in CL-2ME form, and reflect the data of Table 9.1. At any given location and crank angle, instantaneous *NTU* are computed from local Reynolds number *Re* and from steady-flow heat transfer correlations appropriate to local flow passage geometry. In view of the low values of Mach number *Ma* anticipated, Kays' and London's correlations are used from the 1964 compendium.

All computation is carried out in the 'cumulative volume' coordinate system introduced earlier. Diagrams to scale (e.g., maps of particle motion history) are converted back to 'real' ($\omega t - x$) coordinates at the plotting stage. For the specimen particle trajectory map of Fig. 9.5 (only) differences in free-flow area have been exaggerated relative to those of Table 9.1 to highlight the effect of the coordinate transformation.

Figure 9.6(a) and (b) are reliefs ('carpet' plots) of gas temperature. The jagged features interrupting the discontinuities in the temperature surface are in proportion to integration step size, and may be made arbitrarily small. For Fig. 9.6(a) piston motion is discontinuous for correspondence with the 'ideal' situation of Fig. 9.1. The start-up cycle is illustrated. The initial temperature profile is swept leftward. Fluid entering the regenerator at the

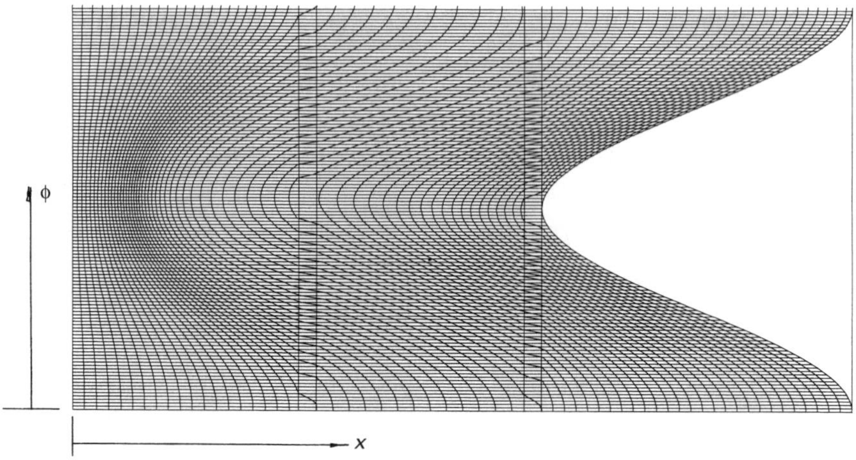

9.5 Particle trajectory map. Discontinuities in free-flow area have been exaggerated relative to those of Table 9.1 to illustrate the effect of the transformation from cumulative volume *v* back to 'real' ($\omega t - x$) coordinates.

point of peak temperature attains a temperature close to that of the matrix. With the piston stationary at inner dead-centre gas temperature in the pulse tube 'relaxes' towards the (linear) wall temperature, accompanied by a slight readjustment of fluid distribution.

Reverse movement of the piston away from inner dead-centre causes a flow of gas into the pulse-tube section. It enters at regenerator exit temperature T_E, carrying the elevated temperature distribution to the pulse-tube exit, i.e., to heat sink entry. The small hydraulic radius of the heat sink ensures that temperature entering the cylinder is close to T_C. The temperature profile in the pulse-tube section relaxes somewhat following expansion, and there is a corresponding redistribution of working fluid.

There is conversion of heat to net indicated work consistent with expansion temperature exceeding mean compression temperature. In Fig. 9.6(b) the same cycle is recomputed assuming simple-harmonic piston motion.

Earlier it was suggested that a way of appreciating the pulse-tube cycle is in terms of an asymmetry between two gas temperature distributions – that during compression and that during expansion. The asymmetry may be highlighted by plotting local *difference*, $\Delta T = T_g - T_w$, between the temperature of the gas and that of the adjacent solid surface. Figures 9.7(a) and (b) plot ΔT corresponding to Figs 9.6(a) and (b) respectively.

A plot of pressure against cylinder volume over a complete crankshaft revolution yields an indicator diagram and a numerical value of indicated work/cycle. Under conditions of limited heat transfer, several complete cycles are necessary for equilibrium to be obtained. Eventually, successive *p-V* loops superimpose, leading to Fig. 9.8 for the data of Table 9.1.

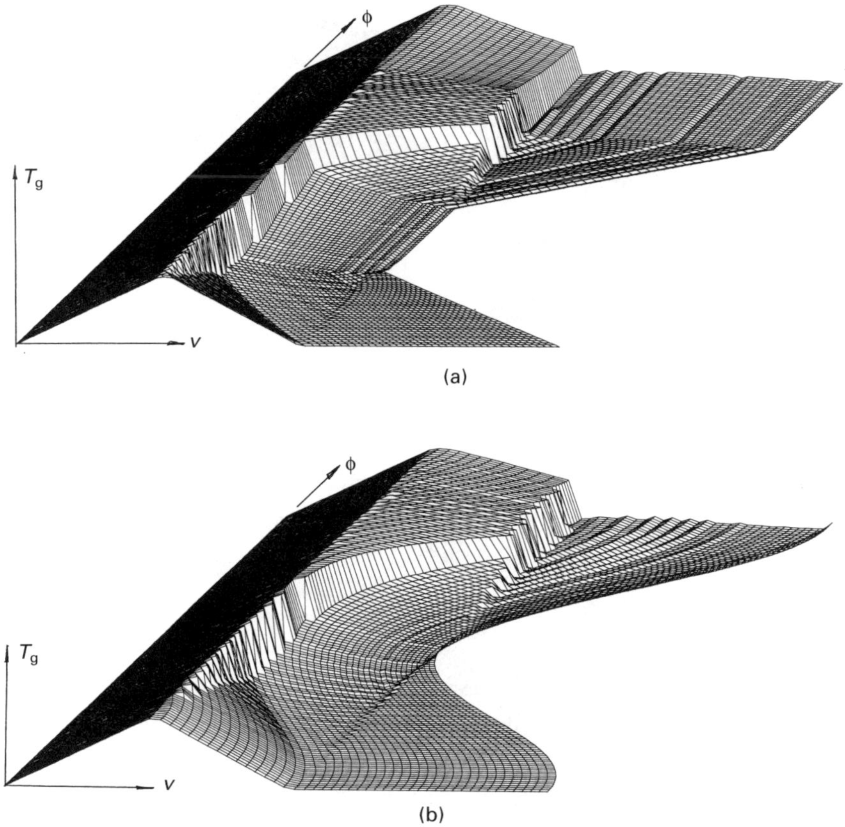

9.6 Temperature reliefs ('carpet' plots) for simulated thermal lag engine defined by data of Table 9.1 (a) piston motion discontinuous (cf Fig. 9.1). (b) piston motion simple-harmonic, but same swept volume as for (a) above.

9.6 Flow losses

The investigation to this point bears out the view that pulse-tube devices function by processes which, in other energy converters, have 'second order' effect. This being so, it is to be expected that pressure gradients associated with movement of the gas will have a strong influence. It is convenient to refer to a gas process model which accounts for flow losses as well as defective heat transfer as a '3 conservation law', or CL-3 model.

In the present study momentum conservation has been implemented differently in the three formulations – implicit, quasi-explicit and explicit.

9.6.1 Incorporation into implicit scheme

Pressure change Dp when following a particle (Lagrange view) is:

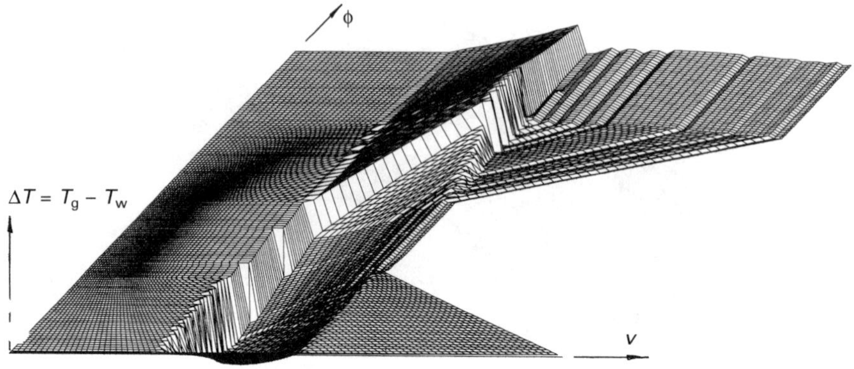

(a)

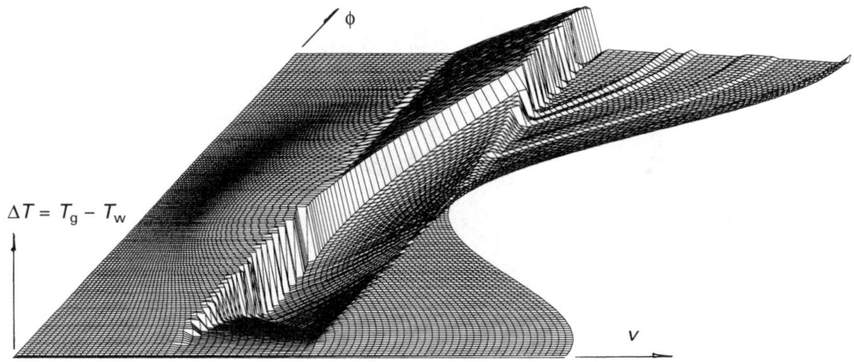

(b)

9.7 Reliefs ('carpet' plots) of local, instantaneous temperature *difference* $\Delta T = T_w - T_g$ for simulated thermal lag engine defined by dimensions and operating parameters of Table 9.1 (a) piston motion discontinuous (cf Fig. 9.1); (b) piston motion simple-harmonic, but same swept volume as for (a) above.

$$\mathrm{D}p\ |_{\text{path}} = (\partial p/\partial t + \underline{u}\partial p/\partial x)\mathrm{d}t \qquad 9.17$$

It will be assumed that the steady-flow term of the $\partial p/\partial x$ component dominates at the modest Mach numbers anticipated:

$$\partial p/\partial x \approx -(\tfrac{1}{2}\rho\underline{u}^2 C_f/r_h)\mathrm{sign}(\underline{u}) \qquad 9.18$$

In Eqn 9.18 C_f is (dimensionless) friction factor and r_h is hydraulic radius (m). The term sign($\underline{u}$) is evaluated as $\underline{u}/|\underline{u}|$, and is present to ensure that friction opposes flow.

Substituting Eqn 9.18 into Eqn 9.17:

$$\mathrm{D}p\ |_{\text{path}} = \{\partial p/\partial t - \underline{u}(\tfrac{1}{2}\rho\underline{u}^2 C_f/r_h)\ \underline{u}/\ |\ \underline{u}\ |\}\mathrm{d}t \qquad 9.19$$

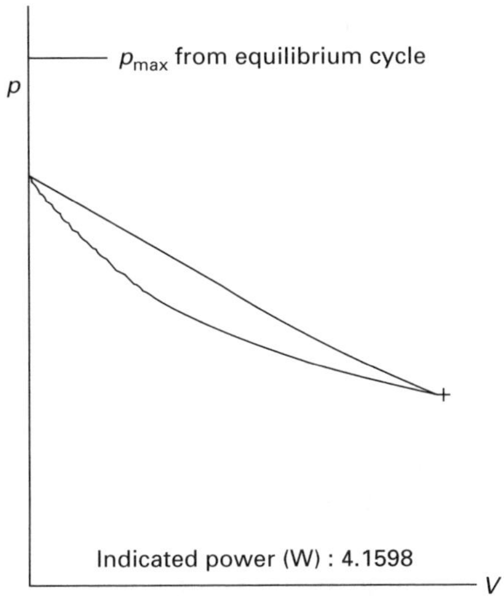

9.8 Indicator diagram for the parameters of Table 9.1. With limited heat transfer it is necessary to compute a number of cycles before successive indicator diagrams superimpose, and thus before the value of indicated work is meaningful.

Embodying Eqn 9.19 in the matrix of coefficients means expressing $\partial p/\partial x$ in terms of the unknowns dv. Figure 9.9 indicates how this is achieved at mid-time step Δt as a proportion of the variable path line increments dx_j, dx_{j+1} either side.

Equation 9.19 is non-linear in $\underline{u}$, and an element of sleight-of-hand is necessary in preparing it for numerical solution: the isolated $\underline{u}$ is represented as $dx_j/\Delta t$ (in fact, by the equivalent in $dv_j/\Delta t$) and the term in $\underline{u}^2$ evaluated conditionally using a previous (stored) value of $\underline{u}$. Iteration corrects the approximation. Velocity $\underline{u} = dx_j/\Delta t$ is based on mid-spatial interval.

There are insufficient equations to define p at all n particle paths. However, velocity, and thus $\partial p/\partial x$ are zero at the left-hand face, and p_2 accordingly equals p_1 for all t. Adding the coefficients for the $n - 1$ unknown dp_j to the array for the dv_j doubles the size of the matrix and quadruples storage requirements and processing time. With the impact on optimization to be considered, other ways of tackling flow loss are worth considering.

9.6.2 Incorporation into any scheme – implicit, quasi-explicit, explicit

The slender indicator diagram of Fig. 9.8 is characteristic of the thermal lag engine. It suggests that end-to-end pressure difference must be a small fraction of absolute pressure for the machine to function. This may justify treating

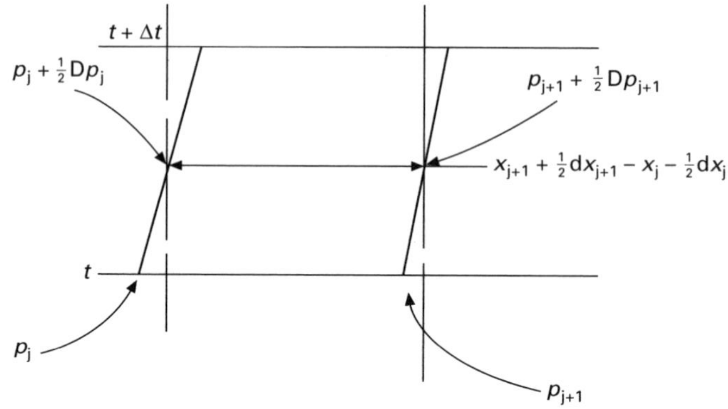

9.9 The $\partial p/\partial t_j$ are expressed in terms of the dv_j by pro-rating in terms of the values either side at $j-1$ and $j+1$.

local pressure gradients at any instant as perturbations of a notional mean pressure at that instant, computing them iteratively from the uniform-pressure treatment.

The $\partial p/\partial x$ computed from Eqn 9.19 are available at all locations j. Adjacent values of p_j follow in terms of cumulative volume v (rather than conventional x):

$$p_{j+1} \approx p_j + \partial p/\partial v\,|_j(v_{j+1} - v_j)$$

For the element at the extreme left:

$$p_2 \approx p_1 + \partial p/\partial v\,|_1(v_2 - v_1)$$

$$p_3 \approx p_2 + \partial p/\partial v\,|_2(v_3 - v_2)$$

$$\approx p_1 + \partial p/\partial v\,|_1(v_2 - v_1) + \partial p/\partial v\,|_2(v_3 - v_2)$$

Thus, pressure p_{nj} at the piston face ($j = n$) is available in terms of p_1 (at the left-hand end) from the recurrence:

$$p_n \approx p_1 + \Sigma\partial p/\partial v\,|_j(v_{j+1} - v_j) \qquad 9.20$$

Summation is carried out over the range $j = 1, n - 1$. Pressure p_1 (or p_n) might be thought of as a constant of integration – except that, to this point, neither has a known value: none of pressures p_j is known until p_1 (or p_n) is determined.

The latter may be computed from the assumption of invariant total working fluid mass M. At start-up, pressure is uniform at specified p_{ref} (typically 1 bar). Gas and wall temperature distributions are initially identical, and compression space volume is V_{sw}. For these conditions, the mass content of each distinct volume V_d in the gas path may be calculated by application of the ideal gas law as per the Schmidt (1871) approach. Individual values are summed to yield M as $p_{ref}/R\Sigma(V_d/T)$:

$$M = p_{ref}/R\{V_{dxe}/T_E + V_{dr}f(N_T)/T_C + V_{pt}f(N_T)/T_C$$
$$+ V_{dxc}/T_C + V_{sw}/T_C\}$$
$$f(N_T) = (\log_e N_T)/(N_T - 1)$$
$$N_T = T_E/T_C$$

Determining the distribution p_j at any instant during the cycle relies on noting that M remains constant under any axial distribution of pressure $p(v)$ with cumulative volume v:

$$M \approx \Sigma p_j(v_{j+1} - v_j)/RT_j \qquad 9.21$$

The summation is carried out over the entire gas path, i.e., from $j = 1$ to $j = n - 1$.
The first two terms of the summation are

$$\{p_1 + \partial p/\partial v_1(v_2 - v_1)\}(v_2 - v_1)/RT_1$$
$$+ \{p_1 + \partial p/\partial v_1(v_2 - v_1) + \partial p/\partial v_2(v_3 - v_2)\}(v_3 - v_2)/RT_2$$
$$+ \ldots . = M$$

This may be rewritten:

$$p_1(v_2 - v_1)/T_1 + \{\partial p/\partial v_1(v_2 - v_1)\}(v_2 - v_1)/T_1$$
$$+ p_1(v_3 - v_2)/T_2 + \{\partial p/\partial v_1(v_2 - v_1)$$
$$+ \partial p/\partial v_2(v_3 - v_2)\}(v_3 - v_2)/T_2$$
$$+ \ldots . = 2MR$$

More generally:

$$\Sigma_1^{nj} p_1(v_{j+1} - v_j)/T_j + \Sigma_1^{nj}\{\Sigma_1^j \partial p/\partial v_j(v_{j+1} - v_j)](v_{j+1} - v_j)/T_j\}$$
$$\approx MR \qquad 9.22$$

The value p_1 at location $j = 1$ which leads to the non-linear pressure distribution consistent with the value of M is

$$p_1 \approx \frac{MR - \Sigma_1^n\{[\Sigma_1^j \partial p/\partial v_j(v_{j+1} - v_j)](v_{j+1} - v_j)/T_j\}}{\Sigma_1^n(v_{j+1} - v_j)/T_j} \qquad 9.23$$

With p_1 known, the p_j follow by successive application of Eqn 9.20.
Figure 9.10 again reflects the data of Table 9.1, and for the start-up cycle plots pressure distribution *minus* the underlying uniform pressure $p(t)$. The magnification factor involved in displaying the small gradients is partly responsible for the irregularities on the pressure surface.

At any given instant the pressure gradient is confined largely to the regenerator, where particle speeds are lowest, but hydraulic radius is smallest. Gradients there increase rapidly from the closed end rightwards due to increasing particle speed and to the increase in coefficient of dynamic viscosity with rising temperature.

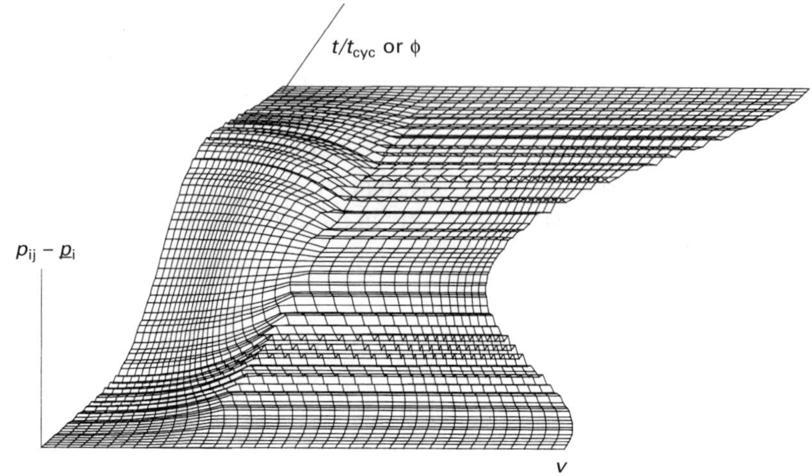

9.10 Pressure distribution *minus* underlying pressure. From data of Table 9.1.

CPU times for the explicit CL-3 formulation are little changed from those of the CL-2. This may be the result of including some of the summation arithmetic within existing computational loops. The author codes in FORTRAN and uses the Lahey-Fujitsu 95 compiler v5.6. The extent of code optimization carried out during the multiple compilation passes is not known.

9.7 A practical 'thermal lag' engine

It is anticipated that indicated power will be small, and brake power negligible. With little as background beyond descriptions of table-top models such as those of Warbrooke (2005), the engine illustrated at Figs 9.11 and Plate 3 (between pages 134 and 135) was designed with elimination of mechanical friction as top priority. Internal proportions are essentially those of Table 9.1.

A quantity of flanged, imperial size miniature ball-races was available, which pre-determined $\frac{1}{8}$-inch (≈ 3 mm) as the diameters of main journal and crank-pin. Crank-case and heat sink were machined from off-cuts of aluminium alloy, material specifications not being available. Pulse-tube and cylindrical regenerator were machined from solid to approximately 0.1 mm wall thickness from low-carbon steel bar stock.

Cylinder bore was determined by the availability of a proprietary Airpot™ cylinder set, comprising a carbon piston running in a pyrex tube. (This technology routinely turns up as 'snubbers' in photocopiers.)

The manufacturers of the Airpot™ units do not guarantee the end faces of the pyrex tubes to be perpendicular to the axis to better than 1 degree. Misalignment outside the plane of motion of the connecting rod shows up as

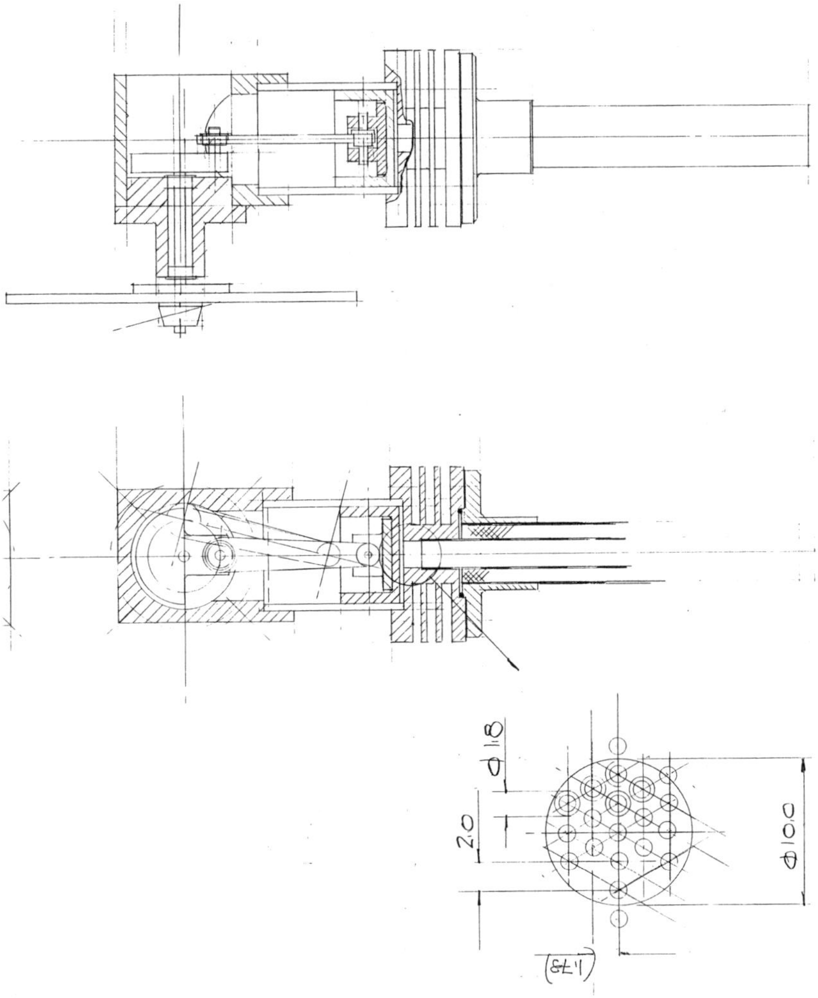

9.11 Arrangement drawing of 20cc experimental 'thermal lag' engine.

unnecessary friction. O-rings (not illustrated in Fig. 9.11) provide sealing between the end faces of the pyrex tube and the crank-case at one end, and the heat sink at the other. Misalignment is dealt with by differential tightening of the four M4 securing studs and accommodated by uneven deformation of the O-rings. The interested reader designing from scratch would be advised to provide for positive alignment.

9.8 Preliminary operating experience

The open flame of a methylated-spirit lamp gives an expansion-end temperature of about 480°C as measured by a portable thermocouple. At this temperature, no-load rpm are about 240. With the extra heat of a gas flame, rpm exceed the maximum scale reading (1000 rpm) of the available tachometer by an amount estimated to be 200 rpm. If the estimate is correct, some 1200 rpm are achieved.

For exploratory power measurements the gas path has been temporarily modified: pulse-tube and regenerator sections have been 'straightened out' and accommodated in a continuous length of pyrex tube. The regenerator consists of a stack of Retimet™ metal foam plugs accurately cut by a numerically-controlled water-jet machine. An electrical resistance coil located internally replaces the external heating and allows input power to be accurately set. A simple friction dynamometer measures torque.

Figure 9.12 shows specimen curves of brake power as a function of rpm. The *lower* of the two values of electric power input (43W) corresponds to the *higher* values of torque, brake power and rpm. The apparent contradiction draws attention to the effects of heat dissipation from the outside diameter of the pyrex tube: with no thermal insulation, input power of 64W was required to yield the lower set of performance figures. A radial thickness of approximately 25 mm of standard glass-fibre insulation resulted in the improvement illustrated.

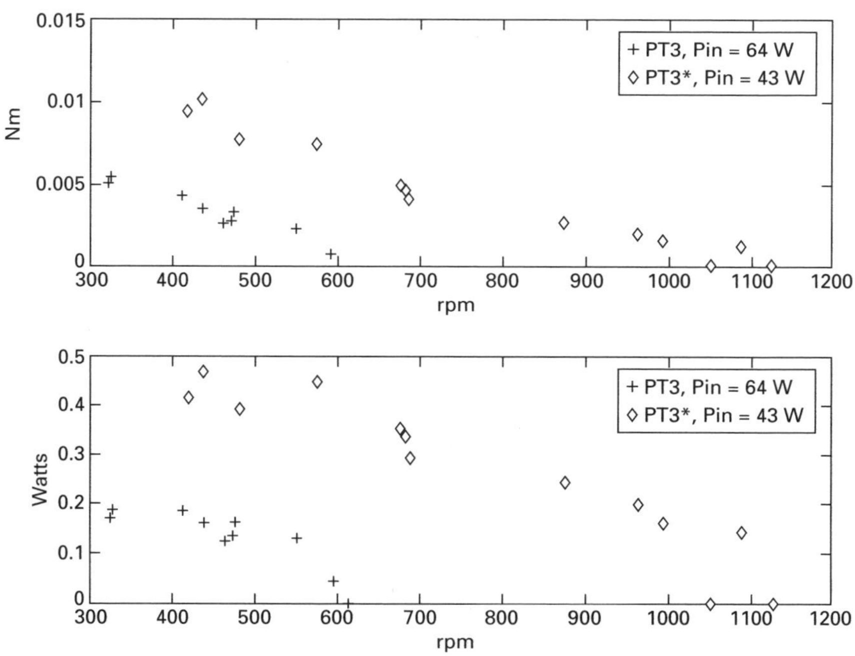

9.12 Torque and brake power as functions of rpm. Engine modified for internal electrical resistance heating. Measurements taken in collaboration with R G James. After van Rijn (2005) with permission.

Hydraulic radius r_h (= free-flow area/wetted perimeter) of the pulse-tube section is readily varied whilst keeping volume and free-flow area constant. Experiment and simulation confirm that the value of r_h is critical. Other things being equal, too large a value means too little heat transfer capacity, and the engine does not run. Too small a value gives excessive heat transfer capacity, too little 'thermal lag', and again the engine does not run. (It also gives high flow resistance.)

Samples of Retimet™ metal foam were secured in three grades, 20 (coarse), 45 and 80 (fine). The results of Plate 3 were achieved with regenerator plugs of 45 grade. If these are substituted by plugs of the finer (80) grade the engine does not function.

9.9 Afterthought

It is noteworthy that the slender indicator diagram of the computer simulation straddles the single line (zero-area curve) of the corresponding Schmidt algebra. An alternative approach to constructing a gas process model would be to formulate it as a 'perturbation' relative to the Schmidt solution. Perturbation offers scope for stabilizing the numerical solution – but above all for *linearizing* the analytical steps.

In the meantime, the three conservation law implicit formulation has been recast in the form of an 'object function' for calling by an optimization routine. At 2000 rpm and a charge pressure of 10 atm abs. (9 atm gauge) the best match of r_h for pulse-tube section and regenerator predicts 25W (indicated) from 20cc displacement.

Before dismissing prospects for the thermal lag engine, it may be worth reflecting that devices operating on the reversed-cycle, namely, pulse-tube coolers, challenge the Stirling counterpart in a small but high-value market for cryogenic refrigeration. Any eventual applications for the engine will almost certainly be of the 'niche' variety, since the type does not compete with any established prime mover.

There are two further reasons why these devices should not be written-off (1) their unique mechanical simplicity and (2) the high probability that exploration of the cycle – and thus of power-producing potential – has barely commenced. All other things being equal, increased size and increased pressure level both reduce *NTU*. The concept *demands* restricted *NTU*. It is seldom that a design requirement plays into the hands of the designer. Indeed it might be said that we owe it to the challenge to respond!

To remain optimistic about prospects for the thermal lag concept it is necessary only to compare the technology, performance (and price) of the first, rustic compression-ignition engines for model aircraft propulsion (Laidlaw-Dickson 1946-7, Bowden 1947) with those of the sophisticated, twenty-first century counterpart.

Part III

Working with the reality of compressible flow

Plate 1 Full-size replica furnace, flue and displacer cylinder. The static structure supporting the cantilever portion of the masonry doubled as scaffolding during construction.

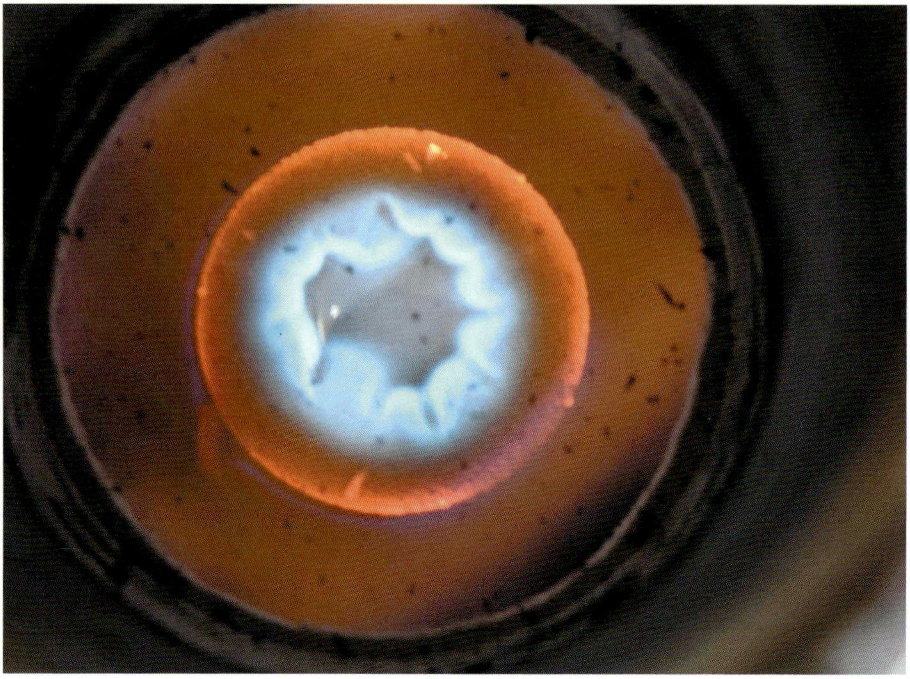

Plate 2 Combustion reaction as seen through a Pyrex window in the dummy cylinder.

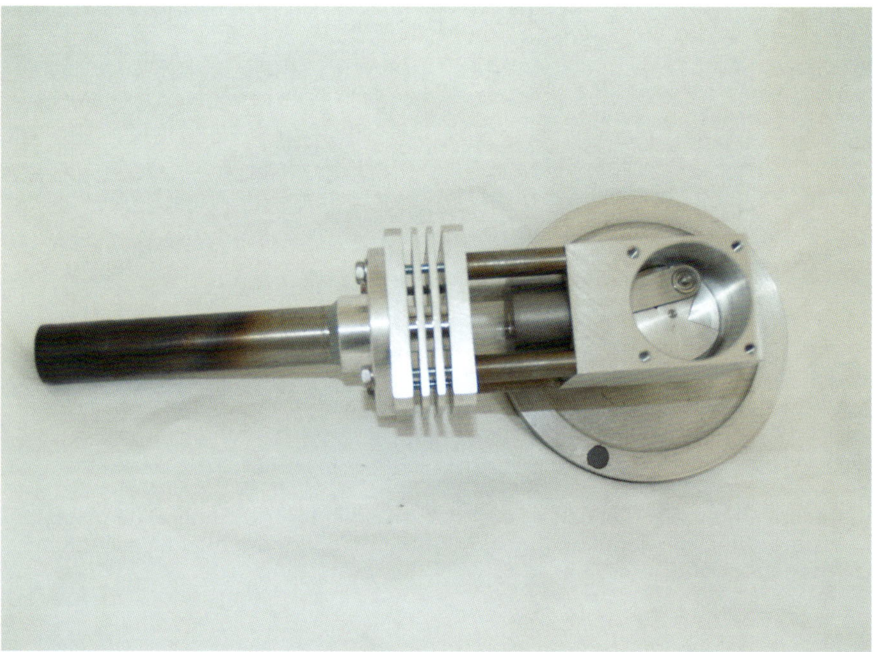

Plate 3 'Thermal lag' engine of 20cc displacement. Machining work by Ian Larque.

10

New correlations for old

10.1 Right data – wrong application

The substantial academic literature on the thermal regenerator occupies itself hardly at all with experimental data. It focuses on the analytical challenge of predicting transient thermal response to arbitrary heat transfer intensity. Flow loss is ignored.

By contrast, design and study of Stirling and like machines calls for prediction of 'real' heat transfer under a realistic estimate of pumping penalty.

Half-a-century ago the US Office of Naval Research supported a 'Heat Transfer Research Project' at Stanford University (Coppage and London 1956). There resulted one of the most highly-regarded and widely-quoted engineering reference works of our time *Compact Heat Exchangers* by Kays and London (1964). The *tour de force* compendium of flow and heat transfer data is the more remarkable for the fact that it flourishes in print in a third edition (1998).

The application which the US Navy had in mind for their heat transfer and flow data is not known – but it is a safe to assume that it was not the air-charged Stirling engine. The fact has not deterred the unquestioning appropriation of these data for use in the alien context. Nor, indeed, has it prevented the rectangular-weave gauze from assuming the status of definitive material for matrix use – a status from which even its high cost appears not to detract.

A new technology is required. Frequent reference will be made to the old, and on the strength of its currency, Kays' and London's classic work will serve as the point of departure.

10.2 The misleading C_f – Re correlation

Consider the definitions of Reynolds number *Re* and Mach number *Ma*:

$$Re = 4\rho|\underline{u}|r_h/\mu \qquad 10.1$$

$$Ma = |\underline{u}|/\sqrt{\gamma RT} \qquad 10.2$$

Referring both definitions to a common location allows velocity $\underline{u}$ to be eliminated, leaving the ratio:

$$Ma/Re = \tfrac{1}{4}\mu(RT/\gamma)^{1/2}/r_h p \qquad 10.3$$

This may be the most important equation the author has yet entered in the word-processor! *Local, instantaneous Ma/Re is independent of particle velocity* $\underline{u}$! Numerical values of γ and μ are fixed by specifying the gas constant. At any given temperature T the relationship between Ma and Re is determined by the numerical value of the simple product pr_h – (absolute) pressure p (Pa) with hydraulic radius r_h (m).

High Re can be achieved from high $\underline{u}$ and/or from large r_h. *Correlations to traditional format do not distinguish.* Kays and London, for example, achieved the higher end of their range Re range by use of large hydraulic radius r_h ($\frac{3}{8}$ inch, or 9.5 mm) rather than by high flow speed.

Their results cannot be reproduced in the fine wire screens of the modern high-power-density regenerator – as is now readily illustrated:

Figure 10.1(a) is the graphical equivalent of Eqn 10.3 for air (or N_2) at ambient temperature. It can be used to read off Ma corresponding to peak Re of the standard correlations, namely, $Re = 10^5$. Take $r_h = 0.04$ mm (0.04×10^{-3} m), $p = 10$ atm (10^6 Pa) and read Ma/Re from the figure, giving $Ma/Re = 2.6 \times 10^{-5}$. So for $Re = 10^5$ $Ma = 2.6$, calling for particle speed to be more than two-and-a-half times the speed of sound!

The result can be confirmed with the hand calculator on noting that air has $R = 287$ J/kgK, coefficient of dynamic viscosity $\mu = 0.017 \times 10^{-3}$ Pas at ambient conditions ($T = 300$K). Figure 10.1(b) facilitates conversion between Re and Ma at 600°C – e.g. at the expansion end of the regenerator, where μ for air (and N_2) is 0.039×10^{-3} Pas.

The finding raises doubts about the applicability to Stirling engines of *any* heat transfer and flow data correlated to Reynolds number Re. Further probing (see later in this chapter and Chapter 14) suggests that, far from being ideal, the woven wire screen is a potential limitation on the performance of Stirling engines charged with heavy gases (air or N_2).

These issues determine the content and direction of the rest of this book: are Mach number effects a feature of routine operation of the air-charged Stirling engine? Are they a limitation on performance – and have they always been there, undetected? Can they be designed around – or even put to advantage?

10.3 Flow data acknowledging *Ma*

The Kays and London correlations are composites from various sources, believed to include the PhD thesis of Coppage (1952) whose stacks were 3,

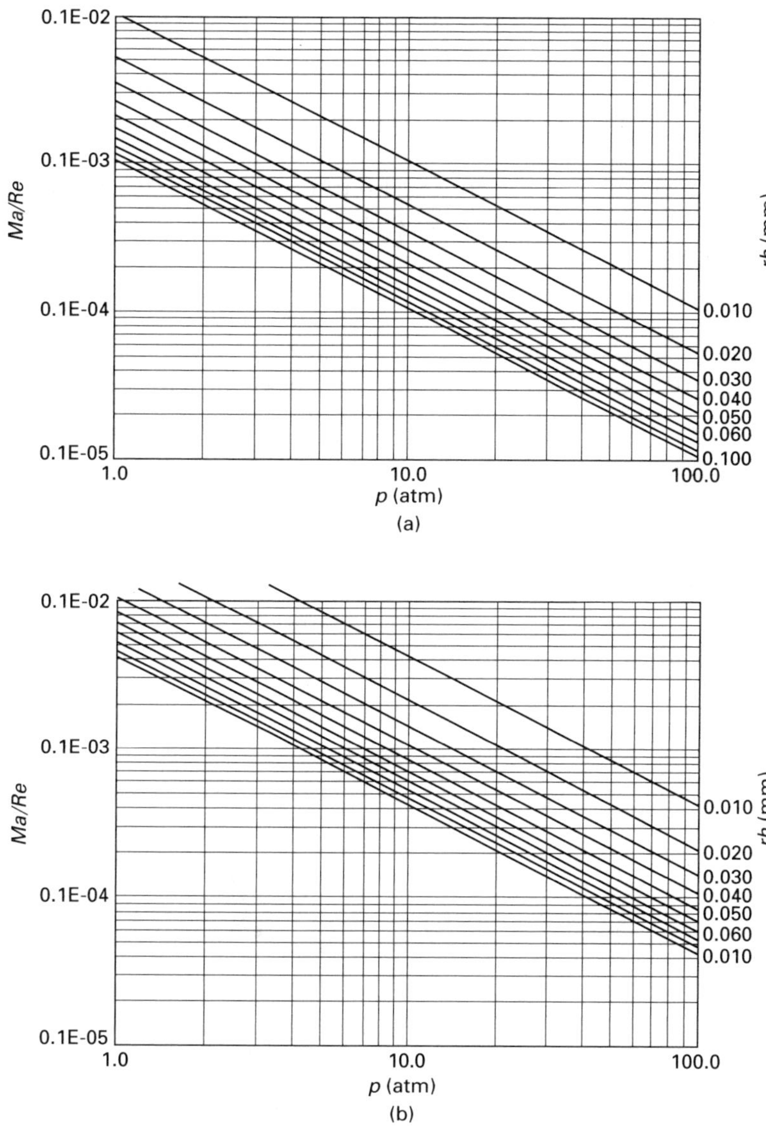

10.1 Graphical equivalent of Eqn 10.3. Numerical values of *Re* and *Ma* are calculated on the basis of particle speed $\underline{u}$ at a common location; (a) air (or N_2) at ambient temperature; (b) air (or N_2) at 600°C.

5, 10, 20, 30, 40, 65 screens deep. Experiments to bring flow data into the twenty-first century must be based on matrices having porosity and hydraulic radius representative of Stirling engine practice. Su (1986) tested stacks of 4, 8, 15, 27 screens, plotting pressure coefficient f against approach Mach number Ma_{app} for a range of constant Re. His results for the 8-deep stack are reproduced in Fig. 10.2. At approach $Ma_{app} \approx 0.16$ flow chokes (pressure coefficient infinite) regardless of *Re*.

Pressure coefficient *f* is defined:

$$f = |\,dp\,| / \tfrac{1}{2}\rho \underline{u}^2 \qquad 10.4$$

Noting that a different aperture ratio from Su's $\alpha = 0.365$ yields a different map, the results may be seen as confirmation of the functional relationship anticipated in Chapter 2:

$$f = f(Re, Ma, \alpha, n_g \ldots) \qquad 10.5$$

In Eqn 10.5 n_g is the number of gauzes in the stack, i.e., 8 for the measurements of Fig. 10.2. Nominal thickness of the individual screen is $2d_w$ in the flow direction. This allows inter-conversion between friction factor C_f and the 'per screen' value of *f*, namely, $|\,dp\,|_{screen} / \tfrac{1}{2}\rho u^2$. C_f is expressed in terms of pressure drop d*p* measured over length d*x* of the stack:

$$C_f = |\,dp/dx\,|\, r_h / \tfrac{1}{2}\rho u^2 \qquad 10.6$$

Stack length element of $dx = 2n_g d_w$. For the square weave (only), hydraulic radius r_h relates to wire diameter d_w in terms of flow passage geometry, itself defined when volume porosity $¶_v$ is specified (author's 1992 text again):

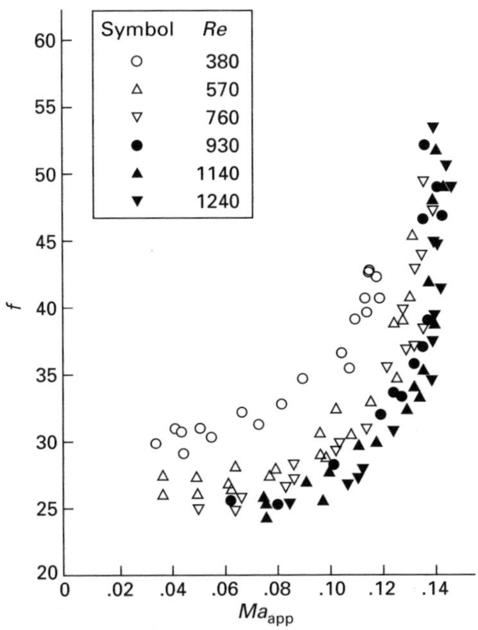

10.2 Su plots *f* vs *Ma* with *Re* as parameter for an eight-screen stack of gauzes having aperture ratio 0.365. A different value of α yields a different chart. Reproduced from the PhD dissertation (1986).

$$r_h/d_w = \tfrac{1}{4}\P_v/(1 - \P_v)$$

It follows that:

$$C_f/f = \tfrac{1}{8}\P_v/(1 - \P_v) \qquad 10.7$$

Chapter 12 will show that aperture ratio α converts to volume porosity $\P_v$:

$$\P_v = 1 - \tfrac{1}{4}\pi(1 - \sqrt{\alpha}) \qquad 10.8$$

With the aid of Eqns 10.7 and 10.8 Su's results may be transformed:

$$C_f = C_f\,(Re, Ma, \P_v \ldots) \qquad 10.9$$

Figure 10.3 shows the result of the transformation, and should remove any doubt about the vulnerability of C_f to *Ma*.

The lower the value of *Re*, the greater is the adverse effect on C_f of a given *Ma*. All other things being equal, minimizing hydraulic radius r_h maximizes surface-to-volume ratio – and potential for heat exchange. However, it minimizes *Re* and thus exacerbates the effect of *Ma*. If thermal recovery were the sole consideration, gauzes in the smallest commercially-available r_h would be specified for Stirling machine use. They are not, and Fig. 10.3 now offers an explanation.

The specimens from Su's tests are not 'freak' results: 12 screen stacks were tested, some with successive adjacent screens given a relative rotation – 0°, 30° and 45°. All showed similar vulnerability to *Ma*. Computer simulation

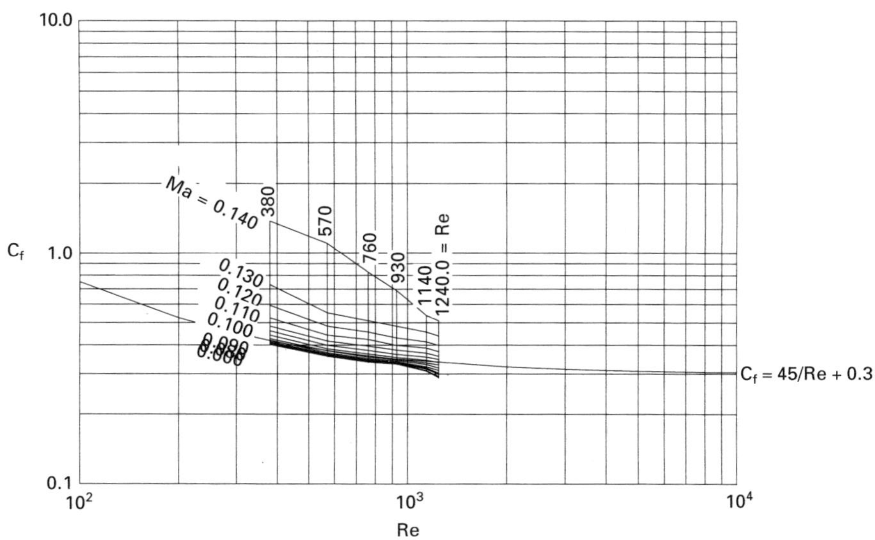

10.3 Su's experimental results of 1986 transformed to C_f – *Re* format. The isolated curve of C_f = 45/*Re* + 0.28 corresponds to Kays and London's consolidated curve for incompressible flow (pp. 7–9 of the 1964 edition).

gave results in agreement with the respective experimental counterpart. All gauzes had aperture ratios α close to 0.365. Pinker's and Herbert's 1967 tests on single gauzes included a gauze with $\alpha = 0.31$. Reference to their Fig. 4, reproduced in Chapter 2 as Fig. 2.2, strongly suggest that the sensitivity of Su's tests to *Ma* would have been even greater at this aperture ratio.

The $C_f - Re$ correlation is at the heart not only of computer simulation, but of scaling as a complementary 'back-of-the-envelope' approach to gas path design (Organ 1997). Discarding it threatens a technology vacuum which can only be filled by a substantial programme of experimental work to acquire flow (and heat transfer) data to a new format.

An already daunting prospect raises a further difficult decision: to reacquire correlations to the enhanced form $C_f = C_f\ (Re, Ma, ¶_v \ldots)$, or to take the opportunity to break with friction factor once for all. In times of greatest difficulty there is only one thing to do: resort to the most powerful tool in the repertoire.

10.4 Dynamic Similarity to the rescue

The 'friction factor' concept originated with the study of flow in pipes and ducts, where, as indicated in Fig. 10.4, shear stress τ_w at the wall acts over peripheral area *perim.*d*x* to give a resisting force aligned with flow direction. This permits the setting-up of a force balance with steady-flow pressure difference $-\mathrm{d}p$ acting on free-flow area A_{ff}:

$$\tau_w perim.\mathrm{d}x = -\mathrm{d}pA_{ff} \qquad 10.10$$

Flow through the matrix is not pipe flow: shear stress doubtless arises, but cannot be separated either experimentally or theoretically from the (significant) component of pressure drop associated with flow separation and wake

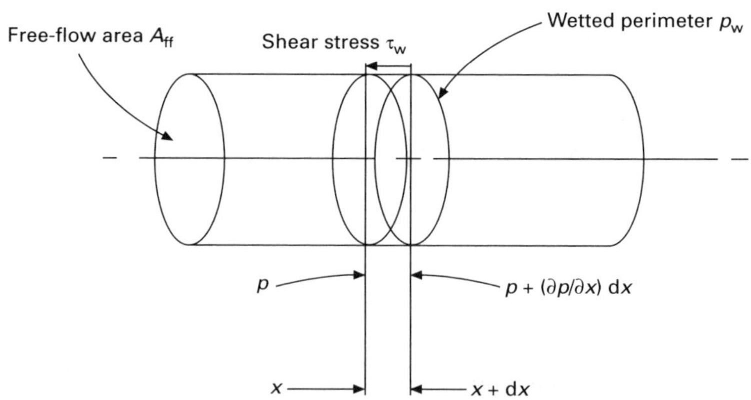

10.4 Element of duct with notation for definition of friction factor C_f in pipe flow.

formation. This is the context in which Kays and London pronounce (p. 8 of 1964 edition):

Whether this equivalent shear force is a true viscous shear, or is primarily a pressure force, as in the case of tube banks, is of no consequence. For most of the surfaces considered it is a combination of viscous shear (skin friction) and pressure force (form drag), but there is no object in design work in attempting to separate these effects.

No object, that is, until it becomes necessary to face up to the fact that flow data correlated on this principle are deficient. But if the notion of an equivalent friction stress does not adequately represent the net force resisting flow, what approach to this complex flow phenomenon will serve better? For all its faults, the Forchheimer law is not predicated on an elementary analytical force balance.

Why not start with the phenomenon of direct interest, which is not friction factor, but pressure drop per screen Δp? Parameters on which Δp may be expected to depend are: p itself (Pa), R (J/kgK), T (K), μ (Pas), $\underline{u}$ (m/s) and r_h (m). Ideal gas behaviour is assumed, so density ρ does not appear as well as p, T and R. Parameters already dimensionless are not considered at this stage – isentropic index γ, number of screens n_g and geometric descriptor $d_w m_w$.

The 'power product' approach (White 1979) allocates an index to each variable:

$$\Delta p = \Delta p\{p^a, R^b, T^c, \mu^d, \underline{u}^e, r_h^f\}$$

Each variable is replaced by its dimensions – M, L, T, K:

$$\mathrm{ML^{-1}T^{-2}}$$

$$= \Delta p\{[\mathrm{ML^{-1}T^{-2}}]^a, [\mathrm{L^2T^{-2}K^{-1}}]^b, [\mathrm{K}]^c, [\mathrm{ML^{-1}T^{-1}}]^d, [\mathrm{LT^{-1}}]^e, [\mathrm{L}]^f\}$$

Equate indices of like dimensions:

$$\mathrm{M}: a + d = 1$$

$$\mathrm{L}: -a + 2b - d + e + f = -1$$

$$\mathrm{T}: -2a - 2b - d - e = -2$$

$$\mathrm{K}: -b + c = 0$$

There are six unknowns and four equations, requiring four of the unknowns to be expressed in terms of two others. The choice is arbitrary, and one choice results in:

$$\Delta p/p = \Delta p/p\{p^{1-2c+e} R^c T^c \mu^{2c+e} \underline{u}^e d_w^{-(2c+e)}\}$$

The expression in parentheses is a 'power product' (White 1979), so the isolated $\underline{u}^e$ may be split into the product $\underline{u}^{2c+e}\underline{u}^{-2c}$. Grouping variables according to indices in common:

$$\Delta p = p\{pr_h/\mu\underline{u},\ \underline{u}/\sqrt{RT}\}$$

Restoring the parameters originally excluded for already being dimensionless

$$\Delta p/p|_{screen} = \Delta p/p\{Sg,\ Ma,\ \gamma,\ d_w m_w\} \qquad 10.11$$

$$Sg = pr_h/\mu\underline{u}$$

$$Ma = \underline{u}/\sqrt{RT}$$

10.5 Farewell to friction factor

Use of C_f in the context of gauze stacks provides a reassuring link with the better-understood pipe flow, but hardly represents the same phenomenon: shear stress τ_w cannot be measured directly, but where free-flow area A_{ff} and wetted perimeter *perim* are clearly defined, as in pipe flow, may be extracted by inverting Eqn 10.10:

$$\tau_w = -(\mathrm{d}p/\mathrm{d}x)A_{ff}/perim \qquad 10.12$$

C_f is defined as τ_w divided by the dynamic pressure head $\frac{1}{2}\rho u^2$, namely $C_f = \tau_w/\frac{1}{2}\rho u^2$. With $A_{ff}/perim$ abbreviated to r_h (hydraulic radius)

$$C_f = |\ \mathrm{d}p/\mathrm{d}x\ |\mathrm{r_h}/\tfrac{1}{2}\rho u^2 \qquad 10.13$$

For the porous matrix A_{ff} and *perim* are averages of some sort. The r_h resulting from taking the ratio serves for defining *Re* – as would any characteristic linear dimension – but has lost the meaning it had for pipe flow. A rough analogy comes from materials testing, and would be the setting up of a force balance in terms of uni-axial tensile stress as a basis for characterizing notch impact sensitivity. Some sort of correlation will doubtless result, but the anomalies of the derivation may be expected to compromise subsequent usage for retrieving values of impact strength.

Friction factor C_f tends to infinity as *Ma* tends to zero. The $\Delta p/p$ of the new format are finite and meaningful over the full range $0 < Ma < Ma_{choking}$, since Eqn 10.11 applies whether flow is compressible or incompressible. The values of *Ma* which divide experimentally-measured Δp to yield $C_f - Re$ correlations are variable, arbitrary – and unknown. Acquiring $\Delta p/p$ (or $\mathrm{d}p/\mathrm{d}x$) from C_f involves multiplying up by *Ma*. If the flow being computed is of different 'mode' from that of the correlation (i.e., is compressible vs 'creeping'), then the computed $\Delta p/p$ is meaningless.

The proposed replacement format reinforces the link between two ostensibly contrasting approaches to gas path design – simulation and scaling. The authority of the latter approach will be increased. The $\Delta p/p$ display will allow fractional pressure drop to be read off directly from numerical values of any pair of *Ma*, *Re* and *Sg*. Wouldn't that be a step in the direction of making regenerator design a less inscrutable business? The new correlations are compatible with the unsteady gas processes of a new generation of closed, regenerative cycles – wave engines and coolers. Design by trial and error is particularly inappropriate to the type, making simulation based on valid and pertinent data the only realistic option.

10.6 The new format

A specimen correlation to the new format is obtained by remapping data from Su's dissertation. Figure 10.5 shows the result. The range of *Re* covered in the original experiments was limited, yielding a plot in the form of a narrow 'ribbon'. Regenerator design calls for a wider coverage of *Re* – and a separate chart for each candidate gauze geometry (i.e., for each $d_w m_w$).

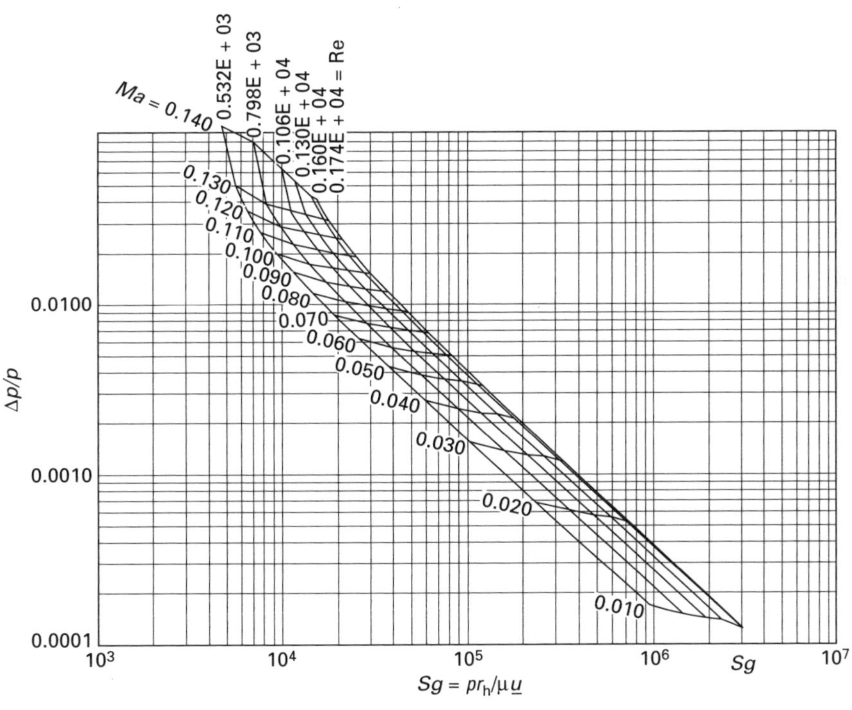

10.5 Specimen $\Delta p/p$ chart constructed by reworking Su's experimental data from the 1986 dissertation. Su covered a limited range of *Re*. Stirling engine design calls for wider coverage. The value of $d_w m_w$ was 0.396. A significantly different value of $d_w m_w$ leads to a different chart.

On the other hand, the limited coverage is more than sufficient to show that achievement of minimum $\Delta p/p$ requires Ma to be minimized and Sg to be maximized. From the definition of Sg (Eqn 10.11) the design target, all other things being equal, is maximum p and maximum r_h. Feulner (2005) reaches the same criterion on the basis of extensive experimentation.

A final look at friction factor against the foregoing background will justify the last rites and a decent burial.

10.7 What the new format reveals about 'incompressible' flow data

Algebraic conversion of the Forchheimer law so as to become a function of Sg and Ma was demonstrated in Chapter 2.

$$\Delta p/p = \{cSg^{-1} + dMa^2\}\mathrm{d}x/r_h \qquad 10.14$$

The result is the precise equivalent of the empirical 'fit' to the $Cf - Re$ correlation, namely to $C_f = c/R_e + d$, as may be seen by dividing through by Ma^2, and recalling that $Re \propto SgMa^2$:

$$\Delta p/p = Ma^2\{cSg^{-1}/Ma^2 + d\}\mathrm{d}x/r_h \qquad 10.14(a)$$

This amounts to the Forchheimer law and/or the traditional 'fit' ($C_f = c/Re + d$), but with two parameters, Sg and Ma exposed rather than combined into the single parameter Re. Specifying numerical values for c, d and $\mathrm{d}x/r_h$, allows back-calculation of the two-parameter form. Figure 10.6 illustrates for $c = 40.0$, $d = 0.3$ and $\mathrm{d}x/r_h = 16.0$.

Over the range of Re common to both, the maps of Figs 10.5 and 10.6 differ fundamentally. The Forschheimer law and the empirical fit to the 'incompressible' correlations are both based on supposed linear addition of viscous and inertia terms. Consistent with the arbitrary assumption, curves of $\Delta p/p$ vs $\mu u/pr_h$ in Fig. 10.6 simply replicate indefinitely as $\Delta p/p$ increases with increasing Ma. Comprehensive measurement demonstrates this to be at odds with reality.

When maps become available for a range of values of the geometric parameter $d_w m_w$ it will be found that lowest vulnerability to Ma is at minimum $d_w m_w$ – i.e., at maximum volume porosity and maximum aperture ratio. All things being equal, minimum $d_w m_w$ is therefore a design target.

10.8 Epitaph

When the material of this chapter was undergoing review for possible publication in *Cryogenics*, a somewhat tetchy referee repeatedly pressed the question 'What experimental evidence do we have that the friction factor at a given Reynolds [sic] and porosity really depends on Mach number?' Well,

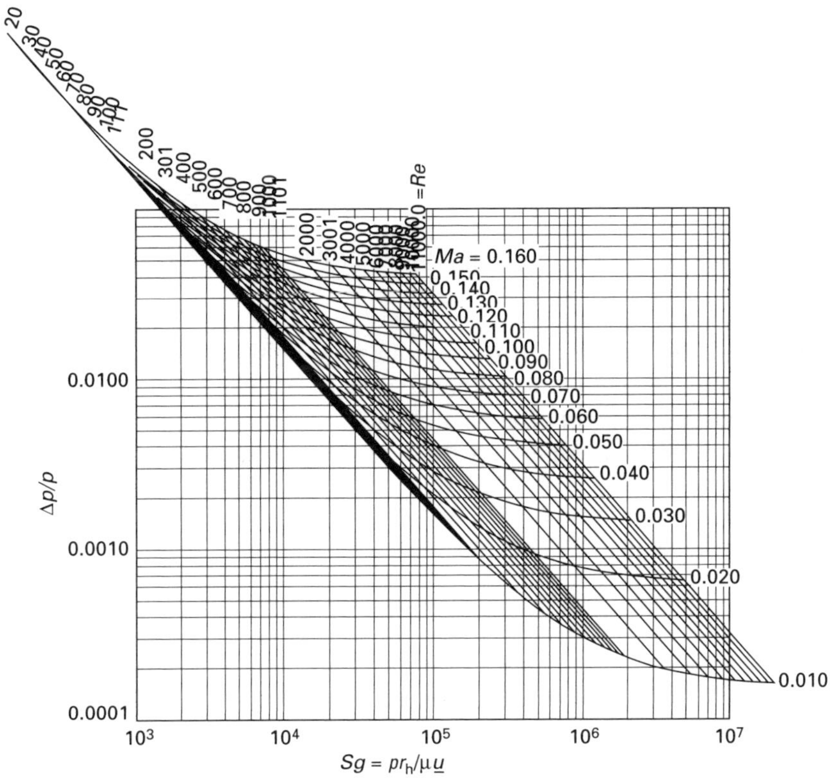

10.6 Forchheimer law and/or tradition 'fit' $C_f = c/Re + d$ displayed by back-calculation using $c = 40.0$, $d = 0.3$ and $dx/r_h = 16.0$. *Not to be relied upon for design*, but to be compared with Fig. 10.5, the data for which were acquired under experimental conditions respecting Dynamic Similarity principles.

only 50 or so plots acquired by Su under the author's supervision, some for f vs Ma with Re as parameter; some with α as parameter; some for $\Delta p/p$ vs Ma with Re as parameter. A range of different gases was covered – air, argon and helium. Some of the curves sets are replicated by computer simulation with good correspondence. All show the features of the specimen reproduced here (Fig. 10.2).

That such body of evidence should fail to impress the referee of a highly-regarded journal has to be taken into account. A factor may be a long tradition in the fluid mechanics of steady flow in pipes, where the rule-of-thumb for the threshold of compressibility effects is $Ma = 0.3$. In the context (the parallel pipe) Shapiro asserts (1954) that friction factor C_f remains unaffected at Mach numbers up to Ma = unity.

It happens that Oswatitsch believes otherwise (see 1956 translation of his 1952 text – bottom of p. 60). But flow through the gauze is not pipe flow: The free-flow area of the gauze aperture varies between inlet and outlet planes by as much as 0.3:1. If there is a relevant point of reference it is not the pipe but the convergent-divergent nozzle, where orthodox engineering teaching emphasizes the potential for *choking* as *Ma* approaches unity in the vicinity of the throat. On the evidence of Su's comprehensive investigation, pipe flow is no criterion for flow through the gauze, where the threshold Ma_{app} at which compressibility compounds C_f decreases with *Re and may be without lower limit!*

Chapter 13 will return to first principles, and derive $\Delta p/p$ correlations to the format of Fig. 10.5 by computer simulation of flow through a detailed geometric model of the individual gauze aperture. Doubtless this will fail to satisfy the referee, but will offer an alternative 'take' on the thermodynamic potential of the Stirling engine for those wishing to explore.

11
Regenerator thermal analysis – unfinished business

11.1 Regenerator design in context

Regenerator design calls for two essentially independent tools: *analysis* and *data*. The *analysis* stage involves manipulating the differential equations of flow and thermal conduction. It can be brought to a satisfactory conclusion – and convincingly verified – using arbitrary data. Predicting the gas processes under specified operating conditions (charge pressure, cyclic speed), requires the analysis, duly verified, to be fed with *data* (flow and heat transfer correlations) appropriate to flow passage geometry. Acquisition, reduction and presentation of data are matters totally independent of analytical formulation and solution.

Improving the commercial prospects for the air-charged Stirling engine means extending its performance envelope. Appropriately formulated, the analytical model is unlimited, in the sense of being valid within the experimentally-demonstrated performance bounds of the engine – and beyond. This is not necessarily true of the data:

- The widely used steady flow and heat transfer correlations for wire screen stacks were acquired *and displayed* under the assumption of *incompressible* flow.
- It may been shown (Organ 2000d) that, at the peak rpm of the Philips MP1002CA air engine, Mach number at the compression end of the regenerator exceeds the threshold (Su 1986) at which compressibility inflates friction factor C_f over the incompressible value at the same Reynolds number.

Before performance potential can be probed by simulation, flow and heat transfer correlations need to be reacquired with due regard for dynamic similarity principles – i.e., in function of both *Ma* and *Re*. This gives the opportunity to address a further concern, namely the extent to which flow in the test section represents that in the functioning regenerator: it is generally assumed in both contexts that flow is 'one-dimensional', i.e. that particle velocity is uniform across the entire matrix flow area. In most situations entry (and exit) conditions do not favour this untested assumption.

That this central matter lies unexplored may reflect the intimidating nature of the Conservations Laws in two (or three) dimensions. On the other hand, a formulation is possible which yields a revealing picture of the axi-symmetric (three-dimensional cylindrical) case for only modest increase in computational intricacy relative to the 'slab'-flow counterpart.

Of necessity, the numerical solutions to be discussed will reflect the use of existing correlations (i.e., will apply to incompressible flow only). Moreover, it will be necessary to assume that the relationship between C_f and Re defining flow parallel to the plane of the gauze is the same as that which applies in the principal (axial) direction.

In the author's experience it is rare for an investigation of this nature to have an outcome favouring the analyst by retrospectively corroborating a necessary starting assumption – in this case, that of one-dimensional flow. Even more rare is that the exploratory analysis should turn into a flexible visualization tool with potential to aid first-principles design. Full exploitation awaits the availability of friction and heat transfer correlations for the two orthogonal directions – perpendicular and parallel to the plane of the gauze. But this chapter opened by pointing out that these will merely become replacement 'plug-ins'.

11.2 Assumptions

The scale of the control volume of Fig. 11.1 exceeds that of a single pore of the matrix. This allows pressure and inertia forces to be balanced by a net resistance to motion defined empirically in a relationship between 'friction factor', C_f, and local Reynolds number, Re. Differential coefficients – $\partial/\partial z$, d/dz, etc., arise, but the customary implication of the infinitesimal limit does not apply. It is now supposed:

- that flow in a given coordinate direction, z or r, experiences uniform volume porosity $¶_v$ and uniform hydraulic radius r_h, although these defining properties may differ between the two directions, allowing the anisotropic annular 'bandage' of wire gauze to be modelled. Dimpled foil, extruded ceramic, corrugated flame-trap, etc., have r_h = zero in the r-direction. This extreme case does *not*, in general, reduce to the conventional, one-dimensional flow model.
- pore size is less by an order of magnitude than that of any control volume used in analysis. The bounding faces of the element of porous matrix coincide with those of the enclosed fluid.
- temperature of the solid phase of the matrix varies linearly from T_E at the left-hand end to T_C at the right.
- the flowing fluid is an ideal gas obeying the equation of state $p/\rho = RT$.

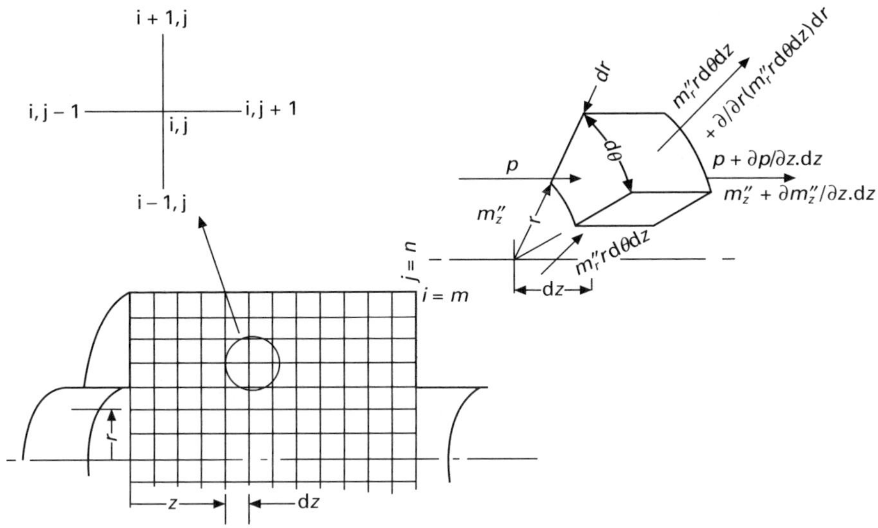

11.1 Regenerator enclosure assumed axi-symmetric. Solution algorithm applied to one symmetrical half of a plane through the axis.

11.3 Modified diffusion law

Upstream pressure, p_0, is assumed to vary cyclically. For the eventual graphical illustrations the variation is simple-harmonic, but the method is not limited to simple functional variations.

The equation of mass continuity in cylindrical polar coordinates is:

$$\partial(\rho u)/\partial z + r^{-1}\partial(r\rho v)/\partial r = -\partial\rho/\partial t \qquad 11.1$$

A condition for the *energetic similarity* of the gas processes between two regenerators is that the cyclic distribution of *fractional mass* shall be the same in both cases. Eqn 11.1 is re-expressed in terms of mass rate per unit area, $g = \rho u$ to capitalize on similarity:

$$\partial\, g_z/\partial z + r^{-1}\partial(rg_r)/\partial r = -R^{-1}\partial(p/T)\partial t \qquad 11.1(a)$$

The steady momentum term is assumed to dominate, and will illustrate the extent of the digression between one- and two-dimensional formulations.

$$\partial p/\partial z = -\tfrac{1}{2}\rho u^2 C_f/r_h \qquad 11.2$$

In terms of g_z:

$$\partial p/\partial z = -\tfrac{1}{2}g_z^2\rho^{-1}C_f/r_h \qquad 11.3$$

For (incompressible) flow perpendicular to a stack of rectangular-woven wire screens, friction factor C_f can be fitted approximately to Reynolds number Re:

$$C_f = c/Re + d \qquad 11.4$$

Re may be expressed in terms of g:

$$Re = 4 \mid g \mid r_h/\mu \qquad 11.5$$

For the purposes of illustration, coefficient of dynamic viscosity μ is assumed invariant. In reality, there is a marked variation over the temperatures spanned by the Stirling cycle regenerator. Modification of the algebra to accommodate is elementary.

For values of overall pressure drop at which regenerator operation is viable:

$$\rho \approx p_0/RT \qquad 11.6$$

Substituting Eqns 11.4, 11.5 and 11.6 into Eqn 11.3 yields:

$$\partial p/\partial z = -cg_z\mu RT(z)/8p_0r_h^2 - \tfrac{1}{2}g_z^2 d\,\mathrm{sign}(g_z)\,RT(z)/p_0r_h \qquad 11.7$$

The term sign(g_z) ensures that this part of the net frictional resistance always opposes flow. The equation is of the form:

$$T(z)^{-1}\partial p/\partial z = -bg_z - a_z g_z^2 \qquad 11.8(a)$$

$$a_z = \tfrac{1}{2}\mathrm{sign}(g_z)dR/p_0r_h$$

$$b_z = c_z\mu R/8p_0\,r_h^2$$

Momentum in the r direction is provisionally as for plane, two-dimensional flow:

$$T(z)^{-1}\partial p/\partial r = -bg_r - a_z g_z^2 \qquad 11.8(b)$$

$$a_r = \tfrac{1}{2}\mathrm{sign}(g_r)dR/p_0r_h$$

$$b_r = c_r\mu R/8p_0\,r_h^2$$

At any given location, z, r, and thus for given local temperature $T(z)$, Eqns 11.8(a) and (b) are quadratic in g_z and g_r respectively and thus soluble by the standard formula:

$$g_z = \tfrac{1}{2}a_z\{-b_z +/- \surd[\,b_z^2 - 4a_zT(z)^{-1}\partial p/\partial z]\} \qquad 11.9(a)$$

$$g_r = \tfrac{1}{2}a_r\{-b_r +/- \surd[\,b_r^2 - 4a_rT(z)^{-1}\partial p/\partial r]\} \qquad 11.9(b)$$

From Eqn 11.3 the g are zero when the pressure gradient term is zero, so the root required is the positive option.

Equation 11.9(a) is differentiated with respect to z. Eqn 11.9(b) multiplied by r is differentiated with respect to r and divided by r. The resulting expressions are added and the sum set equal to the right-hand side of the mass conservation law (Eqn 11.1(a)). For practical purposes the multipliers $-b/2a_z$ and $-b/2a_r$ are independent of z and r respectively, and their differentials therefore vanish. Temperature gradient $\partial T(z)/\partial r$ in the r-direction is close to zero, consistent with which $\partial T(z)/\partial z$ is now written $dT(z)/dz$. With abbreviation R_z for $\sqrt{(b^2 - 4a_zT^{-1}\partial p/\partial z)}$ and R_r for $\sqrt{(b^2 - 4a_rT^{-1}\partial p/\partial r)}$:

$$\partial^2 p/\partial z^2 - (T^{-1}dT/dz)\partial p/\partial z + R_z/R_r\partial^2 p/\partial r^2$$
$$- R_zT/r[(-b_r + R_r)/2a_r] = - R_z\partial p/\partial t \qquad 11.10$$

11.4 Numerical solution

This quasi-explicit equation for p is converted to finite-difference form. In the subscript notation of Fig. 11.1 the first term is represented:

$$\partial^2 p/\partial z^2 \approx (p_{i,j-1} - 2p_{i,j} + p_{i,j+1})/\Delta z^2 \qquad 11.11$$

The remaining terms are dealt with in similar fashion and sorted according to the subscript of p. The resulting difference relationship relates to a single mesh point of Fig. 11.1. Applying to all mesh intersections yields $(m - 1)(n - 1)$ equations linear in the $m \times n$ unknown values of $p_{i,j}$. At flow inlet $g_r = 0$, and the distribution of the z component of mass rate determines the boundary condition. Inserting values of $g_z(r)_{z=0}$ into the first two terms of the right-hand side of Eqn 11.8(a) gives values of pressure gradient as boundary conditions across the entry face. At exit ($z = L$) pressure is assumed uniform with z. At vertical enclosure faces boundary conditions are of the form $\partial p/\partial z \approx (p_{i,j} - p_{i,j-1})/\Delta z = 0$ (flow zero normal to the face). At the horizontal enclosure face, $\partial p/\partial r \approx (p_{i,j} - p_{j-1,j})/\Delta r = 0$. The flow field is assumed symmetrical above and below the axis (horizontal centre-line), as achieved by setting $p_{m+1,j} = p_{m-1,j}$.

With boundary conditions included, the total number of finite difference equations equals the number of unknown $p_{i,j}$.

Solution is by any one of a range of computer library routines for dealing with linear algebraic equations. For an array size up to about $m \times n = 15 \times 30$, a simple, pivotal condensation scheme such as that embodied in SIMQX solves for some 90 subdivisions of cycle time (or of a crank rotation) in a few seconds on a PC having a Pentium 3 processor. Larger numbers of subdivisions justify the extra programming effort of routines storing only those matrix elements taking finite values. The NAG™ routine F04BSF (following a one-off call to the decomposition routine F04BRF) has been found to run particularly fast.

11.5 Parameters of operation

For coding purposes Eqn 11.10 is reworked in terms of normalized variables and normalized parameters. One of several benefits is the resulting link to the generalized parameters of operation of the practical regenerator or Stirling cycle machine.

Normalized pressure, ψ, is defined as p/p_0. Velocities are normalized by $\omega L_{ref} = \omega V_{sw}^{1/3}$, mass rates per unit area g_z and g_r by $p_0 \omega V_{sw}^{1/3}/RT_C$. Operating parameters which emerge from normalization include temperature ratio, $N_T = T_E/T_C$, which is set to 3.0 for Figs 11.2–11.5. The computational sequence is stable and robust over a wide range of speed parameter, $N_{MA} = \omega V_{sw}^{1/3}/\sqrt{RT_C}$, of Stirling number $p_0/\omega\mu$ and of flush ratio N_{FL} (see Notation on page xiv). All curves shown are for infinite thermal capacity ratio N_{TCR}.

11.6 Pressure and velocity fields

Figure 11.2 is a specimen relief of pressure difference p–p_0. Pressure gradient is most intense along the axis, (uppermost trace) where z-velocity u is greatest, and less so along the horizontal wall (lowermost trace) where u is generally less (but *not* as for pipe flow!). Pressure gradients in r direction are consistent with finite r-velocities v, i.e., radially outwards near inlet, and radially inwards near outlet.

Quantitative values of local, instantaneous velocities u, v are obtained from:

$$u = g_z/\rho(z), \quad v = g_r/\rho(z) \qquad 11.12$$

Velocity immediately at entry to the matrix is enhanced by a factor equal to the inverse of porosity, namely, in the ratio $¶_v^{-1}$ (Kays and London 1964). In Fig. 11.3 the fluid in the manifold is assumed to approach with fully-developed (paraboloidal) velocity profile:

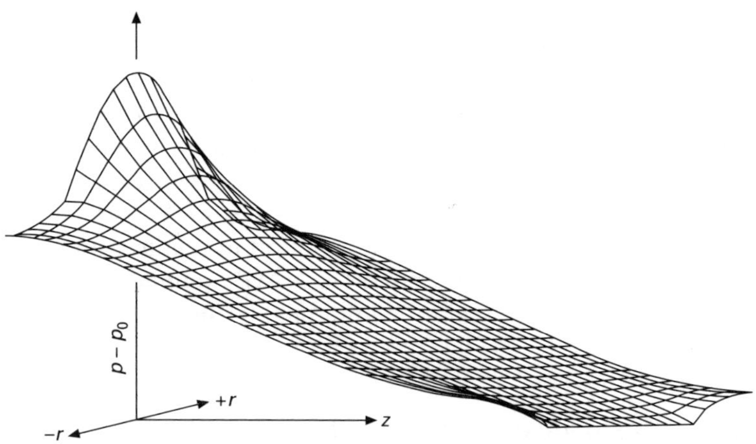

11.2 Relief of pressure difference $p - p_0$. Difference exaggerated for clarity. Note contrast in pressure gradient between axis (uppermost trace) where z-velocity u and pressure drop are highest, and wall (lowermost trace) where u is reduced. Pressure gradients in r direction are consistent with magnitude and sense of r-velocities v.

$$g_z(r)_{z=0} \propto 1 - 4r^2/D^2 \qquad 11.13$$

Arrows indicate the magnitude and direction of local velocities. Manifold free-flow area is less than matrix frontal area. The inflow redistributes rapidly after entry to become essentially 'slab' flow in the main body of the matrix.

As far as the *manifold* (only) is concerned, g_z and u adjacent the wall are zero. If flow rate is allowed to remain identically zero at this point of entry to the matrix it goes nowhere. For Fig. 11.3 an arbitrarily small component has been added to u to 'kickstart' flow at the singularity (the convex corner).

For Fig. 11.4, cross-sectional area of inlet duct is assumed equal to matrix frontal area. Approach velocity is parabolic. It might be expected that purely 'slab' flow within the matrix would follow. Not so! The low velocity adjacent to the inlet manifold wall means there is little inflow to the peripheral pores of the matrix. The core flow accordingly spreads out to occupy matrix free-flow area, giving a strongly two-dimensional velocity field at entry.

11.7 Inevitable asymmetry of flow cycle

Integrating the velocities yields maps of particle displacements. In Fig. 11.5(a) the rectangular grid is scribed on the axial temperature distribution of the matrix, from T_E at the left-hand end to T_C at the right. Manifold free-flow area here is a fraction of matrix frontal area.

Inlet velocity profile is parabolic, so particles near the centre-line have a head start. This means they achieve lower temperature – and thus reduced specific volume – more rapidly than the slower particles at the outboard

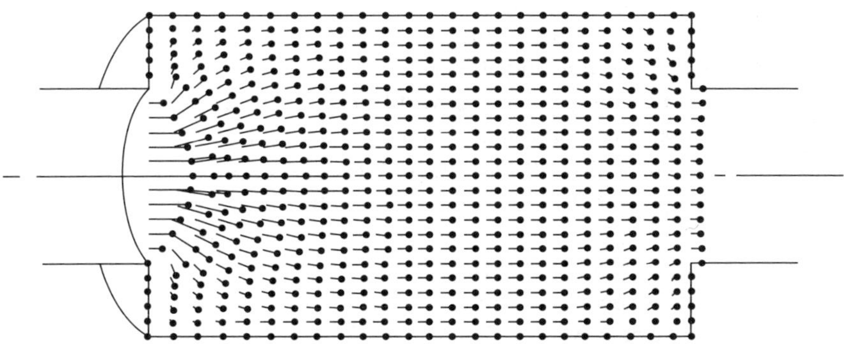

11.3 Approach velocity distribution parabolic within manifold. Matrix free-flow area exceeds that of manifold. Inflow redistributes on entry to become essentially 'slab' flow in the main body of the matrix. Note exit velocity distribution: velocity close to the wall exceeds that on the axis, consistent with the need for outgoing flow to accommodate the radially inward component immediately prior to exit.

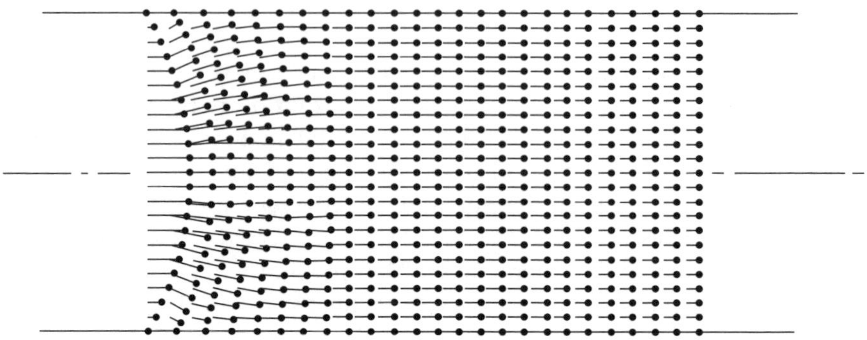

11.4 Even when cross-sectional area of duct and matrix housing are identical, parabolic distribution of approach velocity results in two-dimensional flow at matrix entry: velocity adjacent to wall in manifold is zero, so there is no inflow to the periphery of the matrix. Core flow accordingly spreads out to occupy matrix free-flow area.

locations. This tends to negate the higher velocity at inlet. A temperature gradient which is 'negative' relative to particle velocity might thus be thought of as having a 'stabilizing' effect.

The hot flush (down the temperature gradient) occupies a half-cycle. Inlet mass rate is simple-harmonic. The construction shows the progress, left-to-right, of the 'flow front' – a line of particles which originally (at time $t = 0$) spanned the inlet at $z = 0$. Rightwards progress of the front is marked at selected intervals of the 180° of the half-cycle.

Flow right-to-left *up* the temperature gradient shows the reverse effect: the flow-front near the centre-line enters at the highest velocity and climbs the temperature gradient fastest, expanding ahead of particles at the periphery. The increased displacement is cumulative. The destabilizing effect ensures that precise 'slab' flow does not form.

The picture of regenerator flow now differs in a crucial respect from that of one-dimensional formulation: *on the left-to-right pass the flow front takes a path which is different from that of flow in the reverse direction.*

Availability of particle tracks allows computation of temperature histories. Calculation is along the z–y surface (see Fig. 11.5) on the basis that heat transfer effects act on individual fluid particles. For Fig. 11.6 vertical 'legs' mark the local difference in temperature ΔT between matrix and instantaneously adjacent particle. Local *St* calculated from *Re* lead to small ΔT, and have been arbitrarily reduced to clarify the illustrations.

By contrast with the classic one-dimensional case, temperature histories on the hot-to-cold flush are not symmetrical images of those on the reverse flush (c.f. Organ 1997).

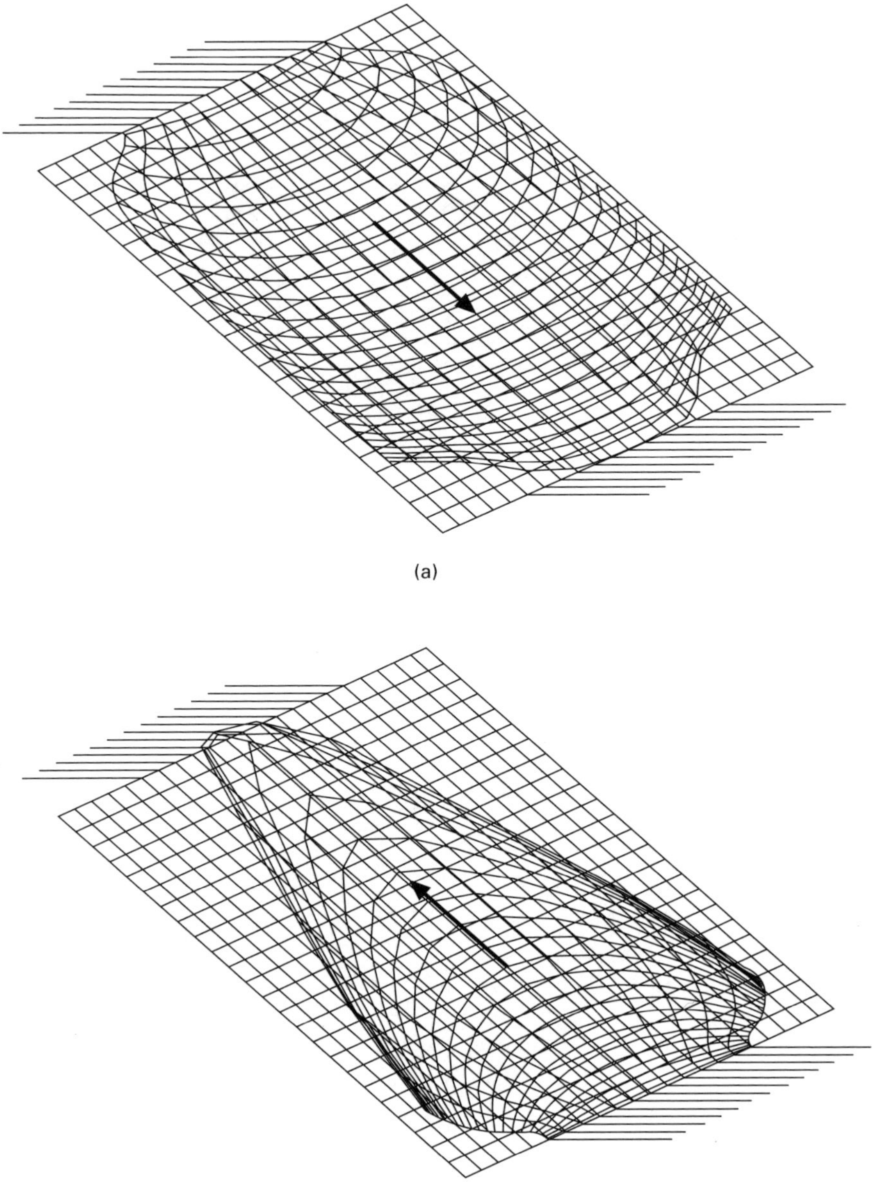

(a)

(b)

11.5 Cumulative displacement of flow front (NB: Not velocity profiles) (a) flow down temperature gradient – flow front decelerating. Faster progress at the core means earlier cooling and greater rate of reduction in specific volume than at the peripheral flow; (b) flow *up* temperature gradient – flow front accelerating. Rate of increase of specific volume at core greater than that at periphery.

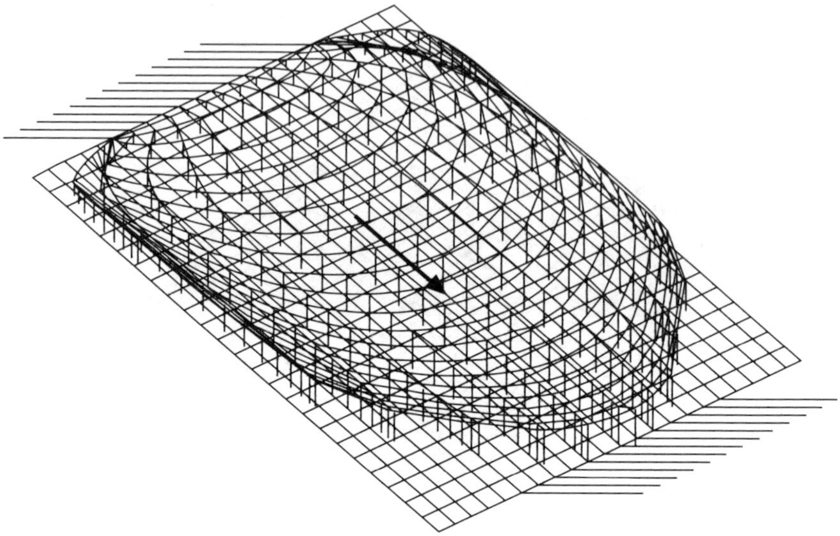

(a)

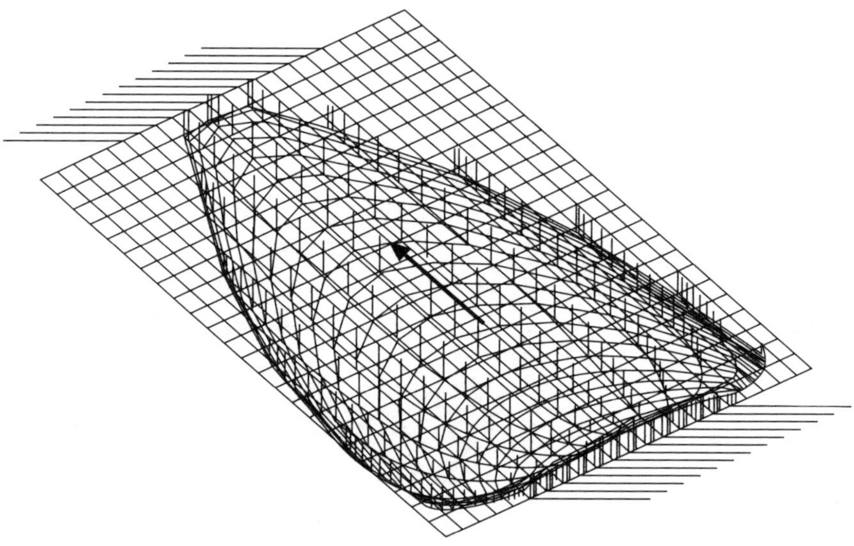

(b)

11.6 Specimen temperature reliefs superimposed on particle tracks. Fluid temperature exceeds matrix temperature on hot-to-cold flush, falling below on reverse flush – as for corresponding one-dimensional models (local temperature differences enhanced by reducing local Stanton number S_t by a substantial factor).

11.8 Anisotropic matrix

To this point it has been assumed that the matrix is isotropic. For this case the coefficients c_z, d_z of friction factor C_{fz} for flow the z-direction take identical respective values to those for C_{fr} for flow in the radial direction.

Neither volume porosity $¶_v$ nor hydraulic radius r_h of the wire screen stack is normally thought of as being directional. On the other hand, such a stack is clearly anisotropic, and intuition suggests that free-flow area perpendicular to the plane of the screens differs from the transverse value.

For the remaining illustrations uniform temperature is assumed, i.e., N_T = unity. This removes the distraction of velocity vectors reducing in length between hot and cold ends. To replicate the dimpled-foil matrix, coefficients c_z, d_z take values relevant to flow between parallel plates, namely, 24.0 and 0.007 respectively. Friction factor C_{fz} is then calculated locally as $24.0/Re + 0.007$. Coefficients c_r and d_r are set infinitely high to inhibit flow in the r-direction. Figure 11.7 shows the result when inlet manifold area is less than matrix frontal area. With no plenum space provided, flow is purely axial, and there is no interaction with the surrounding annulus of stationary gas.

Figure 11.8(a) illustrates flow in the isotropic matrix under the same (isothermal) conditions. The values for C_{fz}, namely, $50.0/Re + 0.15$, are those for the rectangular-weave wire screen. Radial factor C_{fr} takes the same values. There is now a substantial component of radial flow, although conditions do not settle to slab flow at any point in the stack.

For Fig. 11.8(b) axial factor C_{fz} is arbitrarily exaggerated to $250.0/Re + 1.15$, while radial factor C_{fr} remains as $50.0/Re + 0.15$. The enhanced resistance to flow in the axial direction promotes rapid radial redistribution of the

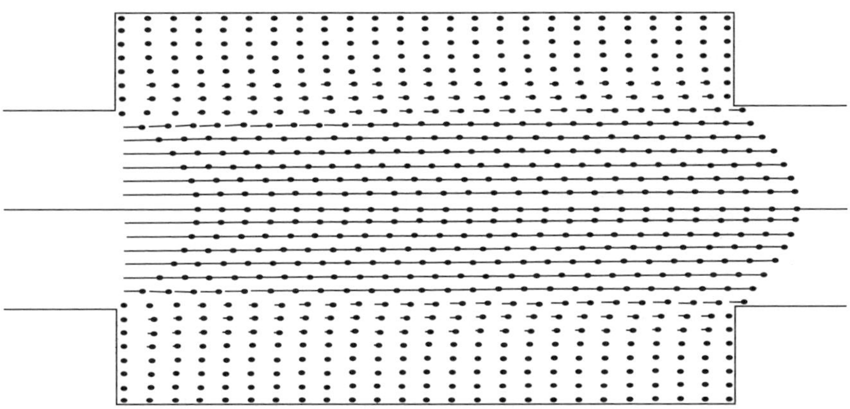

11.7 Inlet and outlet temperatures equal (N_T = unity). Dimpled foil packing replicated by setting C_{fz} for the z-direction as for flow between parallel plates and C_{fr} (for radial direction) to arbitrarily high value to inhibit radial flow component.

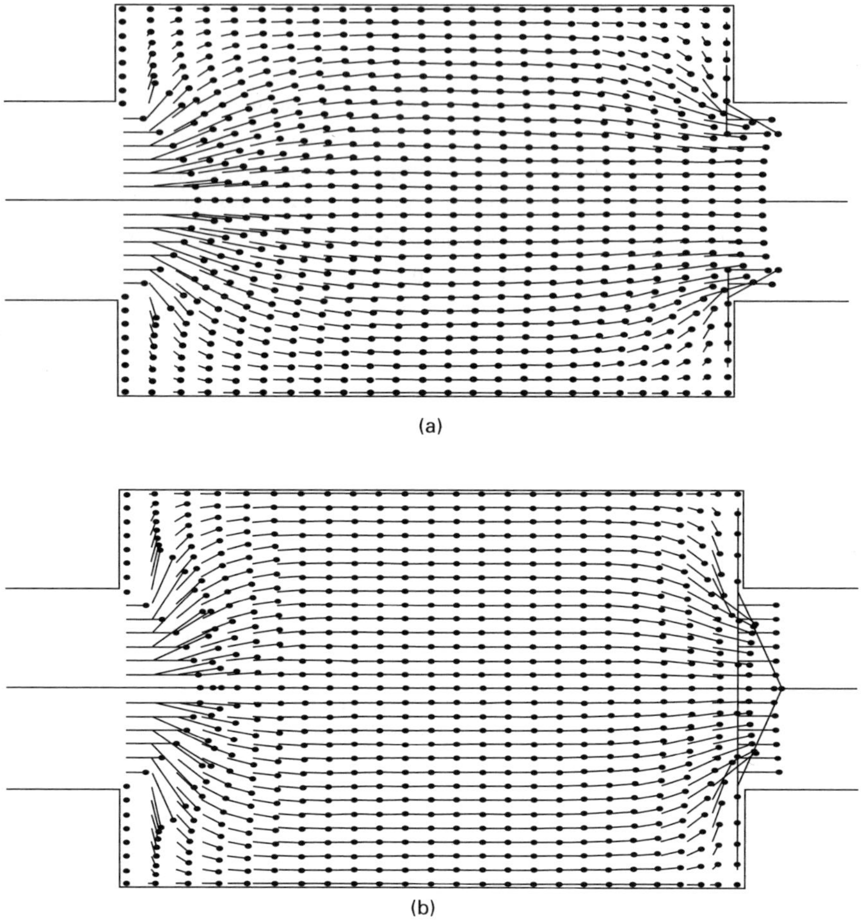

11.8 Specimen anisotropic matrices. Temperature ratio unity in both cases (a) parameters c_r and d_r of radial friction factor set equal to c_z and d_z of axial factor, as for the ideal isotropic matrix. There is now a radial component, although insufficient for net flow to settle to 'slab' flow; (b) axial friction factor C_{fz} arbitrarily exaggerated to $C_{fz} = 250.0/R_e + 1.15$. Radial factor as for Fig. 11.8(a) above. The differential flow resistance is evidently conducive to establishment of slab flow in the middle section.

flow, and in the central region of the stack there is a tendency towards slab flow.

(a) Flow down temperature gradient – flow front decelerating. Faster progress at the core means earlier cooling and greater rate of reduction in specific volume than at the peripheral flow.
(b) Flow *up* temperature gradient – flow front accelerating. Rate of increase of specific volume at core greater than that at periphery.

11.9 Discussion

The investigation pursues the author's current interest in gas process *visualization* as a complement to performance simulation. Reasons for this emphasis include an increasing concern about the adequacy of standard heat transfer and flow friction data for the purposes of first-principles design and optimization.

The outcome to this point resolves under three headings.

11.9.1 Features of the new formulation

It affords:

- a two-dimensional flow model making negligible demands on computing resources while affording evident sensitivity to the geometry and operating conditions of the 'real' gauze stack
- for the axi-symmetric stack the two-dimensional pressure and velocity fields are identically those of the three-dimensional solution
- an urgently-needed alternative to the over-simplification of 'slab'-flow formulations and to the over-kill of the proprietary CFD package
- for conceivable entry conditions, flow in the ideal, isotropic (uniform porosity) matrix differs fundamentally from that of other candidate materials such as extruded ceramic (Cercor™ by Dow-Corning), dimpled foil, corrugated flame-trap etc., for which the *r*-velocity component *v* is zero.

Formulating so that pressure *p* is the sole unknown affords substantial computational economy. The method takes in its stride any degree of anisotropy between axial and radial directions, any desired inlet flow velocity profile, any ratio of inlet area to matrix free-flow area. Any prescribed axial distribution of matrix temperature is handled.

11.9.2 Relevance to reacquisition of correlations for compressible flow

It might have been anticipated (indeed, this writer expected) that the *u*-velocity distribution at matrix mid-length ($z = L_r/2$) would tend to the parabolic distribution of laminar pipe flow. Reality, according to the present computations, could hardly be more different: the tendency is for slab flow to establish itself and to prevail throughout. (No doubt at a microscopic level the Navier–Stokes equations apply, but within the control volume assumed here, local frictional resistance in the *z* direction varies not with $\mu\partial u/\partial r$ but with u and u^2.)

There could hardly be a more reassuring finding: the slab flow assumption is justified in the conversion of experimentally-measured steady-flow pressure drop to correlation form. A slab-flow model of the regenerator is justified, and the correlations are the appropriate input data.

11.9.3 Wider implications

If this – or any other – two-dimensional model is to serve for first-principles design, it becomes imperative to know not only the axial, but also the transverse flow and heat transfer characteristics of the stack. Magara *et al.* (2000) have made a start on acquiring this information. It may be necessary to rethink the reliance on $d_w m_w$ as geometric parameter, since flow passage geometry in the z direction is not that of the r direction.

Downstream outlet temperatures and enthalpy rates have been shown to be inherently non-uniform and unsteady. This does not preclude a numerical estimate of regenerator 'efficiency', but the complex picture which arises prompts questions as to how '*an efficiency of 95 percent and more*' claimed by Rinia and duPré (1946) could have been ascertained by either computation or measurement, given the resources of the era.

12
Flow passage geometry

12.1 Scope

Why an entire chapter on the isolated gauze aperture? Because this geometric feature alone forms the boundary conditions determining the dynamics of the entire flow.

There is no evidence that the performance of air engines yet owes much to detailed attention to regenerator design. Power-producing potential being in terms of heat-rate in *minus* heat rate out, the 'thermal bottleneck' evidently lies elsewhere than with the regenerator. As and when this restriction is eliminated, small changes in regenerator specification may be expected to result in a marked change in performance. Geometric design will then assume its due importance. If the wire-gauze regenerator prevails, it may be anticipated (see Chapter 10) that the trend will be towards high values of aperture ratio.

Within a class of matrices (say, gauzes of given weave) a single numerical parameter suffices to distinguish between flow passage geometries, and so serves for the acquisition of experimental flow data and its use in support of gas path design. Perfectly-packed spheres do not require so much as a single parameter, since pore geometry is unique. It goes without saying that the descriptor is dimensionless in each case.

The literature on the Stirling regenerator generally relies on volume porosity $\P_v$ for this function. A linear measure is also required so that Reynolds number Re may be defined. This text standardizes on hydraulic radius ratio r_h, introduces the dimensionless ratio r_h/d_w, and demonstrates that r_h/d_w and $\P_v$ are inter-convertible.

A literature on inviscid compressible flow through the single gauze enjoys a separate existence, and uses aperture ratio α rather than $\P_v$ as the geometric parameter. For the wire screen, inter-conversion between α and $\P_v$ is via a simple algebraic expression. On the other hand, reconciling the two flow characterizations – viscous incompressible (friction factor vs Re), and inviscid compressible (pressure coefficient vs M_a) – is not a simple transformation: it will require a comprehensive geometric specification of the individual flow aperture to run from inlet to beyond the point of minimum flow area.

In certain classes of material such, as woven wire gauze (one of several candidates), each of $¶_v$, r_h/d_w and α is determined in terms of the technical specification of the weave – diameter d_w of the individual wire and mesh number m_w. Metal foams such as Retimet® and Recemat® have an 'organic' structure for which the geometric description involves a different approach. The former class is considered first.

12.2 Symmetrical gauze – flow perpendicular to plane of weave

The algebra applies to the isolated screen. It extends to a stack of n_g screens if adjacent screens are in close contact ('though not so perfectly aligned as to be 'locked' or 'engaged' (Organ and Maeckel 1996)). Figure 12.1 gives the notation.

Even the most basic approach to performance prediction relies on a numerical value for regenerator dead space ratio δ_r (= V_{dr}/V_{sw}). This derives from volume porosity $¶_v$, defined as the ratio of void volume to envelope volume.

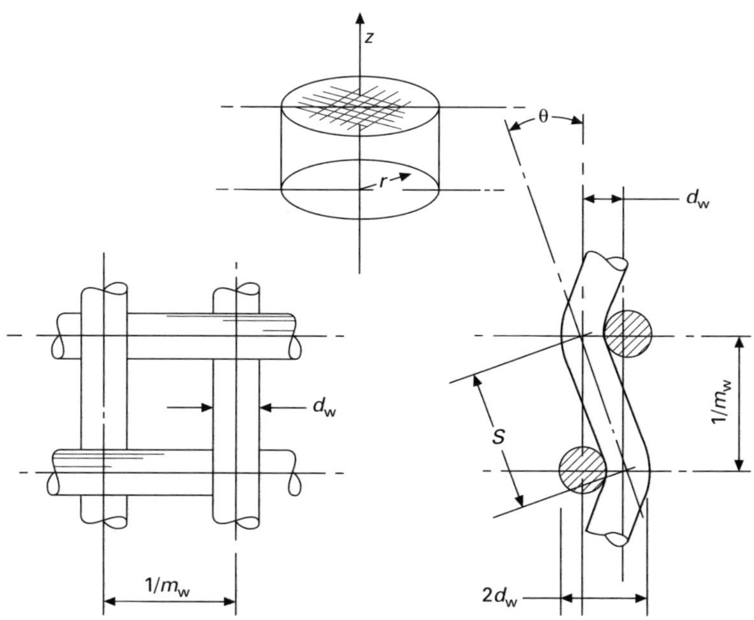

12.1 Notation for flow passage geometry of the square-weave gauze. Flow perpendicular to the plane of simple, square-weave gauze, together with coordinates for axi-symmetric flow in the cylindrical or annular stack.

12.2.1 Geometric parameters

An expression for $¶_v$ may be deduced by isolating the rectangular box $m_w^{-1} \times m_w^{-1} \times 2d_w$ of volume $2m_w^{-2}d_w$ which precisely bounds a representative cell (Fig. 12.1). The volume occupied by solid material is $2(\frac{1}{4}\pi d_w^2 S)$. S is the length of the representative element of wire, for which various approximations are available. Assuming S to be given by application of Pythagoras' theorem (right-hand sketch of Fig. 12.1):

$$¶_v \approx 1 - \tfrac{1}{2}\pi d_w m_w \sqrt{[1 + (d_w m_w)^2]} \qquad 12.1$$

Commercially available gauzes (G. Bopp and Co. undated) have numerical values of $d_w m_w$ ranging between 0.3 and 0.5 (but see Section 12.2.2). Evaluating the square root term of Eqn 12.1 reveals that $1.044 \leq \sqrt{[1 + (d_w m_w)^2]} \leq 1.118$, so that:

$$¶_v \approx 1 - \tfrac{1}{2}\pi d_w m_w \qquad 12.1(a)$$

Flow in the typical matrix differs fundamentally from that in the parallel duct, where flow resistance and forced convective heat transfer may be explored in terms of 'one-dimensional' or 'slab'-flow. The algebra of that straightforward case throws up the ratio free-flow area A_{ff} to wetted perimeter, p_w. The ratio A_{ff}/p_w is routinely denoted hydraulic radius r_h (m). This the preferred characteristic length for use in Reynolds number $Re = 4\rho \underline{u} r_h/\mu$. In the porous medium the fluid–solid interface is not one-dimensional, and at the local level there is no simple force balance to yield an obvious ratio $A_{ff}/p_w = r_h$. It is nevertheless possible to conceive of a volume element ΔV_w in contact with wetted-area element ΔA_w. The ratio $\Delta V_w / \Delta A_w$ serves for hydraulic radius r_h in the definition of Re (and $NTU = StL/r_h$) for the porous matrix.

It is readily shown (e.g. Organ 1992) that r_h, wire diameter d_w and volume porosity $¶_v$ are related through:

$$r_h = \tfrac{1}{4} d_w ¶_v/(1 - ¶_v) \qquad 12.2$$

Hydraulic radius represents size *scale* but not *geometry*. However, for a given class of matrix, such as the square-weave wire screen, two screens having the same value of the ratio r_h/d_w (equivalently of $r_h m_w$) have the same flow passage geometry:

$$r_h/d_w = \tfrac{1}{4} ¶_v/(1 - ¶_v) \qquad 12.3$$

Equations 12.2 and 12.3 are valid beyond wire screens to *any matrix of uniformly-packed wires of uniform diameter* – but obviously require a numerical value of $¶_v$ for the alternative packing (e.g. from calculation or weighing).

The literature offers various definitions for *aperture ratio* α – the ratio of frontal area to minimum free-flow (throat) area. Considering the matrix to consist of crossed rods gives:

$$\alpha = 1 - d_w m_w \qquad 12.4$$

Basing α on projected area (Fig. 12.1) suggests:

$$\alpha = (1 - d_w m_w)^2 \qquad 12.5$$

Reciprocal formulae are useful when checking commercial availability of gauzes against design requirements:

$$\alpha = [1 - (1 - \P_v)4/\pi]^2 \qquad 12.6$$

$$\P_v = 1 - \tfrac{1}{4}\pi(1 - \sqrt{\alpha}) \qquad 12.7$$

$$d_w m_w = 1 - \sqrt{\alpha} \qquad 12.8$$

$$\approx 1 - 4\P_v / \pi \qquad 12.9$$

High $d_w m_w$ means low aperture ratio α and low volume porosity $\P_v$. High aperture ratio α (close to unity) goes with high volume porosity $\P_v$ (also close to unity).

Both expressions for α (Eqns 12.4, 12.5) are based on an under-estimate of minimum area, since no account is taken of the fact that the wires lie out-of-plane. Concerned over this, Pinker and Herbert (1967) suggest, without derivation:

$$\alpha = 1 - 2/\pi^2\{(1 + 2N^2)(\tfrac{1}{2}\pi - \cos^{-1}N) + 3N\sqrt{(1 - N^2)}\} \qquad 12.10$$

$$N = \tfrac{1}{2}\pi d_w m_w$$

All three parameters $\P_v$, r_h/d_w and α, in all alternative versions, are functions of the product $d_w m_w$ – wire diameter (m) *times* pitch (m^{-1}) *only*. Screens having a common value of $d_w m_w$ have the same flow passage geometry notwithstanding different values of d_w and m_w in isolation. For the purposes of *characerizing* screens in a given class (e.g. symmetrical square weave) any one of the parameters $\P_v$, r_h/d_w or α serves, as does the simple product $d_w m_w$, since its value determines the values of all the others.

12.2.2 Upper limit on geometric parameter $d_w m_w$

On the assumption of perfectly cylindricity in both warp and weft directions, there is a minimum achievable value of $d_w m_w$. This reflects the fact that distance $2d_w$ in Fig. 12.2 cannot be reduced. By Pythagoras' theorem $m_w^{-1} \geq \sqrt{\{(2d_w)^2 - d_w^2\}}$ or

$$d_w m_w \leq 1/\sqrt{3} \qquad 12.11$$

Accounts of tests on flow through gauzes which cite values in excess of this limit may be considered suspect. One of the square-weave gauzes featured in heat transfer and friction factor tests reported by Coppage (1952) is given the value $d_w m_w = 0.66$.

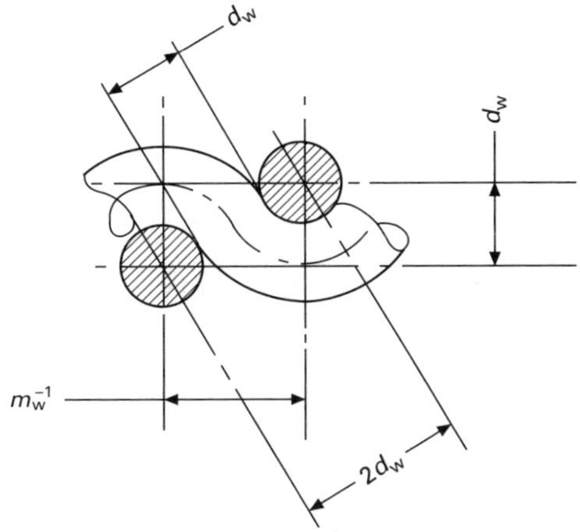

12.2 Upper numerical limit on geometric group $d_w m_w$.

Geometry imposes no lower limit on $d_w m_w$, although commercial practice appears not to go below 0.3.

It can be convenient to calculate $d_w m_w$ from d_w in mm, in which case m_w must be in wires/mm. The same numerical result is achieved by multiplying d_w in inches by m_w in imperial mesh number (wires/inch).

Figure 12.3 plots the three flow-passage parameters $¶_v$, α and r_h/d_w in function of $d_w m_w$. Viable values extend left-to-right as far as the physical limit of $d_w m_w$.

Chapter 2 identified anomalies in the 'incompressible' C_f–Re correlations, Chapter 10 has addressed these, and Chapter 13 will pursue this crucial matter. Possibly concerned about this very problem, Barree and Conway (2004) generated vector plots of two-dimensional flow around rectangular obstructions having a degree of randomness. Denney (2005) summarizes their 2004 paper. Harvey (2003) used the Fluent™ CFD package to achieve a three-dimensional visualization of flow through a geometric model of a single gauze.

The approach taken here will be to analyze the flow in the convergent section of the gauze aperture. Standard concepts from steady, compressible flow are used, including the momentum equation in differential form, one term in which is fractional change of free-flow area dA_{ff}/A_{ff} measured in the flow direction z. Hence the requirement anticipated earlier for an analytical representation of $A_{ff}(z)$.

12.2.3 Insights from proprietary software

The inset to Fig. 12.4 is a 'solid model' of a single gauze created (van Rijn 2006) with the aid of ProEngineer™. The proprietary package offers a ready

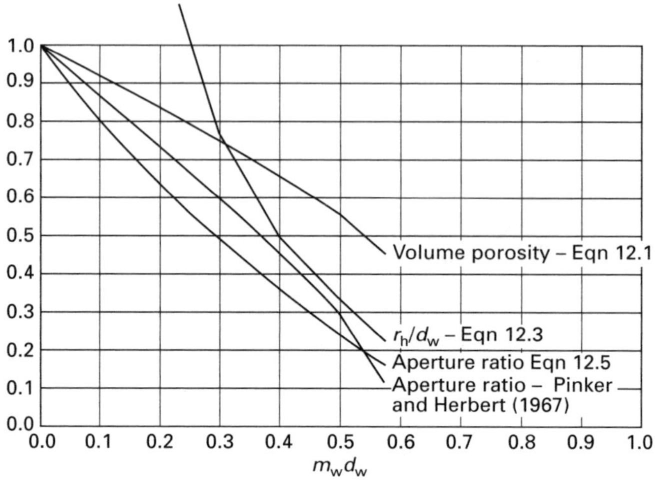

12.3 Volume porosity, ¶v, hydraulic radius ratio, r_h/d_w and area ratios α as functions of geometric parameter $d_w m_w$ for square-weave gauze.

means of taking cross-sections – or 'slices' – through the model. A succession of closely-spaced, parallel slices perpendicular to the flow direction amounts to a definition of the flow passage.

The figure displays slices at z-locations 10 percent, 20 percent, 30 percent, 40 percent and 50 percent of gauze thickness ($2d_w$). The white areas are occupied by the flow, the black by the solid. The final slice is at the mid-plane ($z = d_w$) and white and black areas are predictably symmetrical. The outline of an individual aperture at this point corresponds closely to that deduced by Pinker and Herbert (1967 – their Figs 7 and 8).

The slice at 40 percent highlights a feature which is going to warrant further exploration:

- Focusing on the black areas (representing the wire) gives an initial impression of symmetry. Focusing instead on an individual white area reveals a flow passage outline rotated through an angle relative to the underlying rectangular geometry.
- The angle of rotation of an individual flow passage is equal and opposite to that of any adjacent neighbour.
- Returning to the preceding slices (z < 40 percent of $2d_w$), a white area at any preceding location (z = 20 percent or 30 percent of $2d_w$) can now be interpreted as being rotated relative to the rectangular coordinate system.

Extrapolating to the slice at the 60 percent and subsequent locations (i.e., downstream of the mid-plane), the flow passage must 'unwind' via a mirror image of the initial spiral. The convergent section upstream of the mid-plane

12.4 Geometric model of single gauze. Created by van Rijn (2006) using proprietary solid-modelling package ProEngineer™. The succession of patterns correspond to cross-sections or 'slices' taken perpendicular to the flow direction. Black represents the solid. The gas occupies the white area.

suppresses boundary layer separation, while the divergent section downstream promotes it. If the flow acquires a component of angular velocity or 'swirl' on entry to the gauze, it is likely to retain it downstream of the 'throat' due to relative independence from the wall.

Swirl acquired by flow through any given aperture is equal and opposite to that acquired in all four immediately adjacent apertures. This suggests that net angular momentum downstream of a screen is zero. Moreover, any kinetic energy of rotation gained within a gauze 'stage' is likely to be dissipated before entry to the next.

12.2.4 Analytical definition of flow passage geometry

Figure 12.5 shows the individual flow channel somewhat simplified and in orthographic projection. Each wire element comprises a substantially straight section between two arcs centred on the axis of the circles representing wires in cross-section. Mid-plane X – X is tangent to the hatched cross-sections. In reality, the points of tangency do not coincide with the points of contact with the inclined wire element, but the development will proceed as though they do.

The compressible flow calculations require an expression for fractional area change $dA_{ff}(z)/A_{ff}(z)$ as a function of location z measured in the flow direction.

The equation of an ellipse of semi-axis a in the x' direction, and semi-axis b in the y' direction (Fig. 12.5) is, $(x'/a)^2 + (y'/b)^2 = 1$ (Hodgman 1954), permitting y' to be expressed in terms of x:

$$y' = +/- b\sqrt{[1 - (x'/a)^2]} \qquad 12.12$$

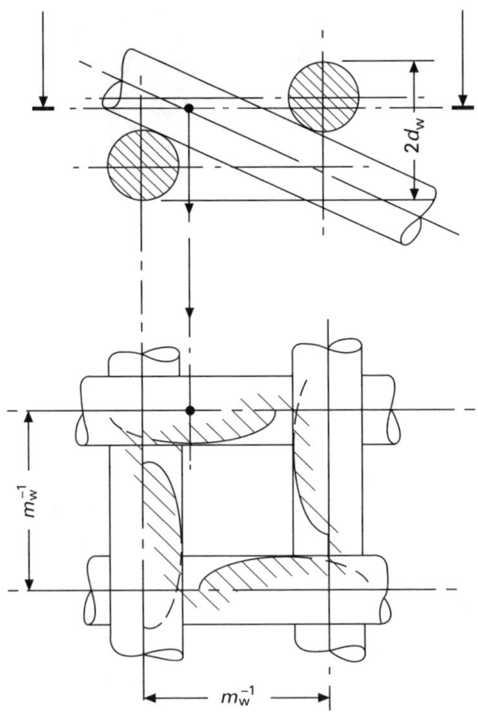

12.5 Individual flow channel somewhat simplified and in orthographic projection.

In Eqn 12.12 $a = \frac{1}{2} m_w$, and $b = \frac{1}{2} d_w$. It is convenient to work in terms of distance x measured from an origin at the 'nose' of the ellipse (i.e. in the (minus) x' direction). The change of coordinate is simply: $x' = a - x$.

The independent variable of the flow problem is z measured from the inlet plane of the gauze aperture. By similar triangles (Fig. 12.5) x and z are related by $x/z = m_w^{-1}/d_w$, or:

$$x = zm_w^{-1}/d_w \tag{12.13}$$

The incremental area of the semi-ellipse (above the horizontal axis) corresponding to increment dx is y'dx or, from Eqn 12.12:

$$dA_{ellipse} = b\sqrt{[1 - (x'/a)^2]}dx \tag{12.14}$$

The outline of the aperture is defined by 4 such, so d$A_{ff}(z)$ = (minus) 4-times d$A_{ellipse}$. The flow calculation is to proceed by small steps in the z direction, so the value of d$A_{ff}(z)$ required at each z is readily calculated as minus $2 \times \Sigma dA_{ellipse}$, where the sum $\Sigma dA_{ellipse}$ is the result of adding d$A_{ellipse}$ at current z to the $\Sigma dA_{ellipse}$ accumulated to that point. Local free-flow area is rectangular entry-plane area m_w^{-2} *minus* the accumulated $2 \times \Sigma dA_{ellipse}$:

$$A_{ff}(z) = m_w^{-2} - 2 \times \Sigma dA_{ellipse} \tag{12.15}$$

Local hydraulic radius $r_h(z)$ is calculated in terms of free-flow area at axial location z and wetted perimeter at the same location. Element ds of the outline of the ellipse is $\sqrt{(dx'^2 + dy'^2)} = dx'\sqrt{[1 + (dy'/dx')^2]}$. Differentiating Eqn 12.12 for y', and anticipating the form of the eventual computer-coded version gives $dy'/dx' = -(b/a)^2 x'/y'$, whence:

$$ds = dx'\sqrt{[1 - (b/a)^2 x'/y']} \tag{12.16}$$

As in the case of the area calculation, curve length $s(z)$ at location z is the sum of all the ds as computation proceeds in increments of dz from the plane of entry to current point z.

At any given z, one symmetrical half of each of 4 ellipses comprises the wetted perimeter. Hydraulic radius at location z is the ratio of local free-flow area $A_{ff}(z)$ to *wetted* perimeter: The area of an ellipse is πab, a and b being respectively the major and minor semi-axes (c.f. $A = \pi r^2$ when $a = b = r$). The minor semi-axis is $\frac{1}{2} d_w$ the major is $\frac{1}{2}/m_w$, making area per ellipse $\frac{1}{4}\pi d_w/m_w$.

Figure 12.6 is an attempt to show the flow passage in perspective between the entry plane at $z = 0$ and the mid plane ($z = d_w$).

The throat aperture at $z = d_w$ is formed by one half in each case of four identical ellipses, and is equal in area to $1/m_w^2 - \frac{1}{2}\pi d_w/m_w$. Frontal area is $1/m_w^2$, on which basis aperture ratio α is:

$$\alpha = 1 - \frac{1}{2}\pi d_w m_w \tag{12.17}$$

Yet again, the defining geometric parameter is $d_w m_w$.

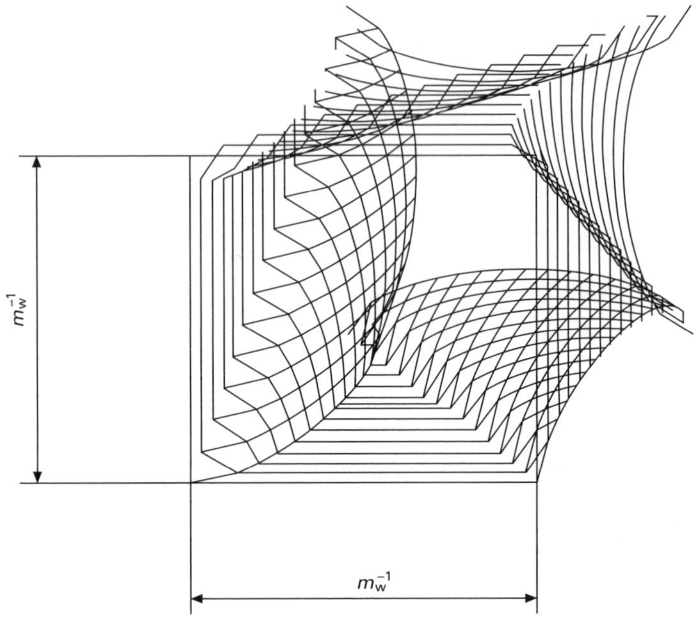

12.6 Flow passage as 'seen' by the gas filament. Planes of flow symmetry are represented by the areas occupied by straight lines.

Figure 12.7 compares aperture ratio α according to Eqn 12.17 with that from previous approximations. Two rectangular-weave gauzes of different wire diameter, but having the same value of $d_w m_w$, have not only the same value of α, but also the same flow passage geometry.

The variation of $A_{ff}(z)$ with z is defined by taking a succession of cross-sectional slices perpendicular to the z direction. Under the assumption of straight-line wire segments, the outline of each slice lying within the $m^{-1} \times m^{-1}$ rectangle defining the aperture is part of an ellipse. $A_{ff}(z)$ at given z is frontal area m^{-2} minus four-times the area of the individual part-elliptical element.

Except at mid-plane X–X, the outline shapes of individual wires do not form a closed boundary. It might be concluded that the fluid path is not bounded – but this is not so: adjacent apertures are mirror images separated by a plane of symmetry. Flow is thus constrained to an outline such as the specimen hatched in the figure.

The outline is not, in general, a simple shape. However, focusing on the individual flow passage, and noting the existence of planes of symmetry, now leads to an even more compelling suggestion of incremental rotation of a substantially rectangular perimeter. The approach highlights a contrast with the more familiar pipe-flow phenomenon. The 'slices' approach implies the 'slab' flow simplification. A 'slab' at location z along the fluid filament

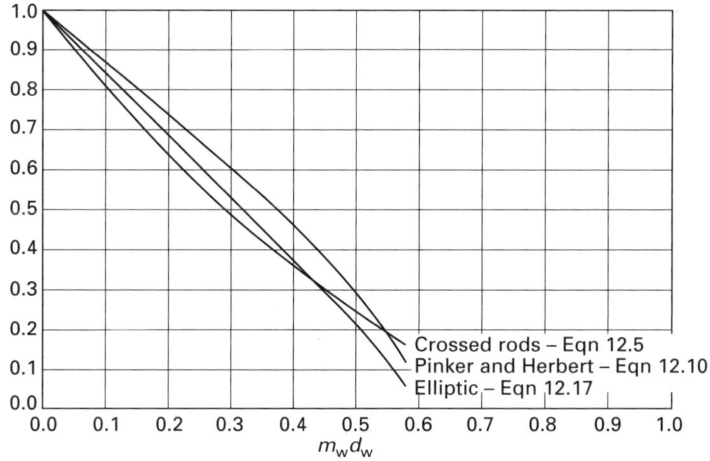

12.7 Aperture ratio α according to Eqn 12.17 compared with values computed from earlier, compromise expressions.

has part (only) of its periphery in contact with a solid surface, and only that part is subject to shear. The rest of the periphery coincides with a plane of symmetry, where shear is zero. This conflicts with the slab-flow concept, of which hydraulic radius and friction factor are an integral part. Pipe-flow thinking does not extend as readily to flow through porous media as is often implied.

Figure 12.1 included a schematic depiction of a cylindrical regenerator. The coordinate direction of net flow is z. As Chapter 11 has demonstrated, the cylindrical gauze stack may be expected to experience a component of flow in the r (radial) direction – and corresponding components of pressure and temperature gradient.

In single-cylinder engines of coaxial (or 'beta') configuration the regenerator tends to occupy an annulus concentric with the cylinder axis. A well-known instance is the Philips MP1002CA. The WhisperGen is an example of a multi-cylinder engine having the same feature. A 'bandage' of wire gauze wound so as to occupy the annulus is considerably more economical of material and easier to fabricate than a stack of gauze washers. On the other hand, the relevant flow characteristics are those in the plane of the gauze (radial direction in Fig. 12.1). The open literature barely refers to the case.

12.3 Flow parallel to plane of weave

Volume porosity is non-directional, and in a sense is a 'scalar' measure. To this juncture, hydraulic radius r_h has been treated in the same way. In its pipe-flow origins, however, the element of wetted area by which shear stress $\mu \partial u / \partial y$ communicates with the volume element is aligned with the flow direction. The gauze stack – and bandage – are anisotropic, and there may be an argument for defining r_h differently between z and r coordinate directions.

There is no doubt that aperture ratio α takes different values in different flow directions within the anisotropic porous medium.

12.3.1 Aperture ratio – perpendicular to net flow direction

There is no counterpart to net (z) flow direction of the isolated gauze:

- the flow path in the r direction is invariably many wire diameters in length – usually some hundreds
- flow edgewise-on (r-direction) to a single gauze in the stack does not generally follow the path to which it would be constrained if the single gauze were sandwiched between two impervious plane surfaces.

For Figure 12.8 a single gauze is shown edge-on. Impervious planes are imagined to be pressed into contact with the upper and lower surfaces, confining flow to the envelope of this isolated gauze. Perpendicular to

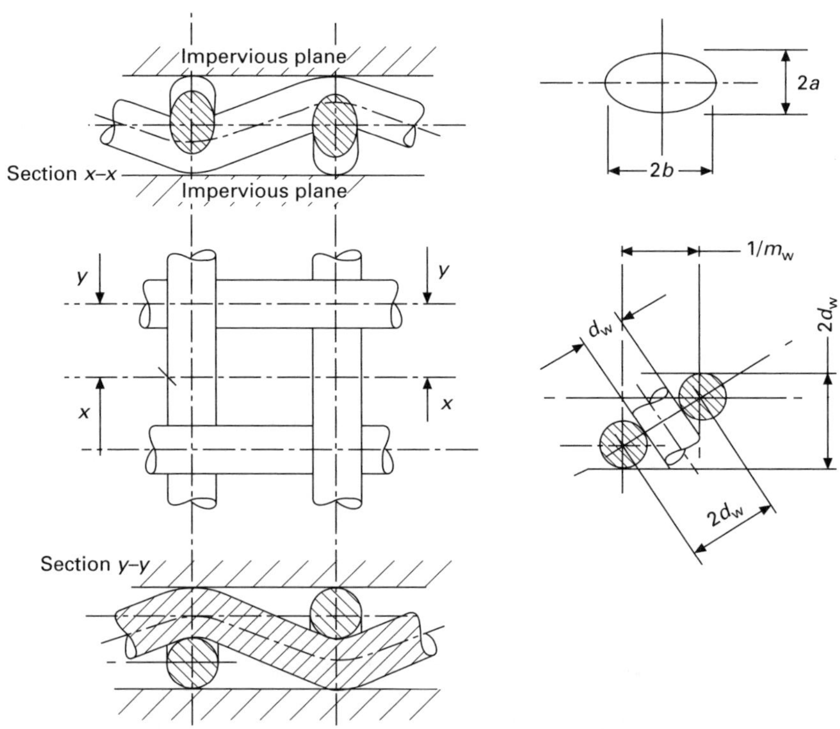

12.8 Notation for flow passage geometry *in the plane of* simple, square-weave gauze – i.e. flow direction perpendicular to that of usual z-direction.

the current flow direction there is a plane *x–x* at which free-flow area A_{ff} is maximum, and given by envelope area *minus* the projected area of the severed wires.

The area of the elliptical outlines is πab, as previously, a and b being minor and major semi-axes respectively. a is equal to $\frac{1}{2}d_w$; b is $\frac{1}{2}d_w/\cos\theta$, where θ is the inclination of the wire element, given in turn by $\cos\theta \approx d_w m_w/\sqrt{[1 + (d_w m_w)^2]}$. Envelope area is $2d_w/m_w$, leaving free-flow area A_{ff}

$$A_{ff} \approx d_w m_w^{-1} \{2 - \tfrac{1}{4}\pi\sqrt{[1 + (d_w m_w)^2]}\} \qquad 12.18$$

At a different plane *y–y* one-half wire pitch $(\frac{1}{2}m_2^{-1})$ in the flow direction (Fig. 12.2), free-flow area for the case under consideration is a minimum. The area obstructed is the sum of the cross-sectional area of a single wire $\frac{1}{4}\pi d_w^2$ plus $d_w S$ of the wire element seen in cross-section. Subtracting from envelope area and simplifying:

$$A_{ff} \approx d_w m_w^{-1} \{2 - \tfrac{1}{4}\pi d_w m_w - \sqrt{[1 + (d_w m_w)^2]}\} \qquad 12.19$$

Taking the ratio of minimum to maximum A_{ff} gives the area ratio through which the flow contracts and expands over each increment m_w^{-1} of distance travelled in the r direction:

$$\alpha_r = 1 - \frac{2 - \frac{1}{4}\pi d_w m_w - \sqrt{[1 + (d_w m_w)^2]}}{2 - \frac{1}{4}\pi\sqrt{[1 + (d_w m_w)^2]}} \qquad 12.20$$

Equation 12.20 is appropriate to any number of layers of gauzes having warp and weft in perfect alignment, since replacing the impervious planes with gauzes stacked in this way has no effect. Flow at an angle other than zero or 90° to warp and weft is subject to a different pattern of interruptions. If the two-dimensionality of the flow is important, attention to this matter is essential. An expression for the general case awaits the attention of an analyst – possibly a statistician – whose analytical skills exceed those of this writer.

In an experiment with flow 'edge-on' to the screen, such as that reported by Magara *et al.*, (2000) flow is subject to a one-off area contraction at entry, and to a one-off expansion at exit, neither of which is addressed by Eqn 12.20. Reference to the illustrations of Chapter 11 will confirm that these entry and exit effects do not affect flow in the *r*-direction when the regenerator is in service.

An alternative to wrestling with the algebra of the wire screen is to take a leaf out of the aircraft designer's book and to follow the principle that a simple design which is amenable to analysis is preferable to a more ambitious one which is not. The principle suggests working with matrix materials chosen for the ease with which flow passage geometry can be specified. In this connection, SES Ltd. are investigating ways of creating matrices to pre-specified geometry using laser remelt technology (Hislop 2006).

12.4 Commercial availability

Figure 12.9 is a selection of gauzes from the standard products catalogue of Messrs. G Bopp and Co Ltd (undated). Values of $d_w m_w$ are tabulated for each entry. Parameters derived from $d_w m_w$ – volume porosity, hydraulic radius etc. – are computed from the definitions of this chapter. Apart from volume

r_h (mm)	d_w (mm)	m_w (Wire/in.)	m_w (Wire/mm)	$m_w d_w$	Volume porosity	Availability – AISI 304/310	Phosphor-bronze	Brass
0.0067	0.0200	635	25.00	0.50	0.56	1	0	0
0.0083	0.0250	510	20.08	0.50	0.56	1	0	0
0.0108	0.0250	445	17.52	0.44	0.62	1	0	0
0.0122	0.0280	395	15.55	0.44	0.63	1	1	0
0.0126	0.0250	405	15.94	0.40	0.66	1	0	0
0.0134	0.0250	390	15.35	0.38	0.68	1	0	0
0.0133	0.0320	355	13.98	0.45	0.62	1	1	0
0.0141	0.0360	325	12.80	0.46	0.60	1	1	0
0.0151	0.0320	330	12.99	0.42	0.65	1	1	0
0.0150	0.0360	315	12.40	0.45	0.62	1	0	0
0.0165	0.0300	320	12.60	0.38	0.68	1	0	0
0.0168	0.0360	295	11.61	0.42	0.64	1	1	0
0.0170	0.0400	280	11.02	0.44	0.62	1	1	0
0.0178	0.0360	285	11.22	0.40	0.66	1	0	0
0.0176	0.0400	275	10.83	0.43	0.63	1	0	0
0.0186	0.0320	290	11.42	0.37	0.69	1	0	0
0.0189	0.0360	275	10.83	0.39	0.67	1	1	0
0.0187	0.0400	265	10.43	0.42	0.64	1	0	0
0.0213	0.0360	255	10.04	0.36	0.70	1	0	0
0.0213	0.0400	245	9.65	0.39	0.68	1	1	0
0.0211	0.0450	235	9.25	0.42	0.65	1	1	0
0.0238	0.0500	210	8.27	0.41	0.65	1	1	0
0.0248	0.0360	230	9.06	0.33	0.73	1	0	0
0.0248	0.0500	205	8.07	0.40	0.66	1	0	0
0.0269	0.0500	195	7.68	0.38	0.68	1	1	0
0.0268	0.0560	187	7.36	0.41	0.65	1	0	0
0.0280	0.0400	205	8.07	0.32	0.73	1	0	0
0.0300	0.0400	195	7.68	0.31	0.75	1	0	0
0.0301	0.0560	174	6.85	0.38	0.68	1	0	0
0.0302	0.0630	166	6.54	0.41	0.65	1	1	0
0.0318	0.0450	181	7.13	0.32	0.74	1	0	0
0.0334	0.0500	169	6.65	0.33	0.72	1	1	0
0.0334	0.0630	156	6.14	0.39	0.67	1	1	1
0.0352	0.0500	163	6.42	0.32	0.74	1	0	0
0.0356	0.0630	150	5.91	0.37	0.69	1	0	0
0.0374	0.0710	139	5.47	0.39	0.67	1	1	1
0.0377	0.0800	132	5.20	0.42	0.65	1	0	0
0.0393	0.0560	146	5.75	0.32	0.73	1	0	0
0.0417	0.0630	135	5.31	0.33	0.72	1	0	0
0.0418	0.0800	124	4.88	0.39	0.67	1	1	1
0.0420	0.0900	118	4.65	0.42	0.64	1	0	0

12.9 Selection of square-weave gauzes from the standard range of G. Bopp and Co. Numerical values of $¶_v$, r_h/d_w and α_{cr} have been calculated from the manufacturer's values of m_w and d_w using Eqns 12.1–12.5, and differ somewhat from official figures (the derived data have not been approved by the manufacturer).

porosity $¶_v$ these relate to flow in the usual (z) direction. Values of aperture ratio differ somewhat from original catalogue entries. (The derived data are not approved by the manufacturer.)

Figure 12.3 may be used for an approximate estimate of parameter values for combinations of m_w and d_w which do not occur in the Fig. 12.9.

12.5 Specimen 'isotropic' material – metal foam

About a dozen producers world-wide now market metal foams (Granta Design 2006). Applications include filtration, gas/oil separation, sound absorption, flame arrest and energy absorption (for crash protection and packaging). Aluminium-based foams are used in the production of light-weight, stiff structures and compact heat exchangers. Of interest in the present context are the nickel and nickel alloy foams, available in a range of pore sizes (Dunlop undated).

There is no clear concept of aperture ratio as there was for the woven screens. Manufacturers' data tend not to help directly in determining what grade of foam – if any – would have potential as a regenerator. Dunlop's literature, for example, cites porosity values, but the company's definition of porosity is number of pores/inch.

With an important reservation (which follows), volume porosity $¶_v$ may be estimated from the ratio of foam density ρ_{foam} to density of the parent (solid) metal, ρ_{parent}:

$$¶_v \approx 1 - \rho_{foam}/\rho_{parent} \qquad 12.21$$

The reservation is that, in some commercial foams, the dendrites are hollow, and to this extent somewhat like bones. Electron microscope images suggest that the hollow cores are not connected, so their surface areas do not contribute to effective hydraulic radius, but invalidate the volume porosity calculation (Eqn 12.21), and compromise the estimate of thermal capacity ratio N_{TCR}. The cyclic history of thermal wave penetration in the hollow element promises to be different from that of the otherwise identical solid – quite possibly reflecting better utilization of a given mass of parent material. If metal foams show promise in this application it may be worthwhile rethinking the analytical description of matrix transient response to get the best out of the material.

An alternative estimate of volume porosity follows on noting that the individual open cell is somewhat like a skeletal sphere. For Fig. 12.10 it is assumed that spheres of identical diameter d_s sit with their origins on at the intersections of a rectangular lattice of pitch distance m_s^{-1} in all three orthogonal directions. The internet page by Thompson (undated) has elegant illustrations in colour.

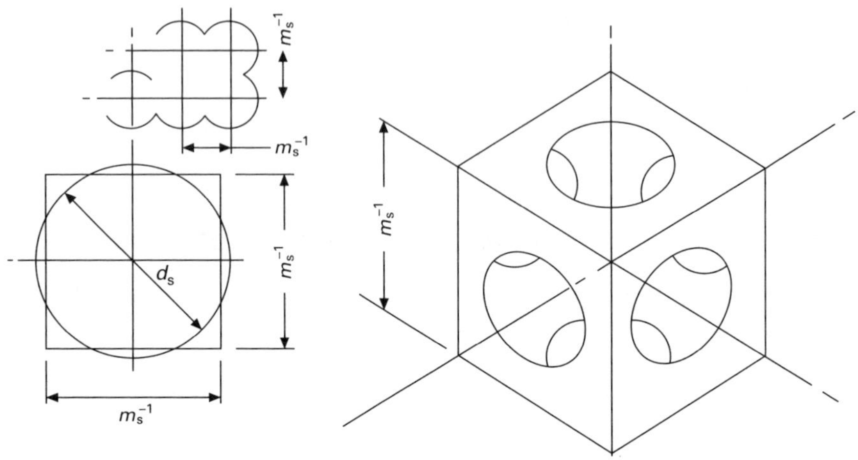

12.10 Notation for estimate of volume porosity $\P_v$ and hydraulic radius r_h of metal foam, assuming the pores to be a regular array of spheres.

Each sphere may be thought of as being centred within a cube of side m_s^{-1}. For values of the product $d_s m_s$ in excess of unity, the sphere breaks though all six plane surfaces of the cube at circular outlines. It protrudes beyond each plane face by the amount h, where $\frac{1}{2}(d_s - m_s^{-1})$. The area A_s of the curved surface of a protruding spherical segment is $\pi d_s h$ (Hodgman 1954). Corresponding segment volume V_s is $\frac{1}{3}\pi h^2(\frac{3}{2}d_s - h)$. The wetted area of the part-spherical pore is the surface area of the complete sphere πd_s^2 minus six-times the individual A_s. Corresponding wetted volume is nominal sphere volume $\frac{1}{3}\pi d_s^3$ minus six-times V_s.

Volume porosity $\P_v$ is the ratio of net pore volume of the cropped sphere to volume of enclosing cube. Making the substitutions:

$$\P_v = \pi(d_s m_s)^3\{\tfrac{1}{6} - \tfrac{1}{4}[1 - 1/d_s m_s]^2[3 - (1 - 1/d_s m_s)]\} \qquad 12.22$$

Hydraulic radius r_h is the ratio of wetted volume to wetted area:

$$r_h/d_s = \frac{\frac{1}{6} - \frac{1}{4}[1 - 1/d_s m_w]^2[3 - (1 - 1/d_s m_s)]}{1 - 3(1 - 1/d_s m_s)} \qquad 12.23$$

The minimum d_s at which adjacent cells connect is $d_s = 1/m_s$. Maximum d_s before the edges of the cube are severed at their mid-point by the sphere is $d_s = \sqrt{2}/m_s$. Equations 12.22 and 12.23 are therefore limited to the following range:

$$\text{unity} < d_s m_s < \sqrt{2} \qquad 12.24$$

As for the wire gauze, the numerical value of a simple dimensionless product – $d_s m_s$ – characterizes $\P_v$ and r_h/d_s for all porous materials for which the regular array of spherical pores is an adequate representation. Figure 12.11 plots $¶_v$ and r_h/d_s over a range of $d_s m_s$. Equation 12.3 expressed r_h/d in terms of volume porosity $¶_v$ for any uniformly distributed packing of wires of diameter d. If reliable value of $¶_v$ is available for the foam, an independent estimate of hydraulic radius comes from noting that adjacent 'spheres' have skeletal arms in common. If the skeletal arms can be considered cylindrical and of uniform diameter d_{arm}, an approximate r_h follows from Eqn 12.3:

$$r_h/d_{arm} = \tfrac{1}{4}¶_v/(1 - ¶_v) \qquad 12.3(a)$$

An alternative view of volume porosity is that the bubbles are spheres packed at maximum (hexagonal) density. Solid spheres occupy $\pi/3\sqrt{2}$ m^3 per m^3 of envelope volume, which converts to a packing density of 0.74048 (Weisstein 2005). The corresponding bubble array thus has $¶_v = 0.74048$. At this value, flow connectivity is zero because the contact between adjacent spheres is a point. Sphere diameter, and thus void volume, must exceed maximum packing density value, making this a 'lower bound':

$$¶_v \geq 0.74048 \ldots \qquad 12.25$$

12.6 Résumé

Much was taken for granted when the steady-flow $C_f - Re$ correlation was appropriated for Stirling engine use. Choice of a replacement format merits

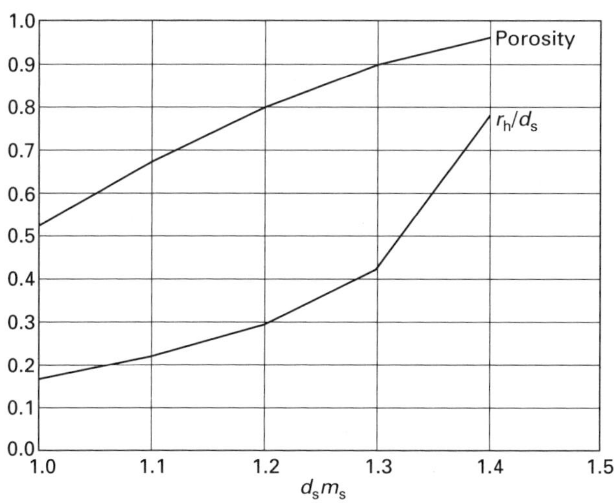

12.11 Volume porosity $¶_v$ and hydraulic radius ratio r_h/d_s as functions of dimensionless geometric parameter $d_s m_s$ for a 'foam' consisting of spherical bubbles of diameter d_s on a square lattice of linear pitch m_s^{-1}.

greater circumspection than went into the earlier choice. With this in mind, Chapter 14 'Bridging the Gap' will generate charts of $\Delta p/p = \Delta p/p\{Sg, Ma \ldots\}$ on the basis of boundary-layer development though a flow passage to the geometry specified in this chapter. Confirmation of all features of the experimental counterpart (see Fig. 10.5) will raise confidence in the new correlations for regenerator design and performance prediction.

The reader nevertheless harbouring misgivings may be prepared to accept that the proposed change threatens *no loss of flow information*: the C_f vs Re correlation is a compromise 'sub-set' version of the more general replacement $\Delta p/p$ vs Sg, Ma … which embodies no pre-conceptions about the interaction between viscous and inertia components.

For the symmetrical, square-weave gauze it is now confirmed that:

- a value for the dimensionless product $d_w m_w$ *alone* fixes respective values of all geometric parameters – $¶_v$, r_h/d_w and several choices for aperture ratio, α – required for the acquisition and application of flow (and heat transfer) correlations
- flow passage geometry is characterized when $¶_v$, or r_h/d_w, or one of the α (or, of course, $d_w m_w$) is specified.

The prospect, examined further in Chapter 14, that flow filaments leaving adjacent apertures with swirl components of opposite sign has a bearing on a long-standing preoccupation, namely the matter of alignment – or mis-alignment – between successive gauzes. The purely geometric location of plane gauze elements in contact is in terms of three degrees of freedom, two of relative translation, one of relative rotation. It now appears that achievement of a specific degree of effective flow passage alignment would have to be achieved taking into consideration the sign of the swirl component of the upstream gauze.

13
Beyond the performance envelope

13.1 Introduction

The underlying principle of Stirling engine operation amounts to biasing a mass of working fluid towards a cold space for a process of overall compression, and subsequently biasing it towards a hot space and allowing an overall expansion.

The biasing process is achieved largely by positive displacement between the two variable-volume spaces – 'largely' because flow resistance causes non-uniformity of pressure distribution, and so influences the distribution of density, and thus of mass.

There is an additional influence: pressure information propagates in the form of waves, and the speed at which it does so is limited. The consequences are minimal in engines charged with light gases such as hydrogen and helium, where acoustic speed is high. However, when the working fluid is air the phenomenon is readily provoked and recorded experimentally (see Organ 1992 text, p. 189 and Rix 1984).

In a reciprocating machine, pressure-waves are largely a response to accelerations. All other things being equal, raising rpm increases the accelerations of piston and displacer. This changes the phase of pressure disturbances relative to the angular position of the crankshaft and generally* increases their amplitude.

There are two implications for the performance potential of the air engine:

1. Power output is the product of *work per cycle* with *cycle frequency*. A modest value of cycle work can amount to high power output if rpm can be elevated.
2. Peak rpm of the Philips MP1002CA air-charged engine coincide (see

* 'Generally' because the cyclic variation of pressure at a given location is the result of the interaction of two waves – one right-travelling, the other left-travelling. The interaction can result in resonance.

Chapter 14) with the threshold of compressibility effects in the regenerator. The 'obvious' deduction is that work/cycle suffers by operating in the compressible region.

Before supersonic flight became a reality there appeared to exist a sound 'barrier'. It is now appreciated that the barrier was imposed by lack of insight rather than by aerodynamic limitations. There is, to the writer's knowledge, no theorem prohibiting the design of closed-cycle, air-charged regenerative engines from *exploiting* compressibility effects rather than being *limited* by them.

No reason, that is, except:

- that the computational tools for flow at elevated Mach numbers have not been much exercised in the present context
- for the paucity of relevant flow friction and heat transfer data – a problem which is now duly recognized, but which these chapters have only just begun to address

The first point can be comprehensively tackled on the basis of arbitrary correlations while the definitive replacements are awaited. This chapter accordingly adapts the numerical modelling tools of classic, unsteady compressible flow to the context (boundary and operating conditions) of the Stirling engine. An air-charged, opposed-piston design is explored having a gas path 1m long and operating at resonance at 12,500 rpm. On the basis of the provisional data, the numerical model predicts that a volume phase angle in the region of 180° is required for achievement of net positive indicated work. At this condition, the traditional contributions of expansion and compression spaces are reversed – compression space work is larger and positive, while expansion space work is smaller and negative. At the rpm called for by operation in this mode, a small value of work per cycle converts to significant indicated power.

13.2 Method of Characteristics

The basic assumption is that flow is 'one-dimensional'. Local (isentropic) acoustic speed is denoted a, and evaluated locally as $\sqrt{\gamma RT}$, where T is instantaneously prevailing temperature. Corresponding particle velocity is u. At any location in a closed duct there is generally a pressure wave travelling leftwards superimposed on a wave travelling rightwards. It is not necessary to resort to analysis to be able to write down the two speeds at which pressure information propagates rightward and leftward:

$$\mathrm{d}x/\mathrm{d}t_{\mathrm{I}} = u + a \qquad 13.1(a)$$

$$\mathrm{d}x/\mathrm{d}t_{\mathrm{II}} = u - a \qquad 13.1(b)$$

In the 'physical' (t–x) plane, the two local speeds are represented by 'characteristic directions' as in Fig. 13.1(a). The right-running and left-running line elements are called I and II characteristics respectively.

A particle, on the other hand, may be travelling in one direction only at any given time and location. The equation of its path line may again be written down directly, and yields a third characteristic direction:

$$dx/dt_{path} = u \tag{13.2}$$

The Method of Characteristics differs from other numerical solutions only in respect of the paths in the t–x plane (the 'Mach' or 'physical plane') along which the integration process steps. Rather than proceeding parallel to the x and y axes, integration follows the three characteristic directions. This calls for 'state equations' defining the changes in states corresponding to an increment in each of the three characteristic directions.

There are various ways of deriving the state equations. This author would be unable to better Shapiro's treatment of 1954, to which the reader is therefore referred (Section 24.7 and pp. 972–976 of the original, or the Appendix to this chapter). The final form of the Shapiro state equations is:

$$du_I = -dp/\rho a$$
$$+ \{-ua\, A_{ff}^{-1}\, dA_{ff}/dx + (\gamma - 1)q'/a + F[-1 + (\gamma - 1)u/a]\}dt_I \quad 13.3(a)$$

$$du_{II} = + dp/\rho a$$
$$+ \{ua\, A_{ff}^{-1}\, dA_{ff}/dx - (\gamma - 1)q'/a + F[-1 - (\gamma - 1)u/a]\}dt_{II} \quad 13.3(b)$$

The path line equation is as follows:

$$dp_{path} = \{[2\gamma/(\gamma - 1)](p/a)da_{path} - \rho[uF + q']\}dt_{path} \tag{13.4}$$

$$F = \tfrac{1}{2}\, C_f u^2 \text{sign}(u)/r_h$$

(and not as misprinted (without the r_h) in Organ (2005a))

For practical use it is necessary to relate heat rate q' (W/kg) to the parameters of the flow problem. In the context of forced convection, heat rate per unit mass q' is:

$$q' = \frac{h p_w\, dx(T_w - T_g)}{\rho A_{ff}\, dx}$$

From the definition of Stanton number St as $h/\rho\,|u|\,c_p$, heat rate q' may be written:

$$q' = St\,|u|\,c_p(T_w - T_g)/r_h \tag{13.5}$$

To keep presentation concise, the term for area variation will be omitted. (It is readily reintroduced retrospectively.) Under the ideal gas assumption, density ρ may be replaced by p/RT, and RT by a^2/γ. Local surface temperature T_w may be replaced by $a_w^2/\gamma R$. Substituting for q' and F leads to state and path line equations in terms of dependent variables dp, da and du:

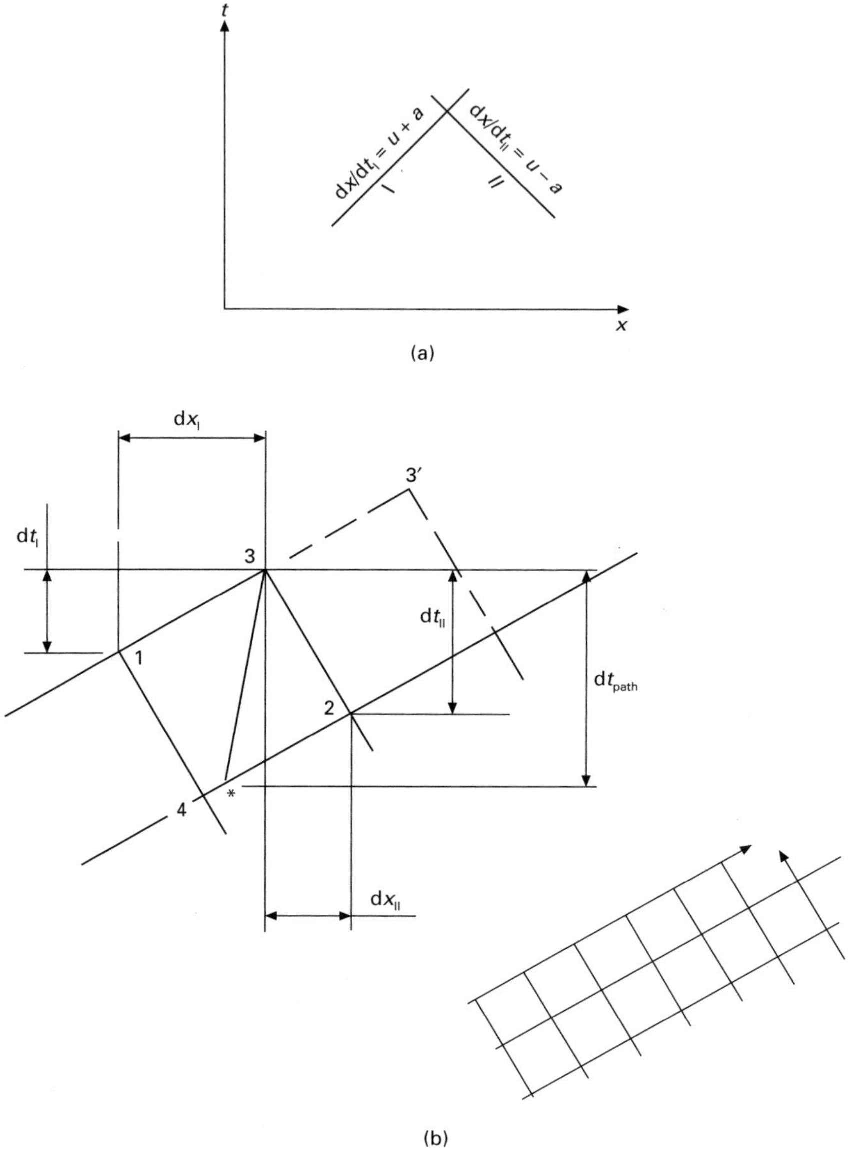

13.1 In Method of Characteristics coordinates the integration grid is not pre-set in terms of a fixed dx and dt. Instead it is constructed step-by-step, alternating with solution of fluid state increments dp, da, du (a) line elements representing velocities dx/dt_I = $u + a$ and dx/dt_{II} = $u - a$ of right- and left-running elements of disturbances of fluid properties in the Mach or physical ($t - x$) plane; (b) typical integration step in the Mach plane.

$$du_{\mathrm{I}} = -\,dp/\underline{p}\;a/\gamma + \{(\gamma-1)NTU[(a_{\mathrm{w}}/\underline{a})^2 - 1] - \tfrac{1}{2}\underline{u}/\underline{a}NFU[1 - (\gamma-1)\underline{u}/\underline{a}]\}dt_{\mathrm{I}}/L_{\mathrm{x}} \qquad 13.6(a)$$

$$du_{\mathrm{II}} = +dp/\underline{p}\;a/\gamma - \{(\gamma-1)NTU[(a_{\mathrm{w}}/\underline{a})^2 - 1] - \tfrac{1}{2}\underline{u}/\underline{a}NFU[1 + (\gamma-1)\underline{u}/\underline{a}]\}dt_{\mathrm{II}}/L_{\mathrm{x}} \qquad 13.6(b)$$

The path line equation is:

$$dp/\underline{p} = [2\gamma/(\gamma-1)]da/\underline{a} - \gamma\{\tfrac{1}{2}NFU(\underline{u}/\underline{a})^2 + NTU[(a_{\mathrm{w}}/a)^2 - 1]\}\,|\,u\,|\,dt_{\mathrm{path}}/L_{\mathrm{x}} \qquad 13.7$$

In which:

$$F = \tfrac{1}{2}C_{\mathrm{f}}u^2\mathrm{sign}(\underline{u})/r_{\mathrm{h}}$$

$$NTU = StL_{\mathrm{x}}/r_{\mathrm{h}};\; NFU = C_{\mathrm{f}}L_{\mathrm{x}}/r_{\mathrm{h}}$$

The underscore (e.g., $\underline{p}$) signifies the mean value between the extremities of the element of integration path.

Shapiro himself does not put these relationships to work as they stand. Instead, he coerces them into the shape of the homentropic counterpart by re-expressing in terms of du and da. The transformation introduces a term for local entropy gradient $\partial/\partial x\{ln[a^{2\gamma/(\gamma-1)}/p]\}$ – Shapiro's Eqn 24.52. The x coordinate direction is *across* path lines in the Mach (t–x) plane, which are loci of potential discontinuities in the gradients of fluid properties. This obliges him to treat the time-dependent terms of Eqns 13.6(a) and 13.6(b) (i.e. those multiplied by dt_{I}, dt_{II}) as adjustments to the original homentropic construction. While a Mach line net may then be built up without *explicit* involvement of the path lines, it is not clear that cumulative 'smearing' of particle trajectories has been avoided in the process.

Plausible account of heat transfer calls for use of the state equations (Eqns 13.6(a), 13.6(b) as they stand. This allows information carried along the path line to feed explicitly into the Mach lines at their point of intersection. This is achieved by reconciling states a, p, u, at the intersection.

13.3 'Unit process' of the integration sequence

Figure 13.1(b) illustrates an integration step in the physical (t–x) plane. It typifies every such step which has to be taken, with the exception of those at left- and right-hand boundaries. It may therefore be thought of as a 'unit' process. Applying the unit process repeatedly and systematically to cover the entire flow field yields the required solution – the coordinates t, x, of all Mach line intersections, and values of p, a and u at each such intersection. The t and x coordinates at 1 and 2 are also known. If values of t, x, p, a and u can be found at point 3, then the process may be repeated to take construction to point 3'.

The construction applies equally to the special case of a stationary fluid. The calculations are trivial, since local I and II directions are simply $dx/dt_I = +a$, $dx/dt_{II} = -a$. The waves are then of 'zero strength', i.e., the disturbances in pressure, temperature (or acoustic speed) and particle are 'virtual'. The construction is also valid where, by the conventional text-book criterion ($a < 0.3$), flow would be thought of as incompressible. The method of characteristics is therefore of wider applicability than is sometimes thought.

Each unit process builds on the previous one. As indicated in the diagram, states p, a and u are assumed known at intersections 1 and 2. The coordinates t and x of point 3 are first located approximately by finding the intersection of a I-line element projected from point 1 at gradient $dt/dx_I = 1/(u_1 + a_1)$ with a II-line element projected from point 2 with gradient $dt/dx_{II} = 1/(u_2 - a_2)$. The construction yields approximate values for time increments dt_I and dt_{II}.

State Eqns 13.6(a) and (b) in conjunction with Eqn 13.4 form a set of three simultaneous equations in the unknown differences dp, da and du, which may respectively be written $p_3 - p_1$ (or $p_3 - p_2$), $a_3 - a_1$, (or $a_3 - a_2$) as appropriate. The path line reaching point 3 will have passed close to point 4, in the process cutting either the I-characteristic between 4 and 2, or the II-characteristic between 4 and 1. Intersection is at auxiliary point t^*, x^*, located by projecting a path line backwards from point 3 at provisional gradient $dt/dx_{path} = 1/u_4$. This yields provisional values of p^*, a^*, u^* and of dt_{path}. Substituting into the three state equations (13.3(a), (b) and 13.4) and solving (e.g. in 3×3 determinant form) yields provisional values of p_3, a_3 and u_3. Tentative values of the respective time increments dt_I, dt_{II} and dt_{path} may now be picked off in the physical plane.

The t–x coordinates of point 3 are now improved by use of mean values of u and a between 1 and 3, and between 2 and 3. The unit process is iterated to convergence.

13.4 High-speed operation – the wave engine

Figure 13.2 shows the opposed-piston engine schematically. A difference from the well-known arrangement is that piston (and thus volume) phase angle is nearer to 180° than to 90°. Geometric parameters and specimen operating conditions are listed in Table 13.1.

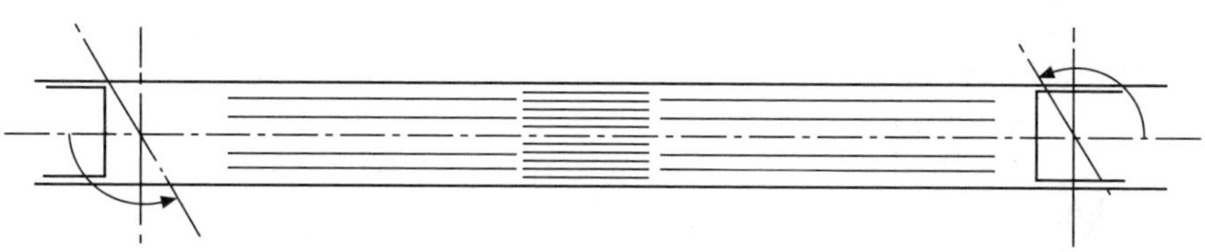

13.2 Schematic representation of opposed-piston engine. Piston phase angle is shown nearer to 180° than to the more usual 90°.

Table 13.1 Specification and operating conditions of opposed-piston 'engine' of Fig. 13.2

Working fluid	air
Charge pressure p_0	10 bar
Gas constant	287 J/kgK
P_r	0.7 –
T_C	330 K
T_E	930 K
Rpm	12,750 min^{-1}
A_{ff} (uniform throughout)	0.8E-03 m^2
Hydraulic radii r_h	
expansion exchanger	0.25E-03 m
regenerator	0.125E-03 m
compression exchanger	0.25E-03 m
Regenerator matrix material	Alternate layers of corrugated and plain foil – as in Amal™ flame-trap construction.
Individual exchanger lengths L_x	
expansion	0.410 m
regenerator	0.125 m
compression	0.375 m
Piston strokes	
expansion	16.0E-03 m
compression	16.0E-03 m
Volume phase angle	180 degrees

Figure 13.3 is a specimen Mach plane, which may be thought of as a plan view of the pressure surface. The solution shown is that immediately following start-up. Starting the pistons from inner (or outer) dead-centre position aids convergence towards steady state, as velocity is instantaneously zero, and no expansion fan is generated. For the purposes of display the number of Mach lines per left-to-right integration sweep has been limited to 75. (A larger number gives a smoother p–V diagram.) The lines which appear to be particle trajectories indicate only the *local* particle path: individual consecutive line elements do not join up precisely to give a continuous history of a specific particle.

Replotting with the Mach plane inclined slightly, as in Fig. 13.4, reveals the pressure surface in quasi-perspective. The lines ruled on the surface from lower left to upper right outline the lateral extent of the regenerator. The corresponding relief of local particle velocity is shown at Fig. 13.5.

The solution scheme is coded in FORTRAN 90 using double-precision arithmetic (REAL*8). For most operating conditions investigated, computing and displaying 3000 sweeps of the characteristics mesh takes about 3 sec. on a PC with Intel D845 chipset.

In the fluid temperature relief (Fig. 13.6) detail is overwhelmed by the massive temperature difference across the regenerator. (Future results will be displayed as local fluid-wall temperature difference, $T_g - T_w$.)

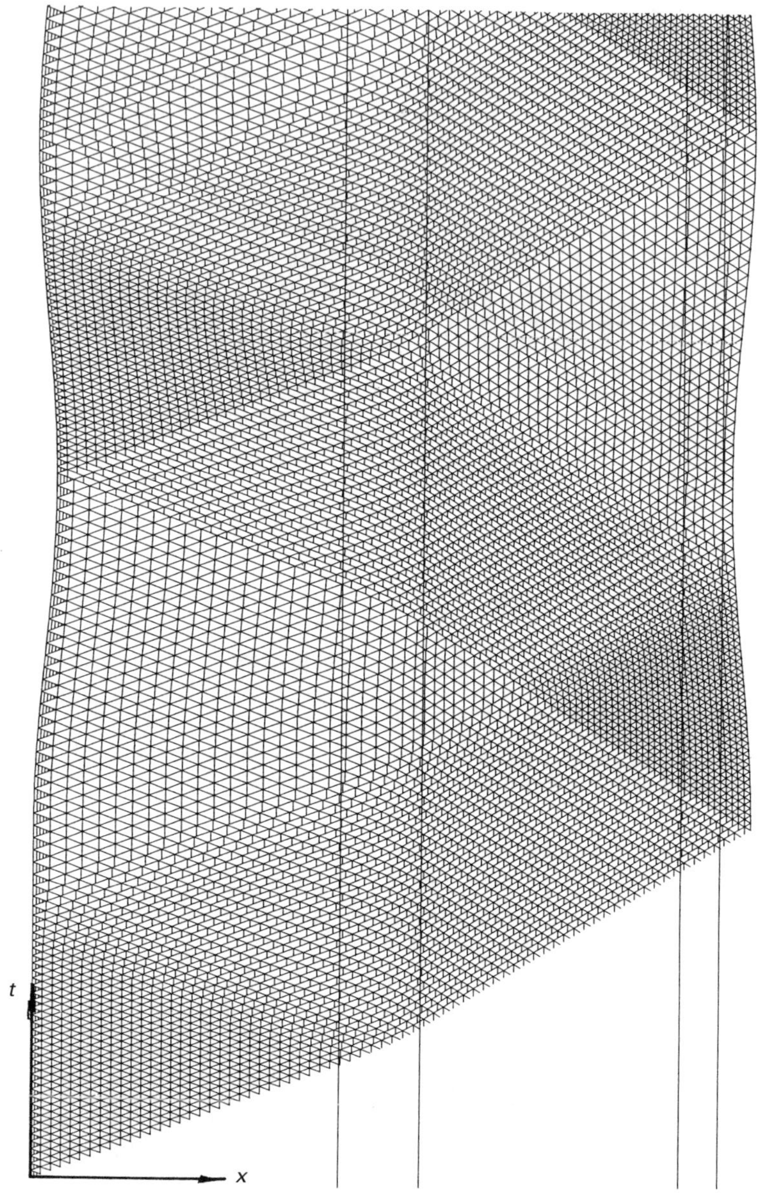

13.3 Specimen $t - x$ (Mach) plane construction. May be thought of as plan view of the pressure (or velocity – or temperature – or acoustic speed) surface.

Figure 13.7 shows the development of p–V loops from start-up. Both sets enlarge as operation gets underway. There is an initial tendency to float upwards, suggestive of an increase in mean pressure. For the parameters so far investigated, this stabilizes after a few more cycles.

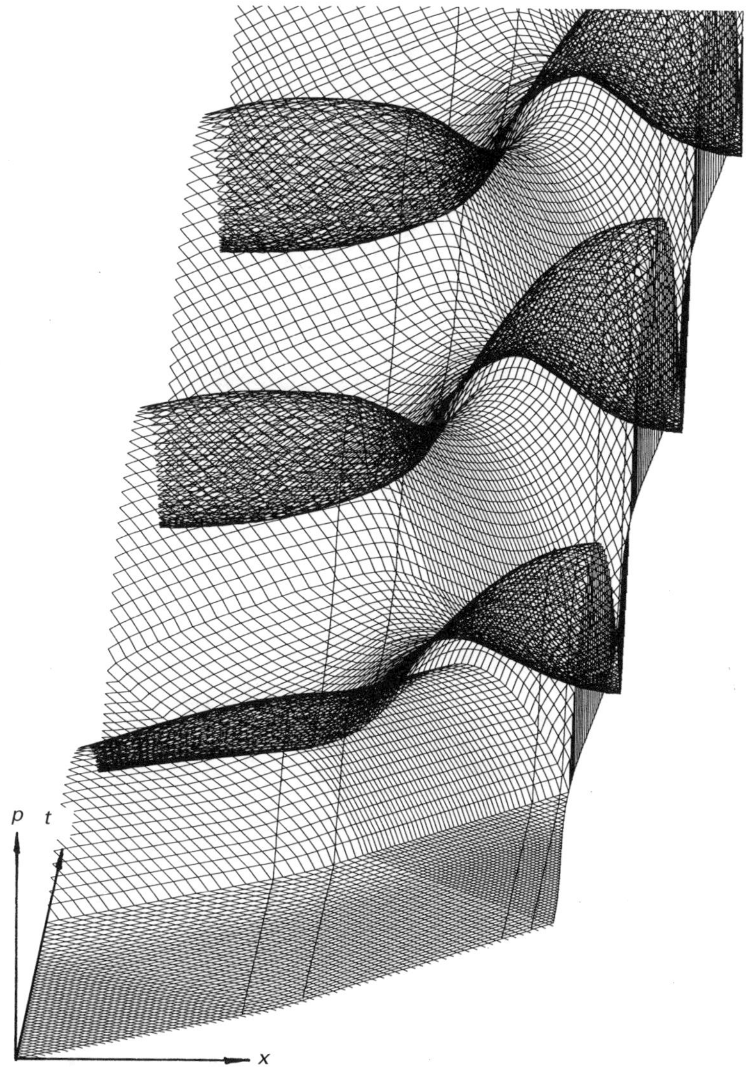

13.4 Pressure surface corresponding to the Mach line solution of Fig. 13.3 outlining the lateral extent of the regenerator. The corresponding relief of local particle velocity is shown at Fig. 13.5.

The loops for the expansion space are traced anti-clockwise, those for the compression space are traced clockwise – contrary in both cases to expectation from study of incompressible operation. On settling down, the area of the compression loop exceeds that of the corresponding expansion loop, consistent with a surplus of indicated work. Constructing an accurate net indicator diagram is less straightforward than for the incompressible case, since the

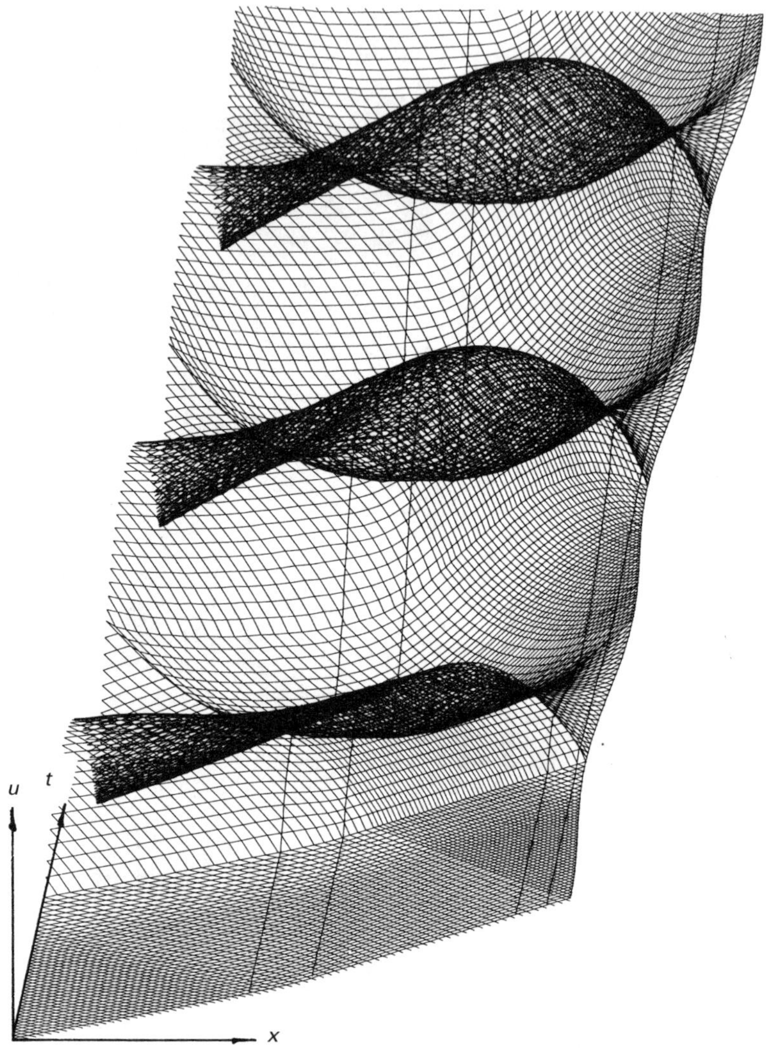

13.5 Particle velocity surface corresponding to Figs 13.3 and 13.4.

individual loops are out-of-phase, and time (and thus crank-angle) increments at the piston faces are variable and generally out of step.

It would not be normal for an engine to produce power if net thermal energy input occurred at some temperature other than at the upper limit of the cycle – i.e., at the expansion end. According to the cycle simulation, the expansion end receives work in addition to heat. For network output to occur at the compression space suggests a large flux of energy from the high-temperature end to the lower. In the context of the complex interactions portrayed in Figs 13.3–13.5, intuition is not generally the best route to an

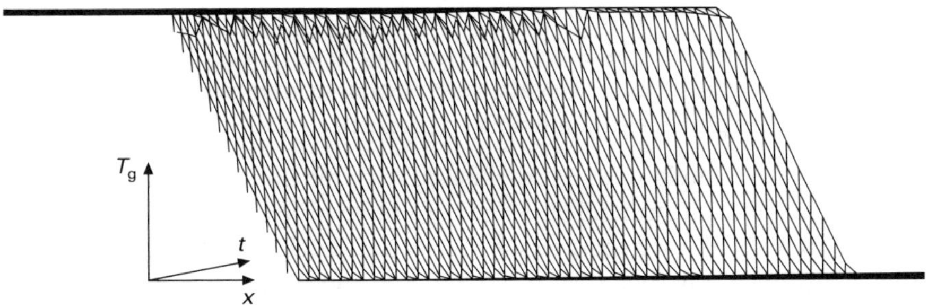

13.6 Fluid temperature relief. Detail is overwhelmed by the massive end-to-end temperature difference across the regenerator (600°C). Future results will be displayed as local gas/wall temperature difference, $T_g - T_w$.

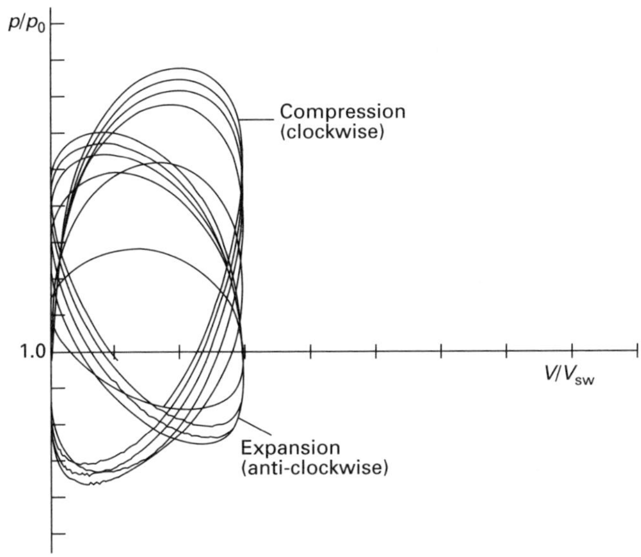

13.7 Development of *p–V* loops from start-up. By contrast with incompressible operation, the expansion loop traces anti-clockwise, the compression loop clockwise. Compression loop area exceeds that of the expansion loop, consistent with a surplus of (indicated) shaft power.

explanation. A study is in progress to integrate local, instantaneous energy exchanges into a net cycle energy balance for the machine, and thus to clarify.

13.5 Discussion

Extending the performance envelope of the air- or N2-charged Stirling engine calls for two things:

1. Heat transfer and flow friction data for regenerator matrices acquired in function of *Re* and *Ma*, and valid beyond $Ma = 0.02$.
2. Tools for gas process visualization which convincingly quantify and portray the physical phenomena of interest.

Item 1 may be thought of as a pre-existing requirement if confidence is to be restored in the first-principles design of conventional N_2- and air-charged engines. It will call for experimental work considerably more comprehensive than Kays and London's classic undertaking. On the other hand, modern methods of real-time data capture and reduction may make this wide-ranging task less onerous than the 1964 original.

As regards requirement 2, the solution algorithms of one-dimensional, unsteady gas dynamics extend readily to the Stirling engine. They offer to facilitate investigation of the gas processes at cycle speeds which might otherwise not receive attention.

The outline design study reported here has temporarily sidestepped the compressibility issues by specifying a regenerator of corrugated metal film as used in flame traps. This does not have the successive enlargements and contractions in free-flow area which are a feature of the wire screen. However, it would not be safe to make assumptions about Mach effects arising, since accessible data do not explore such issues.

It may be anticipated, even without the benefit of simulation, that piston face pressure excursions in the vicinity of resonance of a system such as that of Fig. 13.2 will exceed those predicted by a simple 'static' analysis, provided friction damping is not overwhelming. The expectation is confirmed by the present simulation exercise. However, there are no means to carry out a definitive enquiry until data become available to the format $St = St(Re, Ma, Pr)$ and $C_f = C_f(Re, Ma)$.

If the possibility of work-producing resonant operation is confirmed, and if insuperable problems of friction, sealing, dynamic balance and mechanical integrity do not arise, the simple, air- or N_2-charged Stirling engine may have interesting potential. Indeed, air or N_2 may turn out to be the only realistic choice of working fluid for this mode of operation: achieving with H_e the resonant condition for the engine of Table 13.1 would call for rpm to be increased by a factor of 2.75, i.e., to some 35,000 rpm. To achieve resonance in the same machine charged with H_2 would require 48,000 rpm.

Appendix: State equations – Shapiro's derivation

The First Law of Thermodynamics is written using q' for rate of heat transfer per unit time and per unit mass of working fluid:

$$q'\rho A_{ff}dx = \partial/\partial t\{\rho A_{ff}dx(c_v T + \tfrac{1}{2}u^2)\}$$
$$+ \partial/\partial x\{\rho u A_{ff}(c_v T + p/\rho + \tfrac{1}{2}u^2)\}dx$$

Selected terms are eliminated with the aid of continuity and momentum equations:

$$(\gamma - 1)\, \rho(q' + uF) = \partial p/\partial t + u\partial p/\partial x - a^2(\partial \rho/\partial t + u\partial \rho/\partial x) \qquad 13.8$$

The equations of continuity of properties p, ρ and u, together with those of mass conservation, momentum conservation and Eqn 13.8 (energy conservation) form a set of six simultaneous equations in the unknowns $\partial p/\partial x$, $\partial p/\partial t$, $\partial \rho/\partial x$, $\partial \rho/\partial t$, $\partial u/\partial x$ and $\partial u/\partial t$:

$$\begin{array}{lllllll}
\mathrm{d}x\partial u/\partial x & \mathrm{d}t\partial u/\partial t & & & & & = \mathrm{d}u \\
 & & \mathrm{d}x\partial p/\partial x & \mathrm{d}t\partial p/\partial t & & & = \mathrm{d}p \\
 & & & & \mathrm{d}x\partial \rho/\partial x & \mathrm{d}t\partial \rho/\partial t & = \mathrm{d}\rho \\
\rho\, \partial u/\partial x & & & & u\, \partial \rho/\partial x & \partial \rho/\partial t & = -(\rho u/\mathrm{A}_{\mathrm{ff}})\mathrm{d}A_{\mathrm{ff}}/\mathrm{d}x \\
u\, \partial u/\partial x & \partial u/\partial t & \rho^{-1}\partial p/\partial x & & & & = -F \\
 & & u\partial p/\partial x & \partial p/\partial t & -ua^2\partial \rho/\partial x & -a^2\partial \rho/\partial t & = (\gamma - 1)\, \rho(q' + uF)
\end{array}$$

In momentum and energy equations:

$$F = \tfrac{1}{2} C_{\mathrm{f}} u^2 \mathrm{sign}(u)/r_{\mathrm{h}}$$

$$q' = \frac{h p_{\mathrm{w}}\, \mathrm{d}x(T_{\mathrm{w}} - T_g)}{\rho A_{\mathrm{ff}}\, \mathrm{d}x}$$

The coefficients are laid out in the form of a determinant (rank 6×6). Pursuing Shapiro's approach of 1954 the solutions for $\partial p/\partial x$, $\partial p/\partial t$ etc are not evaluated, but instead are *inspected* to find under what conditions they are indeterminate, i.e. under which the determinant has the form zero/zero. This is equivalent to identifying the loci in the x–t plane of discontinuities in the derivatives.

The process confirms the gradients of the Mach and path line elements in the physical plane (Eqns 13.1 and 13.2) and yields the equations for change of state along all three directions (Eqns 13.3 and 13.4).

14
For the sceptic

14.1 What does it all add up to?

Experimental measurements reported in Chapters 2 and 10 confirm the dependence of fractional pressure drop $\Delta p/p$ on Mach number *Ma* as well as Reynolds number *Re* – and indeed on flow passage geometry as characterized by $d_w m_w$. On the other hand, the *Re* range of experiments reported to date is narrow. If flow through the gauze aperture could be modelled, and if numerical model and experiment were to agree over the limited range of the tests, this might justify tentative use of simulated flow data pending experimental coverage of an extended range of *Re*.

The outcome might also placate those who, like the referee for the material of Chapter 10, remain unconvinced by the accumulating evidence for the influence of *Ma*.

14.2 Flow in the isolated gauze aperture

Su himself (1986) undertook a flow analysis which he coded for computer. For single screens and for stacks of up to 16 gauzes, the resulting simulation predicted the essential features of each corresponding experiment. Su's geometric model of the flow passage was two-dimensional, requiring an arbitrary approximation to the variation in flow direction z of free-flow area $A_{ff}(z)$ and wetted perimeter $perim(z)$. Recourse to the simplification is unsurprising given the apparent geometric complexity of the flow channel, well illustrated in the electron-beam microscope image of Fig. 14.1.

It has since been appreciated (Chapter 12) that planes of flow symmetry can be identified such that a representative flow filament is isolated within a simple square 'box'. Figure 14.2 shows a wooden mock-up of the aperture identified in this way. A square shroud of thin plywood occupies the planes of flow symmetry and thus defines the box which, in the model, serves to support the set of four straight 'wire' elements. Only the symmetrical, semi-circular half of a wire contributes to the wall surface of the flow channel.

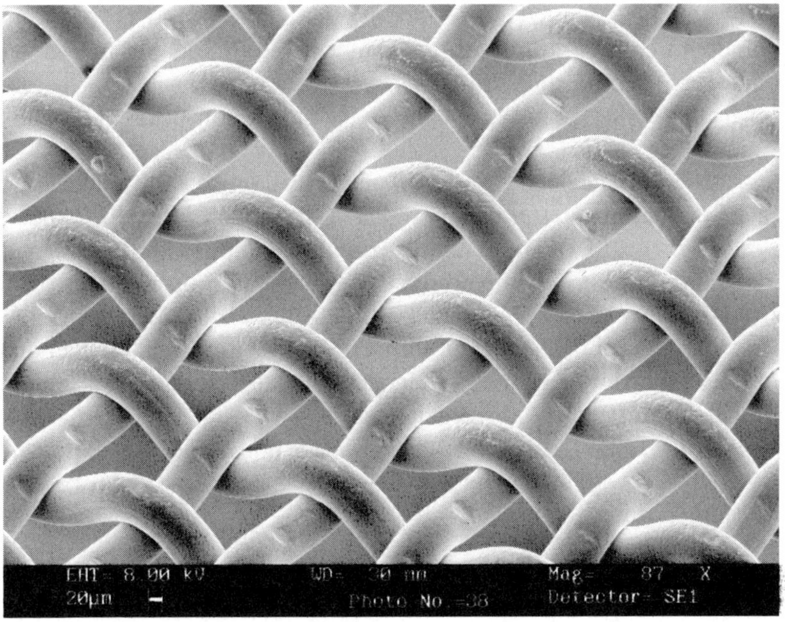

14.1 Electron microscope image of gauze of d_w = 0.071 mm, m_w = 5.46 wire/mm (NB units) and $d_w m_w$ = 0.388. Image recorded by Mr A Heaver of the Engineering Laboratory, Cambridge University from specimen under test in flow rig by van Rijn (2006).

14.2 Wind-tunnel model of gauze aperture (500-times natural size).

The algebra in Chapter 12 derives free-flow area $A_{ff}(z)$ in terms of location z in the flow direction. It also gives fractional free-flow area increment $dA_{ff}(z)/A_{ff}(z)$ and hydraulic radius $r_h(z)$. Over much of the flow passage, the fluid element is in contact with the bounding surfaces over a fraction of its periphery. This important reality is accounted for in the algebra.

It will be assumed that the flow filament comprises an inviscid core subject to compressibility effects. Where the filament is in contact with an enclosing surface, the core is isolated by a boundary layer. The latter builds up in thickness from zero at the entry stagnation point. The build-up is a function of local *Re*.

A first estimate of core flow is obtained by provisionally neglecting boundary layer development (but including a term for wall friction where there is fluid-metal contact). The resulting axial distributions of pressure p and velocity $\underline{u}$ are then used to estimate boundary layer build-up in terms of distance z from the plane of entry. An initial estimate of the z-location of the downstream separation point is also acquired. Core flow is then recalculated on the basis of local free-flow area $A_{ff}(z)$ reduced by local boundary layer thickness. The process is iterated to convergence.

Pressure recovery between separation point and the exit plane of the gauze is dealt with in a single step assuming conservation of the axial component of momentum.

14.3 Defining equations

14.3.1 Inviscid core

Su made use of the standard, steady-flow Mach number relationships between local *Ma* and upstream, critical values. This confined his enquiry to (a) ideal gas behaviour $p = \rho RT$, and (b) isentropic processes. The present, revised development will make provision for handling the behaviour of a real gas by use of van der Waals' equation. This will call for numerical means to integrate the conservation equations. There is then little additional complication in discarding the isentropic assumption and including a wall friction term from the outset. Formulation will be illustrated using the ideal gas assumption.

With A_{ff} for local free-flow area, the steady flow mass conservation law is $\rho\underline{u}A_{ff}$ = constant. Differentiating:

$$\mathrm{d}\rho/\rho + \mathrm{d}\underline{u}/\underline{u} = -\mathrm{d}A_{ff}/A_{ff} \qquad 14.1$$

In earlier chapters the value of friction factor C_f to correlate pressure drop with flow through the *gauze stack* was challenged. Attention is now on the individual aperture, here being treated as a duct, to which the conventional form of the steady-flow momentum equation is appropriate. After subtraction of Eqn 14.1:

$$\underline{u}\mathrm{d}\underline{u} + \mathrm{d}p/\rho = -\tfrac{1}{2}C_f\underline{u}^2\mathrm{d}z/r_h \qquad 14.2$$

The expression for energy conservation in adiabatic flow is written so as allow flexibility to explore the contribution of rotational energy. To this end it includes a variable lead angle λ_{lead}, namely, $c_pT + \frac{1}{2}u^2(1 + \frac{1}{2}/\tan^2\lambda_{lead}) =$

constant. The algebra leading to the term for kinetic energy of rotation is given later. In view of the complex variation of flow channel outline with z, provision is included for λ_{lead} to vary with axial location at variable local rate $d\lambda_{lead}/dz$:

$$c_p dT + \underline{u}d\underline{u}\{1 - (\cot\lambda_{lead}\csc^2\lambda_{lead})d\lambda_{lead}\} = 0 \qquad 14.3$$

For the ideal gas provisionally under consideration, small changes in properties p, ρ, T, are related through:

$$dp/p - d\rho/\rho - dT/T = 0 \qquad 14.4$$

Four unknowns have arisen: dp/p, dT/T, $d\rho/\rho$ and $d\underline{u}/\underline{u}$. Committing to the ideal gas assumption would have permitted the number to be held to three, and led to explicit solution e.g., by 3×3 determinant. To allow eventual substitution of the equation of state for a van der Waals gas, all four unknowns are retained – the penalty of which is a 4×4 matrix. Until ideal gas behaviour is discarded, $\underline{u}/\sqrt{\gamma RT}$ is set equal to Ma and R/c_p to isentropic index ratio $(\gamma - 1)/\gamma$. This leads to the following set of coefficients for the set of four equations – conservation laws plus gas law:

dp/p	dT/T	$d\rho/\rho$	$d\underline{u}/\underline{u}$	RHS
0	0	1	1	$-dA_{ff}/A_{ff}$
1	0	0	γMa^2	$-\frac{1}{2}\gamma Ma^2 C_f (Re)dz/r_h$
0	1	0	$(\gamma - 1)Ma^2 \{1 - \cot\lambda\csc^2\lambda d\lambda\}$	0
1	−1	−1	0	0

14.5

14.3.2 Boundary layer

Both Su and the present author have found the need to approach the final solution by dispensing with successive simplifications incrementally during iteration. Initial calculation of boundary-layer momentum thickness θ, for example, makes use of Thwaites incompressible approximation (1949):

$$\theta^2 = \theta_0^2 + 0.45(\rho/\underline{u}\mu)\underline{u}^{-5} \int \underline{u}^5 dz \qquad 14.6$$

θ_0 is momentum thickness at the front stagnation point, usually taken as zero (White 1979). The term in parentheses evidently converts to local Reynolds number Re, and integration in practice is a simple numerical summation.

Displacement thickness δ^* is obtained from momentum thickness with the aid of shape factor H:

$$H = \delta^*/\theta \qquad 14.7$$

Thwaites defines momentum-thickness variable m:

$$m = -\theta^2(\rho/\mu)\mathrm{d}\underline{u}/\mathrm{d}z \qquad 14.8$$

Su fitted a curve to Thwaites' correlation of H vs m shown in Fig. 14.3:

$$H \approx 24410m^5 + 6958m^4 + 558.3m^3 + 9.75m^2 + 3.21m + 2.61 \qquad 14.9$$

With displacement thickness known as a function of z, the distribution of p, T, ρ and $\underline{u}$ are calculated between entry and the point of separation. The latter occurs under the influence of negative pressure gradient. A commonly-used criterion of separated flow is $m \leq -0.082$. The calculation sequence is iterated to convergence.

The formulation will allow incorporation of some embellishments. In the meantime, Fig. 14.4 anticipates the independent influence on the eventual solution of parameters Re and Ma. The comparison assumes that flow rate m' – and thus the approach value Ma_{app} of Ma – are common to both cases. A low value of Re (left diagram) results in a relatively thick boundary layer, and a correspondingly high core velocity $\underline{u}$. A high value of Re results in a thinner boundary layer. Core velocity $\underline{u}$ and Ma at the throat are correspondingly lower (diagram on right).

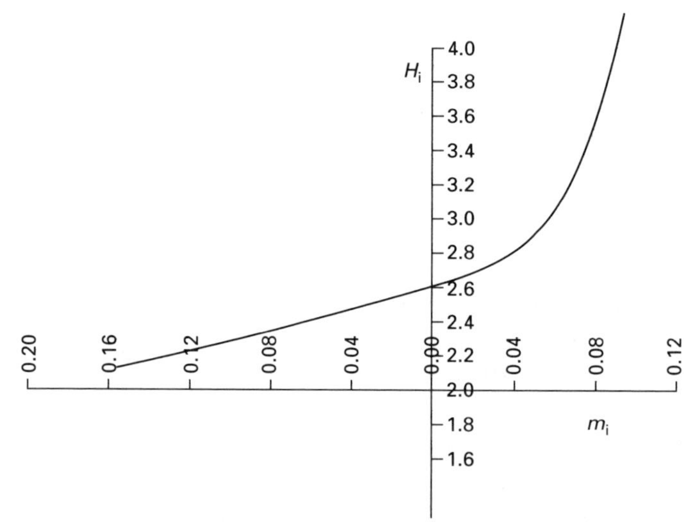

14.3 Thwaites' correlation between m (Eqn 14.8) and shape factor H (Eqn 14.9).

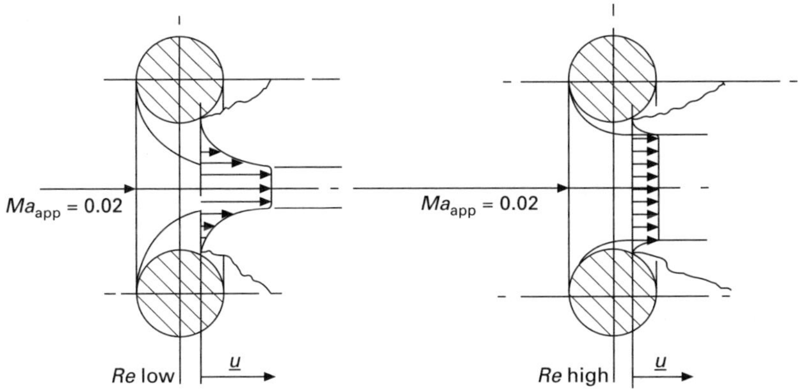

14.4 Stylized representation of effect of extreme values of *Re* at a common value of approach Mach number Ma_{app}.

Supposing Ma_{sep} at the separation point (where the velocity profile has reversed) has reached the choking value of unity for the lower *Re*, then by definition, corresponding Ma_{sep} at the higher *Re* is less than unity. Even before downstream pressure recovery is taken into account, dealing with the effects of *Re* and *Ma* by adding them linearly, as in the Forchheimer law, seems inappropriate.

Under the assumption of 'slab' flow for the core, high values of dA_{ff}/dz at inlet ($z = 0$) suggest a large radial component of velocity. Modification of the energy equation to reflect the extra kinetic energy is achieved by incorporating a simple algebraic factor once the effect has been quantified.

14.4 Radial component of kinetic energy

The approach is illustrated by assuming the flow passage to be axi-symmetric and of local free-flow area $A_{ff} = \pi r_0^2$, where A_{ff} and r_0 are functions of axial location z, as in Fig. 14.5. At entry, rate of change of A_{ff} with z is locally infinite, suggesting high values of the radial component of velocity v (and of corresponding acceleration).

A disc of fluid is considered having outer radius r_0. The assumptions of 'slab' flow and uniformity of pressure p with radial location mean that, as r_0 changes by dr_0 with change dz in location z, so the radial location r of a fluid particle changes in proportion:

$$dr = (r/r_0)dr_0 \qquad 14.10$$

The mass conservation law (Eqn 14.1) is in terms of free-flow area, A_{ff}, inviting dr_0 to be expressed in the same way:

$$dr_0 = dA_{ff}/2\pi r_0 \qquad 14.11$$

Substituting Eqn 14.11 into Eqn 14.10, denoting radial velocity dr/dt by v and noting that $\underline{u}$ = dz/dt:

$$v = \tfrac{1}{2} r\underline{u}(\mathrm{d}A_{\mathrm{ff}}/A_{\mathrm{ff}})\mathrm{d}z^{-1}$$

Total instantaneous kinetic energy in the radial direction of an annulus of axial length dz and radial thickness dr is $\tfrac{1}{2}\rho 2\pi r\mathrm{d}r\mathrm{d}zv^2$. This is integrated between $r = 0$ to $r = r_0$ with respect to = r to give the total radial component of kinetic energy of the disc.

$$KE_{\mathrm{r}} = \underline{u}^2 A_{\mathrm{ff}}(\mathrm{d}A_{\mathrm{ff}}/A_{\mathrm{ff}})^2/8\pi\mathrm{d}z^2 \qquad 14.12$$

The units of the energy equation (Eqn 14.3) are energy per unit mass, L^2T^{-2}. Dividing Eqn 14.12 by the mass $\pi r_0^2\rho\mathrm{d}z$ of the disc yields the term by which the $\tfrac{1}{2}\underline{u}^2$ of the energy equation is augmented by the radial component:

For pedantic consistency with the slab flow assumption, v at entry – and thus KE_{r} – are infinite. With dz set to $0.01d_{\mathrm{w}}$ for numerical processing of Eqn 14.12, KE_{r} is approximately four times KE_z.

It may be of significance for eventual heat transfer computations that entry of the core flow to the gauze aperture is accompanied by a drop in temperature as enthalpy is exchanged for kinetic energy. Boundary layer thickness at this point is zero, and temperature gradient between solid surface and fluid is locally infinite.

14.5 The not-so-square-weave wire gauze

Figure 12.5 isolated a single aperture. The four bounding wire elements all make the same angle $\tan^{-1}(d_{\mathrm{w}}m_{\mathrm{w}})$ with the plane of the gauze. A screw thread having a four-start thread with *lead angle* $\tan^{-1}(d_{\mathrm{w}}m_{\mathrm{w}})$ and engaged in the aperture would advance, per revolution, by linear distance πd_{w} perpendicular

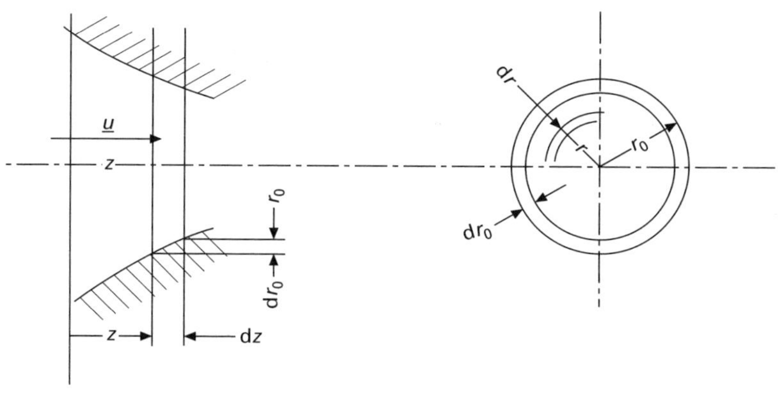

14.5 Notation for algebra of radial component of velocity *v*.

to the plane of the gauze. The kinematic analogy suggests use of notation from screw-thread technology (Faires 1969):

$$\lambda_{lead} = \tan^{-1}(d_w m_w) \qquad 14.13$$

The defining geometric parameter turns out yet again to be the simple product $d_w m_w$. (The helix direction of the mock-up of Fig. 14.2 is opposite to that of the computer-generated profile of Fig. 12.2). Linear scale is approximately 500:1. Tunnel testing has yet to take place, but blowing the entry section from a paint spray gun supplied at shop air line pressure causes a streamer attached downstream to rotate rapidly in an anti-clockwise direction.

The direction of rotation is consistent with a simple experiment which the reader may care to repeat. The scientific apparatus required is a piece of square-weave gauze and a small quantity of putty or plasticene. The gauze should have wire diameter several times larger than that of the typical regenerator: a diameter d_w of 0.025 in. (0.635 mm) in a mesh of 8 wire/in. (312 wire/m) is a common size, and will demonstrate the phenomenon. (A higher mesh number will produce a more spectacular result.) Also required is a spray can of WD-40™.

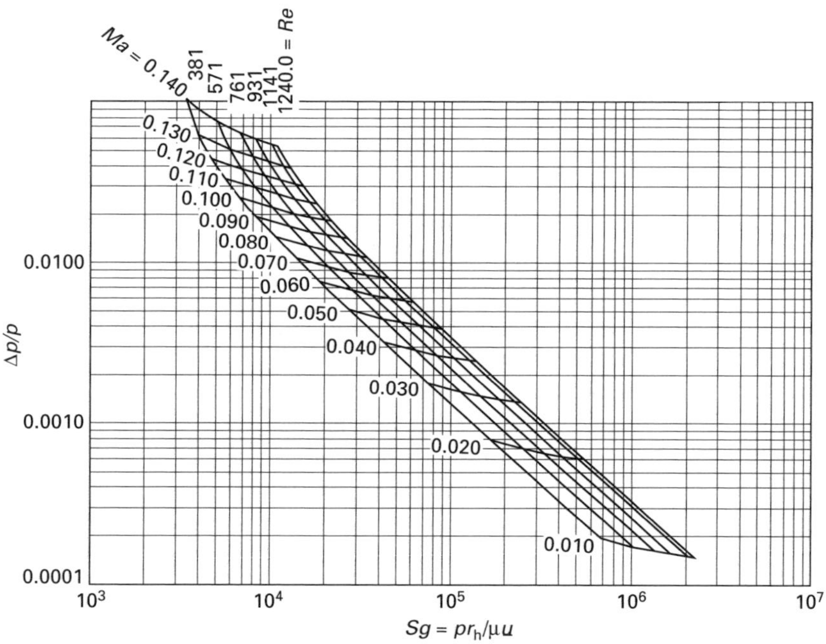

14.6 Specimen map of simulated $\Delta p/p|_{gauze}$ vs *Sg* for an eight-screen stack with Mach number *Ma* as parameter. Alternative parameter *Re* is also displayed over the range of Su's experimental results of 1986. May be compared with the transformed experimental results here plotted (for a second time) as Fig. 14.7.

The putty is kneaded until soft, the screen sprayed with a thin film of WD-40™, and the putty pressed through the gauze by hand. The extruded filaments will be found to be twisted into a shallow helix, the spiral angle depending somewhat on the malleability of the putty. The helix angles of adjacent filaments will be found to be of opposite sign.

14.6 Kinetic energy of rotation

The fluid filament is treated provisionally as a body of revolution. A fluid element of axial length dz and local radius r s considered. Angular velocity ω is related to axial velocity $\underline{u}$ through:

$$r\omega \approx \underline{u}/\tan\lambda_{\text{lead}}$$

$$\approx \underline{u}/d_{\text{w}}m_{\text{w}} \qquad 14.14$$

The mass moment of inertia I of the filament about its axis is:

$$I \approx \tfrac{1}{2}\pi r^4 \rho \mathrm{d}z \qquad 14.15$$

The kinetic energy of rotation of the element is $\frac{1}{2}I\omega^2$. However, the equation for steady, adiabatic flow is customarily expressed on the basis of unit mass. The mass of the element is $\pi r^2\rho \mathrm{d}z$, so that specific rotational energy is $\frac{1}{4}r^2\omega^2$. Steady-flow energy $c_{\text{p}}T + \frac{1}{2}\underline{u}^2 + \frac{1}{4}r^2\omega^2$ is assumed to be conserved up to the point where the rotational component is dissipated. Taking account of Eqn 14.14:

$$c_pT + \tfrac{1}{2}\underline{u}^2(1 + \tfrac{1}{2}/d_{\text{w}}^2 m_{\text{w}}^2) = \text{constant} \qquad 14.16$$

Substitution of $d_{\text{w}}m_{\text{w}} = 0.365$ for correspondence with the geometry of gauzes tested by Su (1986) suggests that the energy of rotation exceeds that of translation by a factor of 3.75. This is improbably high, and the modest helix angle achieved in the experiment with plasticene compounds the doubt. Moreover, the gas does not behave as plasticene, which itself conforms only partially to the shape of the passage.

Except at very low *Re*, flow may be expected to separate just down-stream of the point of minimum free-flow area, i.e., at down-stream of mid-plane at $z = d_{\text{w}}$. An alternative estimate of rotational energy might assume that flow separates as a jet of square cross-section at the apparent angle β_{s} of the flow passage outline. Angular velocity ω might then be estimated as the ratio of β_{s} to time of travel between entry and separation point.

Individual gas filaments emerge from adjacent apertures with opposite swirl, so there is only one realistic conclusion about the net component of rotational kinetic energy per gauze: it is dissipated before or during entry to the next gauze stage.

The literature makes much ado about the effect of alignment – or misalignment – between successive gauzes. Any useful concept of perfect alignment must be judged by its success in ensuring reproducibility of local flow patterns between gauzes. The alternating sign of the swirl of adjacent filaments makes the likelihood of achieving perfect alignment even more remote.

14.7 'Real' (van der Waals) gas

The equation of state for the van der Waals gas is:

$$(p + a/\rho^2)(\rho^{-1} - b) = RT \qquad 14.17$$

$$a = 27/64\ (RT_{crit})^2/p_{crit}$$

$$b = \tfrac{1}{8} RT_{crit}/p_{crit}$$

Differentiating, as for the ideal gas,

$$\mathrm{d}p/p(1 + a\rho^2/p)^{-1} + \mathrm{d}\rho/\rho\{2a\rho^2(1 + a\rho^2/p)^{-1}/p - (1 - \rho b)^{-1}\} - \mathrm{d}T/T = 0 \qquad 14.18$$

The coefficients of the dp/p, dT/T and dρ/ρ replace the unity values for the ideal gas in the fourth line of the matrix (Eqn 14.5).

14.8 Downstream pressure recovery

Achieving a solution between separation and exit plane at $z = 2d_w$ calls for an assumption: gauzes of the size of present interest yield high NTU under the flow conditions and extreme temperature gradients (25×10^3K/m) of Stirling engine operation. This suggests that the temperature at outlet from the single gauze in steady flow will differ little from the inlet value, thereby fixing that latter value. Fractional pressure drop per gauze, $\Delta p/p\,|_{gauze}$ follows as the sum of the individual drops between inlet and separation, and between separation and exit plane.

Stacks are dealt with by using the outlet values of p, $\underline{u}$, and T, from any given gauze as the inlet conditions for the next in the sequence. Under the earlier assumption of complete temperature recovery, overall pressure gradient is negative. Net gradient in particle velocity is positive. For isothermal flow, the highest value of Ma and the highest $\Delta p/p|_{gauze}$ therefore occur at the downstream gauze (but see Section 14.9 dealing with features of regenerator operation *in situ*).

The flow model must embody the conditions for choking, and anticipate the possibility of supersonic flow downstream of the point of minimum effective free-flow area. Reversion to sub-sonic flow is via a normal shock.

For the perfect gas the equations defining states either side of the shock are in terms of a function $f(\gamma)$ of γ.

$$f(\gamma) = (\gamma + 1)/(\gamma - 1) \qquad 14.19$$

Upstream density ρ_{down}/ρ_{up} is expressed in terms of p_{down}/p_{up}. White's 1979 treatment includes the derivation:

$$\rho_{down}/\rho_{up} = \frac{1 + f(\gamma)p_{down}/p_{up}}{f(\gamma) + p_{down}/p_{up}} \qquad 14.20$$

Pressure ratio is available (White 1979) in terms of upstream *Ma*:

$$p_{down}/p_{up} = [2\gamma Ma_{up}^2 - (\gamma - 1)]/(\gamma + 1) \qquad 14.21$$

14.9 Simulated correlation $\Delta p/p = \Delta p/p\{Sg, Ma, \gamma, d_w m_w\}$

Holding nominal Reynolds number *Re* ($= 4\rho \underline{u} r_h/\mu$) constant and calculating simulated $\Delta p/p|_{gauze}$ for a range of upstream Mach number *Ma* is the numerical equivalent of Su's experiments of 1986. The result is the correlation of Fig. 14.6, which is for the eight-screen stack. This is seen to have all the essential features of the earlier remapping of Su's original plots. To facilitate comparison that experimental counterpart is presented here (for a second time) as Fig. 14.7.

At the top left of the diagram, where Stirling number $Sg = pr_h/\mu\underline{u}$ is lowest and $\Delta p/p\,|_{gauze}$ highest, the simulated curves are somewhat less steep than the experimental counterpart. The former are computed using the same aperture

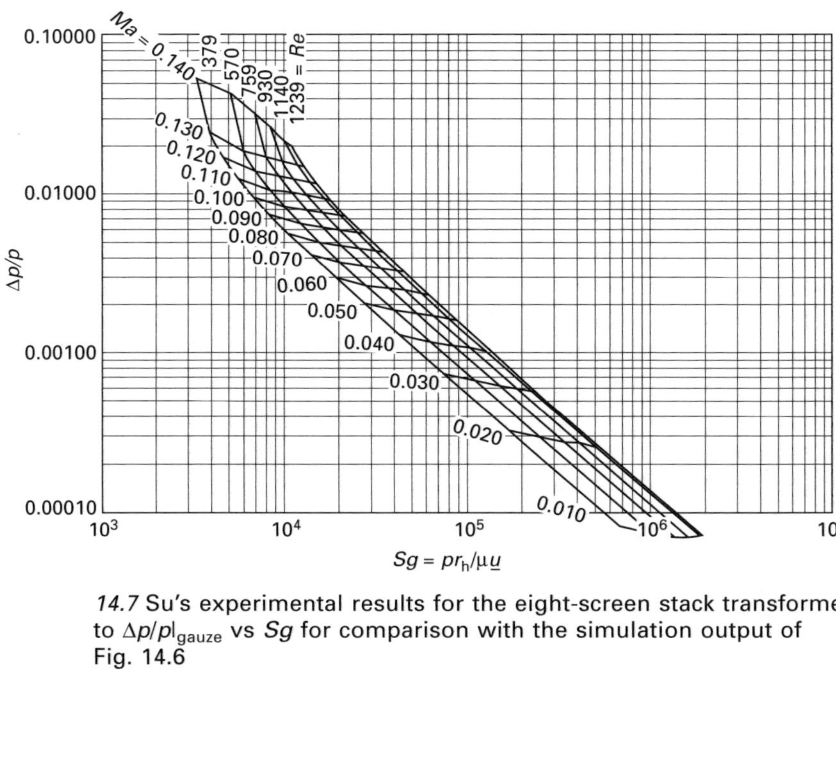

14.7 Su's experimental results for the eight-screen stack transformed to $\Delta p/p|_{gauze}$ vs *Sg* for comparison with the simulation output of Fig. 14.6

ratio $\alpha = d_w m_w$ as the latter. The encouragingly small discrepancy disappears almost completely on 'cheating' by use of a slightly reduced value of α as input data for the simulation, the gradient of the simulated curves almost exactly replicating the experimental values over the entire range.

The increasingly negative gradient of the curves in the top left-hand corners of Figs 14.6 (simulated) and 14.7 (converted from Su's experiments) is a symptom of choking. For upstream conditions in this region ($Ma_{app} = 0.14$, $Re = 381$) the simulation generates values of Ma which peak at values close to 1.3 at the separation point. The phenomenological correspondence between the two sets of curves – experimental and simulated – strongly suggests that peak Ma of this order arose in Su's experiments.

In steady flow, nominal particle speed $\underline{u}$ corresponding to (uniform) mass rate m' is $\underline{u} = m'/\rho A_{ff}$. For the ideal gas this may be expressed in terms of p and T as $\underline{u} = m'RT/pA_{ff}$. In service there are temperature and pressure gradients $\mathrm{d}T/\mathrm{d}x$ and $\mathrm{d}p/\mathrm{d}x$, as suggested in Fig. 14.8. Gradient $\mathrm{d}p/\mathrm{d}x$ is variable, but its sign is always opposite to that of $\underline{u}$.

In terms of the foregoing expression for $\underline{u}$, $Ma = |m'|(\sqrt{RT/\gamma})/pA_{ff}$. Differentiation with respect to flow direction z gives:

$$Ma^{-1}\mathrm{d}Ma/\mathrm{d}x = \tfrac{1}{2}T^{-1}\mathrm{d}T/\mathrm{d}z - p^{-1}\mathrm{d}p/\mathrm{d}z \qquad 14.22$$

Ma here is local nominal Mach number. It is not responsive to flow area variation through the individual gauze. Nevertheless, from screen to screen it will bear a systematic relationship to the Ma_{app} used in analysing flow through the individual aperture.

In the isothermal stack (Fig. 14.8(a)) $\mathrm{d}T/\mathrm{d}z$ is zero. Regardless of flow direction (left-to-right or the reverse) Ma increases down the pressure gradient, peaking at the downstream exit. If any gauze chokes, it will be the exit gauze. The $\Delta p/p$ of that gauze will account for the majority of the $\Delta p/p$ of the stack. The notional pressure distribution for isothermal flow left-to-right is a mirror image of that right-to-left. Both depart from the 'default' assumption of linearity.

In Fig. 14.8(b) temperature gradient is negative: a numerical value for $\tfrac{1}{2}T^{-1}\,\mathrm{d}T/\mathrm{d}z = $ (minus) 40 (m^{-1}) would be representative. This will overwhelm the likely value of 2 (m^{-1}) or so for $p^{-1}\mathrm{d}p/\mathrm{d}z$. In steady flow from expansion space to compression space (left-to-right in the Fig. 14.8(b)) Ma now decreases. If the first gauze in the sequence does not choke, then neither do any subsequent gauzes. The linear pressure distribution indicated is a possibility.

For steady flow in the reverse direction (compression to expansion), pressure and temperature gradients combine to enhance the conditions for choking at exit. Fig 14.8(c) suggests a pressure distribution consistent with this.

Dynamic pressure measurements taken from an engine motored to high rpm (Rix 1984) confirm the dependence of pressure asymmetry on Ma. The increasing asymmetry with increasing Ma is a symptom of the fall-off in

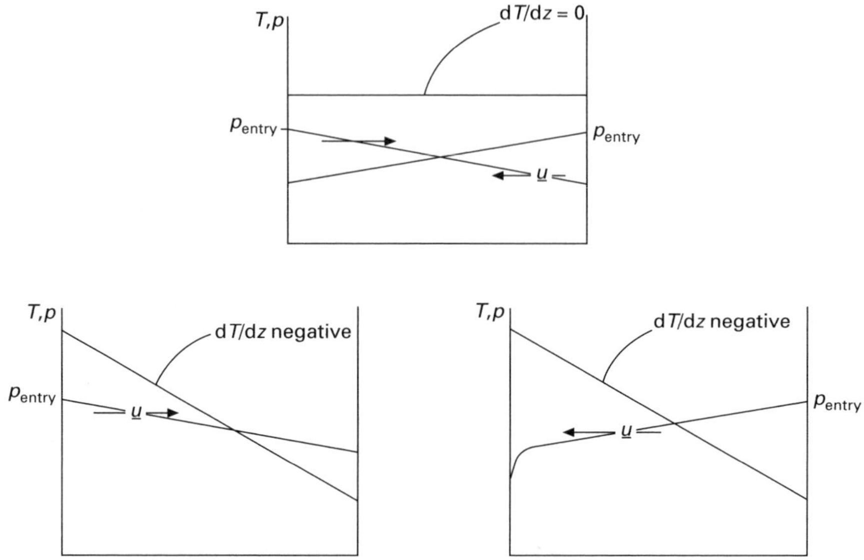

14.8 Temperature and pressure gradients in steady flow under the effect of *Ma* (a) flow left-to-right, isothermal stack; (b) left-to-right, strong negative d*T*/d*z*; (c) flow right-to-left, strong negative d*T*/d*z*.

indicated work/cycle as rated cycle speed is increasingly exceeded. Understanding – *and exploiting* – the phenomena behind asymmetry are essential steps in extending the performance envelope of Stirling engines in general, and that of the air engine in particular.

There is no practical limit to the number of gauzes for which the $\Delta p/p$ plots can be generated by simulation. A practical regenerator might comprise a stack of n_g = 1000 screens. This number takes 30 seconds to simulate and plot. Displays are reduced to $\Delta p/p|_{gauze}$ in each case by dividing computed $\Delta p/p|_{stack}$ by number of screens, n_g. For a stack susceptible to a large overall pressure drop, the value of $\Delta p/p$ read from the map does not represent the $\Delta p/p$ of any particular gauze.

If this is a shortcoming, it is shared by friction factor C_f, which if measured (or calculated) over length L_{reg} representing a large number of screens, also says nothing about local C_f over elemental length d*z*.

The maps reveal phenomena which C_f cannot: Fig. 14.9 is for a single screen of d_w = 0.04 mm and m_w = 11.5 mm^{-1}. The low aperture ratio α is conducive to choking, in evidence at the peak Ma_{app} of the plot, namely, at Ma_{app} = 0.14. Approach Mach number to a second screen added to the stack, Ma_{app}, would exceed 0.14 because of the reduced density and corresponding higher $\underline{u}$. Stacks of gauzes of the $d_w m_w$ in question allow unchoked flow only if Ma_{app} to the first gauze is reduced. Lowering peak Ma_{app} from 0.14 to 0.084 for the entry screen gives Fig. 14.10 for a stack of n_g = 6 screens.

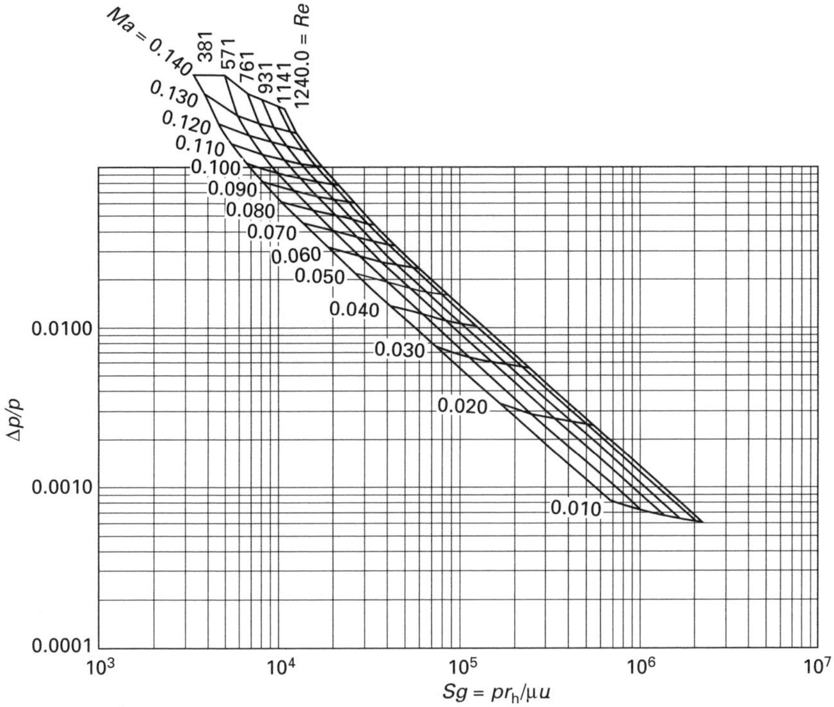

14.9 Single screen (n_g = 1) having d_w = 0.04 mm and m_w = 11.5 mm^{-1} and correspondingly small aperture ratio α.

Stacks of larger numbers of screens function below choking only if Ma_{app} is reduced still further – or alternatively if screens of higher aperture ratio α are substituted. For Fig. 14.11 n_g = 100 screens. Wire diameter d_w remains 0.04 mm, but mesh number m_w is reduced to 4.4 mm^{-1}.

14.10 Implications for first-principles design

- The correlations of Fig. 14.6 (simulated) and that of Fig. 14.7 (remapped from experiment) agree that minimum d*p*/*p* is achieved when Stirling number *Sg* is a maximum. From the definition of *Sg* as $pr_h/\underline{u}\mu$ the condition occurs when pressure *p* and/or hydraulic radius r_h are a maximum, and/ or when particle speed $\underline{u}$ and coefficient of dynamic viscosity μ are a minimum.
- Hydraulic radius r_h is related to wire diameter d_w and ¶$_v$ *via*:

$$r_h = \tfrac{1}{4} d_w \P_v/(1 - \P_v) \qquad 14.23$$

Thus the predictable conclusion (maximum d_w, minimum d*p*/*p*) is confirmed in the new format.

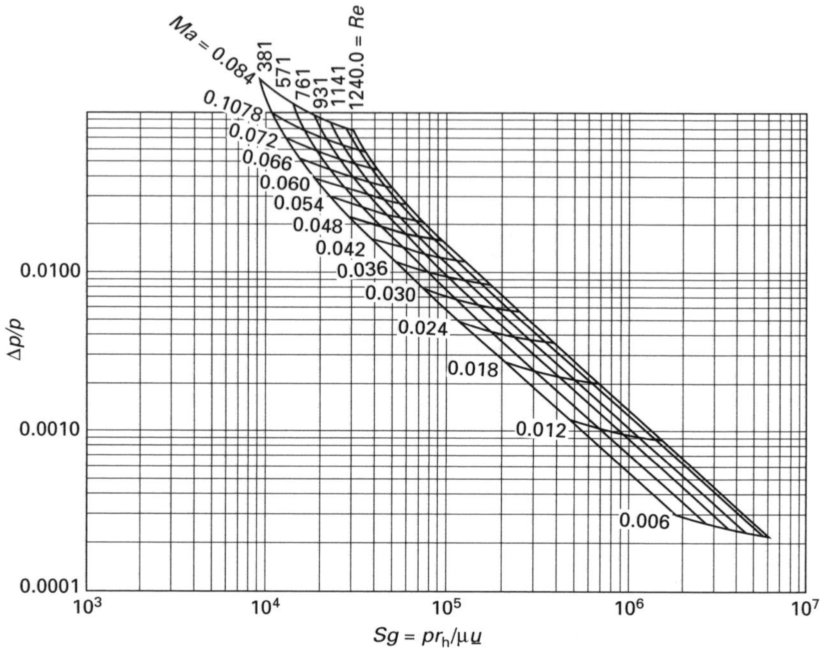

14.10 If number of screens n_g is increased above unity for the aperture ratio of Fig. 14.9, Ma_{app} to the entry screen must be reduced to give unchoked flow.

- Maps simulated for different values of aperture ratio α confirm Pinker and Herbert's 1967 finding from tests on single gauzes, namely, that the adverse effect of compressibility is least when aperture ratio is highest (closest to unity), and greatest when it is lowest. The favourable condition of α coincides with highest numerical value of volume porosity $¶_v$ – and with the lowest value of the geometric parameter $d_w m_w$.
- If the marked anisotropy of the gauze stack is a disadvantage, metal foams do not suffer from it. Indeed, foam technology appears to offer prospects for controlled anisotropy, such as grading of hydraulic radius and free-flow area in the direction of the temperature gradient.
- *Ma* has entered Stirling cycle analysis to stay.

The ideal regenerator combines high temperature recovery with low pumping loss. The conditions for achievement of the latter remain in conflict with those of the former, so design will always be an intricate balancing act.

14.11 Résumé

In appropriate applications of the Bernoulli equation, the term $\frac{1}{2}\rho \underline{u}^2$ (= $\frac{1}{2}p\gamma Ma^2$) effectively normalizes out kinetic energy effects. Bernoulli's equation

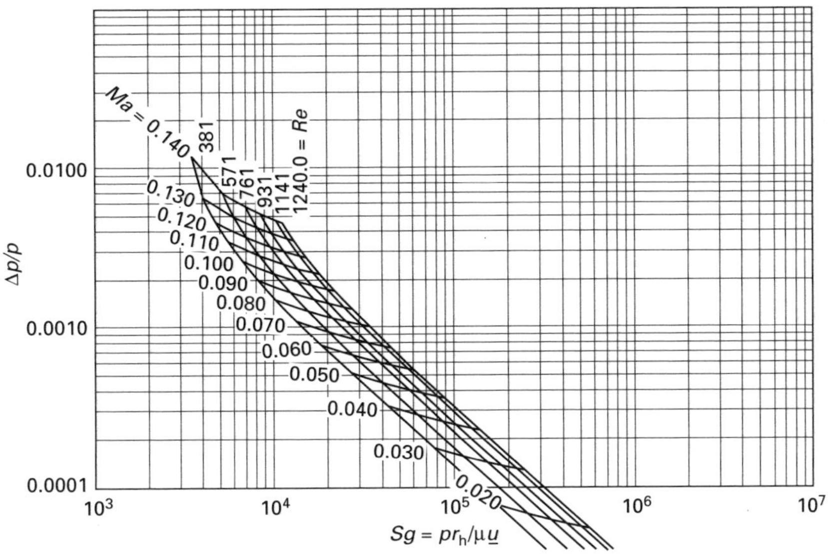

14.11 Number of screens increased to $n_g = 100$ without decreasing Ma_{app} at the entry screen. But avoiding choking calls for increased aperture ratio α, achieved in this case with same wire diameter $d_w = 0.04$, but with mesh number m_w decreased to 4.4 mm^{-1}.

does not apply to dissipative flow through the regenerator matrix – so neither does the simple algebraic expedient.

The flow model of this chapter captures the essential features of relevant experimental measurement. Replacing the C_f–*Re* correlation of the conventional gas process simulation should allow assessment of the benefits to be expected from extending Su's experiments to cover the entire range of *Ma*, *Re* and $d_w m_w$ required by first-principles design.

A new form of correlation is offered as successor to the traditional C_f–*Re* presentation. It is compatible with design and research needs of the modern closed-cycle regenerative air engine. It affords a clear visual distinction between the influence of *Re* and that of *Ma*. Most importantly, it offers instant pictorial display of the factors determining the intensity of pressure drop per gauze.

The time has arrived when Stirling engine design requires heat transfer and flow data in consistent and relevant notation. Studies such as those of Boomsma *et al.* (2003) and of Wirtz *et al.* (2003) show signs of bridging the gap between the science of porous media and the Kays and London (1964) tradition. The material of this chapter has shown how far there is still to go.

The greater the effect of *Ma* on pressure drop, the greater is the likelihood that the convective heat transfer process is also skewed by *Ma* – and the greater becomes the urgency to investigate. The task of experimental reacquisition of the heat transfer correlations with due account of *Ma* threatens to be an order of magnitude more time-consuming than that behind the Kays and London $StPr^{2/3} - Re$ originals. Analytical modelling is no alternative, but remains essential: experimentation which promises to be lengthy and costly will be even more so unless carried out to guidelines mapped out by comprehensive analytical homework.

Part IV

Some design considerations

15

Scaling – and the neglected art of back-of-the-envelope calculation

15.1 The overriding objective

After two centuries of evolution, Stirling engine science still does not offer a step-by-step recipe for the design of a viable prime mover. With the bi-centenary of the original invention just around the corner in 2016, the evolutionary process might appropriately be speeded up. This will involve leaving behind such irrelevances as Carnot efficiency and 'ideal' cycles in favour of relevant and sound engineering science.

It will be necessary to range beyond the academic engineering syllabus. In commercial engineering a guiding principle is that the difference between failure and success lies with attention to detail. A celebrated instance arose during development of the Bristol Aeroplane Company's famous range of air-cooled radial engines. One model was plagued by problems of interchangeability, traced in turn to ovality of the sleeve valves. Problems persisted until a batch of sleeves was finish-machined by grinding using an undressed (i.e. blunt) grinding wheel. There is no explanation in terms of basic principles in the documentary account (Gunston 1998), and isolating one might prove challenging.

Many technological gaps remain to be filled. Meanwhile, the designer looking for assurance of a viable design prior to embarking on the expense of machining may have to settle for 'bracketing' the specification using a range of resources – scaling, simulation and back-of-the-envelope arithmetic.

15.2 Gas path scaling – update

Scaling is an application of Dynamic Similarity principles. It cannot 'design' an engine to any arbitrary level of performance. It offers a means to achieving, in a 'derivative' engine and at an operating point specified by the designer (speed, pressure, working fluid), a cycle of gas processes mirroring that in a 'prototype' engine which, in general, operated at different speed, pressure and working fluid. The scaling process applies *internal* to the gas path only.

Earlier versions of scaling gained some flexibility over 'orthodox' Dynamic Similarity by exploiting the fact that flow friction and heat transfer correlations to traditional format – C_f vs *Re* and $StPr^{2/3}$ vs *Re* – were 'fitted' by expressions of the form $C_f = c/Re + d$. Two things have since changed: (a) it has been demonstrated that pressure distribution depends on *Ma* as well as on *Re*, and (b) the concept of the pressure-wave engine has emerged, which cannot be fully explored using design data which fail to reflect the rôle of *Ma*.

In the changed circumstances, achieving 'similarity' relies on knowing that dp/p (for example) is a function of the form $dp/p\{Re, Ma, \gamma\}$. Dimensionless *Re* is a product of four variables, ρ, μ, $\underline{u}$ and r_h having dimensions. The number of dimensionless variables required fully to specify the cycle of gas process is fewer than that required to define it in terms of absolute variables (p, R, V_{sw} etc.). Conversely, an infinite number of respective numerical values of the individual variables combine to yield a given numerical value of *Re*. Scaling fixes the absolute values – rpm, p_{ref}, V_{sw}, say – to be achieved in the derivative design. These are then substituted into the *dimensionless* parameters of the prototype design and numerical values determined for the absolute variables of the derivative design which keep the *Re*, etc. at the prototype values.

Some stages are elementary. A condition for the cycle history of particle speed to be the same between two engines is that there should be complete geometric and kinematic similarity. The latter condition is imposed by carrying over to the derivative design the thermodynamic phase angle α, the thermodynamic volume ratio κ and all dead space ratios V_d/V_{sw} of the prototype. The T_E and T_C of the derivative must retain the original value of temperature ratio $N_T = T_E/T_C$. Working fluids must be either diatomic ($\gamma = 1.4$) in both cases, or monatomic ($\gamma = 1.66$). At all pressures, temperatures and locations during a cycle the working fluid of the two engines must behave according to the same p, ρ, T relationship. The requirement is automatically satisfied if both working fluids behave as per the ideal gas law $p/\rho = RT$.

In two formally 'similar' engines, fractional pressure drop $\Delta p/p$ across the regenerator at, say, 25 degrees after maximum volume position has the same numerical value in the derivative as in the prototype – as, indeed, does the entire lengthwise distribution of $\Delta p/p$ at that crank angle. The same applies to temperature distribution. The condition requires *Sg* and *Ma* to have the same respective values in both machines at all crank locations. Mach number *Ma* is defined $Ma = \underline{u}/\sqrt{\gamma RT}$, but the γ is not required, since it has been agreed that the value may not change during the scaling process.

$$Ma = \underline{u}\sqrt{RT} = (\underline{u}/\omega V_{sw}^{1/3})(\omega V_{sw}^{1/3}/\sqrt{RT})$$

$$= (\underline{u}/\omega V_{sw}^{1/3})\, N_{MA}/\sqrt{(T/T_C)} \qquad 15.1$$

N_{MA} is characteristic Mach number, or speed parameter, $\omega V_{sw}^{1/3}/\sqrt{RT_C}$. The term $u/\omega V_{sw}^{1/3}$ is dimensionless particle speed.

$$Sg = pr_h / \underline{u}\mu = (p/p_{ref})(p_{ref}/\omega\mu)(\omega V_{sw}^{1/3}/\underline{u})(r_h/V_{sw}^{1/3})$$
$$= (p/p_{ref}) N_{SG} (r_h/L_x)(L_x/V_{sw}^{1/3})(\underline{u}/\omega V_{sw}^{1/3})^{-1} \qquad 15.2$$

N_{SG} is Stirling parameter $p_{ref}/\omega\mu$.

Similarity of fractional pressure drop across the regenerator $\Delta p/p$ calls for $\Delta p/p\,|_{screen}$ to be the same between prototype and derivative – and for there to be the same number of screens. This in turn requires the same respective variations of *Sg* and *Ma* over the cycle, i.e., that Eqns 15.1 and 15.2 should undergo the same respective cyclic variation in numerical values in the derivative design as in the prototype.

The groups in Eqns 15.1 and 15.2 divide into independent parameters (e.g. N_{MA}, r_h/L_x) set by the designer, and dependent (dimensionless) variables such as p/p_{ref} T/T_C. The respective cyclic variations in the latter are determined by the numerical values of the former. Similarity of r_h/L_x, $L_x/V_{sw}^{1/3}$ imposes geometric similarity. This in conjunction with similarity of N_{SG}, N_{MA} (and N_T, γ etc.) ensures similarity of the gas process cycle. Volume porosity and aperture ratio of the regenerator are dealt with by retaining the prototype value of $d_w m_w$.

Scaling will not deliver a viable gas path design for all choices of derivative specification. A numerical example will illustrate.

With General Motors' GPU-3 rhombic drive engine as 'prototype', the viability of an N_2-charged derivative is to be explored. A comprehensive specification of the GPU-3, together with a scale drawing, appears in Organ 1997). Swept volume was 118.63 cm^3. The performance point to be replicated is 8.95 kW, originally produced at 3600 rev/min on H_2 at 69 bar.

The scaling rules, which ensure similarity of the gas processes *only*, call for a rhombic drive, scaled by the linear factor about to be determined for the gas path. There is no need at this stage to cite bore, stroke, the various hydraulic radii, free-flow area etc.: all will scale by the same linear factor. The similarity conditions are $N_{MA}\,|_{deriv} = N_{MA}\,|_{prot}$; $N_{SG}\,|_{deriv} = N_{SG}\,|_{prot}$, together with equality of dimensionless power, as expressed by Beale number, $N_B = 2\pi\,\text{power}/(p_{ref}V_{sw}\omega)$. The individual groups may be expanded somewhat for clarity:

Equivalence of N_{MA}: $\quad \omega V_{sw}^{1/3}\,|_{deriv} = N_{MA}\,|_{prot}\sqrt{RT_C}\,|_{deriv}$
Equivalence of N_{SG}: $\quad p_{ref}\,|_{deriv} = N_{SG}\,|_{prot}\,\mu\omega\,|_{deriv}$

Equivalence of N_B (under the condition of equal power):

$$p_{ref}\omega\, V_{sw}\,|_{deriv} = p_{ref}\omega\, V_{sw}\,|_{prot}$$

All prototype values are known, allowing:

$$\omega V_{sw}^{1/3}\,|_{deriv} = 4.64\ \text{(m/s)} \qquad 15.3(a)$$

$$p_{ref}\,|_{deriv} = 3.62 \times 10^4\ \omega\,|_{deriv}\ \text{(Pa)} \qquad 15.3(b)$$

$$p_{ref}\omega V_{sw}\,|_{deriv} = 2.93 \times 10^5\ (m^3Pa/s) \qquad 15.3(c)$$

Dropping the $|_{deriv}$ labels and substituting Eqn 15.3(b) into 15.3(c):

$$\omega^2 V_{sw} = 2.97 \times 10^5/3.7 \times 10^4 = 8.027\ (m^3/s^2)$$

Eliminating ω between the last equation and Eqn 15.3(a) gives $V_{sw}^{1/3} = 0.573$ m, or $V_{sw} = 0.188$ m^3. Charge pressure p_{ref} is retrieved from Eqn 15.3 as 1.92 (bar). The linear scale factor is the cube root of the ratio of the swept volumes, namely, 11.67. The arithmetic may be checked by calculating the N_{MA}, N_{SG} and N_B of the new design, confirming that they are respectively as for the GPU-3.

The outcome is an unattractive proposition. Expansion exchanger tubes are of 2860 mm total length and 35 mm internal diameter. Bore and stroke are over 800 mm and 360 mm respectively. The massive CRE N_2-charged engine (Dando 1997) is a table-top model by comparison.

This particular scaling exercise may be seen as a dead-end, as an indictment of air and N_2 as working fluids – or as a valuable insight: The GPU-3 gas process cycle has been achieved – and now laid bare! The Stirling engine suffers at least two 'barriers' to unlimited cycle speed: all variants, regardless of working fluid, have moving components which are heavy relative to their counterparts in the internal combustion engine. Air- and N_2-charged types run up against *Ma* effects before the mechanical inertia limit is reached (see Section 15.3). The rpm limit of engines charged with light gases is imposed by *inertia loads.*

Particle speeds in the H_2-charged GPU-3 lie clear of *Ma* effects. Scaling has built this same margin into the specification for the air-charged derivative. Embodying this conditions has imposed a value of rpm which is impracticably low.

The reader may care to carry out a scaling in the reverse sense – say, from the MP1002CA specification to an H_2-charged engine of the same rating. The result will indicate the rpm potential – and thus the potential for specific power – which in the air-charged engine is sacrificed to mechanical limitations.

15.3 Back-of-the-envelope *Ma* and *Re* in the regenerator

The early stages in the modern study of the Stirling engine were preoccupied with the volume ratio κ and phase angle α which maximized ideal, indicated specific performance – the $W_{cycle}/p_{ref}V_{sw}$ of the (lossless) Schmidt cycle. There are no accounts of such exercises being responsible for performance breakthroughs.

Contemporary thinking would probably dismiss the influence of κ and α as insignificant over wide limits relative to that of flow passage geometry –

hydraulic radii, free-flow areas and passage lengths. On this thinking, a design based on volume ratio κ of unity and volume phase angle α of 90 degrees would probably achieve of over 90 percent of ultimate power and efficiency potential. (Different datum values no doubt apply in the case of the pressure wave engine, but a similar argument applies.) Accepting that the temperature limits of the practical engine will not differ greatly from 600°C (expansion) and 60°C (compression) allows dramatic simplification of basic cycle algebra, and brings back-of-the-envelope estimating within the scope of the hand calculator.

Figure 15.1 is the opposed-piston configuration to which the unit gas path of any Stirling engine may be reduced for 'one-dimensional' flow analysis. At the compression end of the regenerator instantaneous mass rate may be estimated by differentiating the ideal gas law $m = pV/RT_C$:

$$dm/dt = m' = (RT_C)^{-1}[pdV_c/dt + (V_c + V_{dxc})dp/dt]$$

With A_{ffr} for regenerator free-flow area, instantaneous particle velocity is available by inverting $m' = \rho \underline{u} A_{ffr}$. Substituting for ρ from the ideal gas law:

$$\underline{u} = A_{ffr}^{-1}[dV_c/dt + p^{-1}(V_c + V_{dxc})dp/dt]$$

Denoting crank-pin offset by r and using A_C for cross-sectional area of the compression cylinder, peak value of the term dV_c/dt is $\omega r A_C = \frac{1}{2}\omega V_C$. ($V_C$ is the *amplitude* of compression space volume variation – not necessarily equal to swept volume V_{sw}.) The combined value of the terms within parentheses will be somewhat greater or somewhat less than $\frac{1}{2}\omega V_C$ depending on whether pressure is rising or falling. Present purposes will be served by working with dp/dt = zero:

$$\underline{u} \approx \frac{1}{2}A_{ffr}^{-1}\omega V_C \qquad 15.4$$

Substituting into the definition of 'isothermal' Mach number:

$$Ma = \frac{1}{2}\omega V_C/A_{ffr}\sqrt{(RT_C)} \qquad 15.5$$

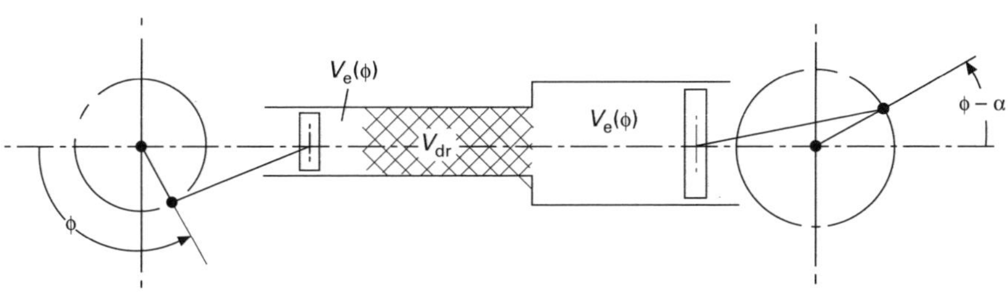

15.1 Opposed-piston configuration to which the unit gas path of any Stirling engine may be reduced for purposes of 'one-dimensional' flow analysis.

For the simple-harmonic volume variations assumed, the relationship between V_{sw} and V_C is in terms of thermodynamic phase angle α and thermodynamic volume ratio κ (Finkelstein 1960):

$$V_{sw} = V_C\kappa^{-1}\surd(1 + \kappa^2 + 2\kappa\cos\alpha) \qquad 15.6$$

If the machine is coaxial or of 'gamma' configuration, Eqn 15.6 is evaluated in terms of *equivalent* α and κ. Finklestein's conversion formulae or charts are appropriate.

Speed parameter – or characteristic Mach number – N_{MA} is $\omega V_{sw}^{1/3}/\surd(RT_C)$. Substituting:

$$Ma = \frac{\frac{1}{2}\kappa}{\surd(1 + \kappa^2 + 2\kappa\cos\alpha)} N_{MA}/\alpha_{ffr} \qquad 15.7$$

In Eqn 15.7 α_{ffr} is dimensionless free-flow area, $A_{ffr}/V_{sw}^{2/3}$. The compound dimensionless group N_{MA}/α_{ffr} will be recognized as *regenerator speed parameter*, numerical values of which are available for a range of benchmark Stirling engines (Organ 1997).

For Fig. 15.2, Eqn 15.7 is evaluated for ten Stirling engines whose specifications, operating conditions and measured performance are documented. The oblique line corresponds to the threshold of compressibility inferred, e.g. from Su's results (1986) namely, $Ma_{(peak)} = 0.02$. By this criterion the engines to the left of the line function wholly within the incompressible

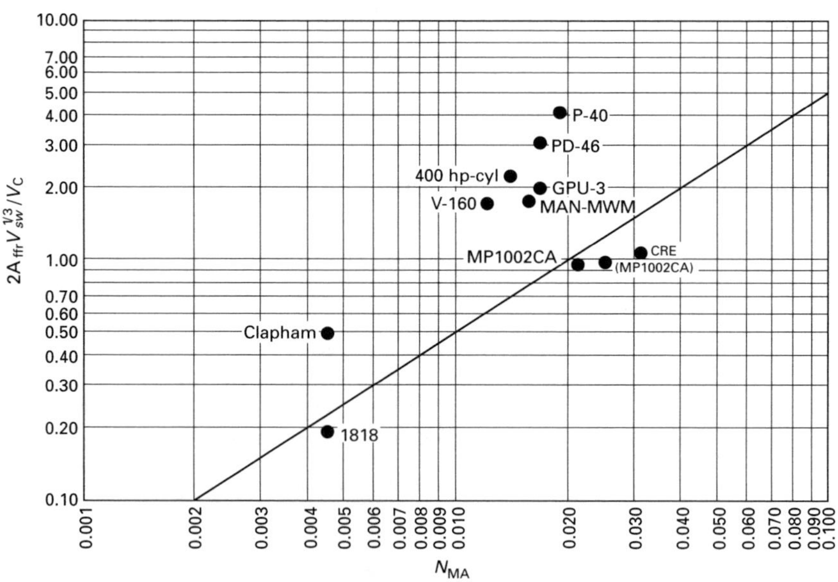

15.2 Equation 15.7 evaluated for 10 Stirling engines. The oblique line denotes the limit of validity of the incompressible flow friction correlations assuming this to be $Ma_{cr,peak} \leq 0.02$.

regime. Those to the right are subject to the (unknown) effects of compressibility.

The regenerators of the MP1002CA and original (1818) engines were wire coils wound *in situ*. The Clapham engine had pre-formed coils inserted. Sufficient information is available to allow an estimate of free-flow area. Some of the data on Stirling's original 1818 engine have been back-calculated from a comprehensive 'forensic' study (Finkelstein and Organ 2001). The operating conditions used for the CRE engine are *design target* values (Dando 1997). The location of the CRE machine is consistent with measured performance falling short of expectation.

15.4 Limiting *Ma*

Chapter 10 proposed working in terms of the ratio of *Ma* to *Re*. For scaling purposes the specific heat ratio term γ may be dropped, allowing the earlier Eqn 10.3 to be slightly simplified:

$$Ma/Re = \tfrac{1}{2}\mu\sqrt{RT}/r_{h}p \qquad 15.8$$

In a regenerator under steady-flow, all terms on the right-hand side have known constant values. In a functioning engine, absolute pressure p is a variable, but the insight on offer is retained if p is replaced by constant charge pressure, p_{ref}. Alternatively, expressing p as an harmonic function, e.g., $p = \underline{p} + C\cos\omega t$ retains all the detail, since the question of phase angle between pressure and crank angle does not arise. Use of p_{ref} gets rid of all temporal variation from Eqn 15.8. Assuming a linear gradient in T pre-defines the spatial variation.

Chapter 10 concluded that *Ma* effects modify the steady-flow C_f–*Re* correlations at *Ma* as low as 0.02. For any critical value of *Ma* Eqn 15.8 offers the corresponding critical value of *Re*.

Figure 15.3 plots *Ma* vs *Re* according to Eqn 15.8 over values of r_h achievable using wire gauze from the standard product range of G. Bopp and Co (undated). To read off the upper bound on *Re* corresponding to an alternative threshold *Ma* note the intersection of the chosen *Ma* value (say 0.05) with the appropriate r_h line.

15.5 Compressibility vulnerability chart

Su's curves of f vs *Ma* for *Re* = constant for the eight-screen stack (Fig. 10.2) are inspected for the *Ma* value marking the threshold of compressibility effects. Along a curve of given *Re* the onset of compressibility effects is taken to be the *Ma* at which the gradient reaches 5 percent. More formally, $\partial C_f/\partial Ma\,|_{Re} = 0.05$. Points from each curve are denoted the Ma_{crit} for that curve, and plotted against respective *Re* as in Fig. 15.4.

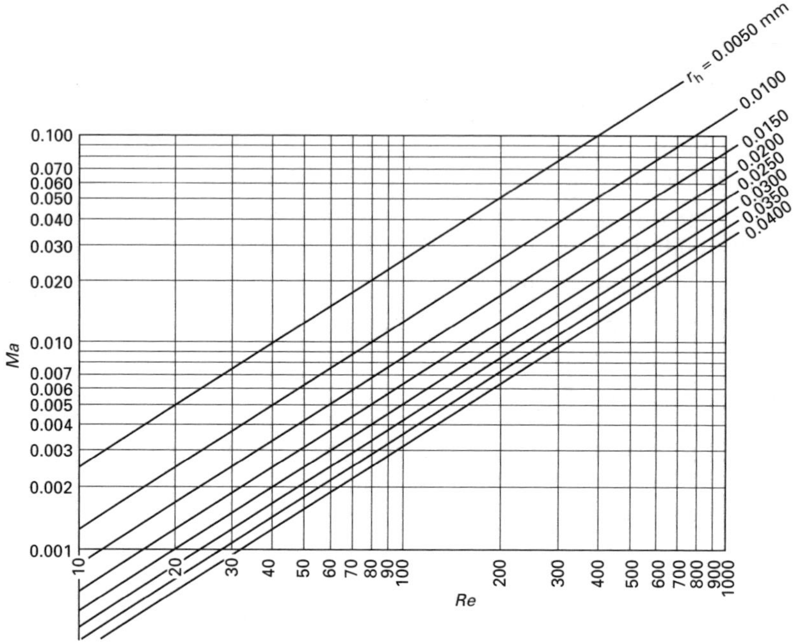

15.3 Ma vs *Re* for values of r_h achievable using wire gauze from the standard product range of G. Bopp and Co. (undated). Plotted from Eqn 15.8.

The heavy curve joining the points separates incompressible from compressible regimes. Inspection of Fig. 15.4 shows that, for *Re* sufficiently high, Ma_{crit} is the choking value 0.14. For the aperture ratio in question, this appears in Fig. 15.4 as a horizontal line at $Ma = 0.14$. Evaluating *Ma*/*Re* from Eqn 15.8 for a range of values of the product pr_h (Pam) and marking these in as the diagonal lines completes the design chart. Figure 15.4 applies to air (or N_2) at $T = 300K$ = i.e. to the compression end of the regenerator.

To steer clear of compressibility effects up to the peak *Re* of the chart, operation must be at $pr_h > 20$. On the basis of $p = 10$ bar (10^6 Pa) r_h should exceed 2.0×10^{-5} m, or 0.02 mm. The value will be looked at later in the light of independent criteria.

Pending availability of correlations sensitive to both *Re* (or *Sg*) and *Ma* it is necessary to proceed using existing data. With this reservation, the major component of pressure gradient d*p*/d*x* is:

$$dp/p \approx -\tfrac{1}{2}Ma^2 C_f L/r_h \qquad 15.9$$

For the wire screen stack of volume porosity $¶_v$, friction factor C_f is fitted to *Re*:

$$C_f = c/R_e + d \qquad 15.10$$

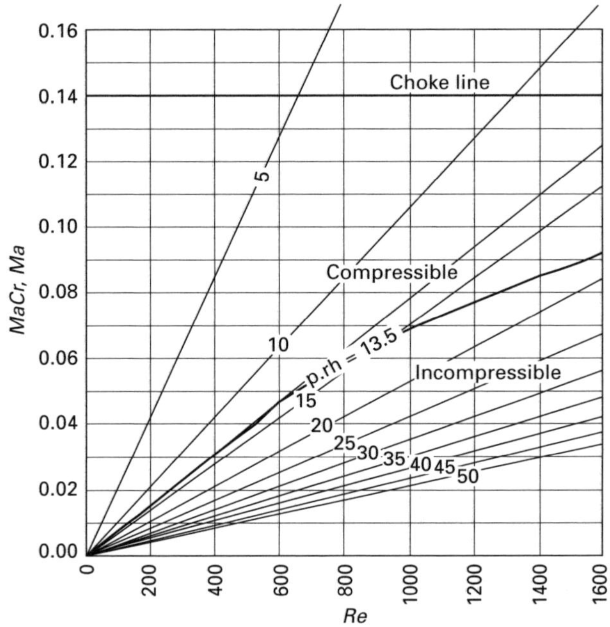

15.4 Compressibility vulnerability chart. Applicable to air and N_2 at ambient temperature and square-weave screens close-packed of aperture ratio 0.365. A combination of *Ma* and *Re* above the heavy curve means that friction factor is enhanced over the incompressible value at the same R_e. Combinations below the line are immune, and data for incompressible flow are appropriate, such as the correlations of Kays and London.

Substituting Eqns 15.8 and 15.10 into Eqn 15.9

$$p^{-1}\mathrm{d}p/\mathrm{d}x \approx -\tfrac{1}{2}\{cMa\mu\sqrt{RT}/4r_{\mathrm{h}}p + dMa^2\}/r_{\mathrm{h}} \qquad 15.9(a)$$

For a candidate design such as that specified by the data of Table 15.1 Eqn 15.9(a) yields fractional pressure gradient $p^{-1}\mathrm{d}p/\mathrm{d}x$ corresponding to specified *Ma*. Over the full exchanger length L_x, and with temperature T and *Ma* assumed uniform with x:

$$\Delta p/p \approx -\tfrac{1}{2}\{cMa\mu\sqrt{RT}/4r_{\mathrm{h}}p + dMa^2\}L_x/r_{\mathrm{h}} \qquad 15.9(b)$$

Figure 15.5 plots $\Delta p/p$ vs *Ma* for the data of Tables 15.1 and 15.2. The parameter is hydraulic radius r_{h}. No specific figure is to hand to serve as a working limit on $\Delta p/p$, although it is hard to image an engine functioning effectively if peak $\Delta p/p$ over the entire gas path reached, say, 20 percent (0.2 bar/bar). Corresponding to this, the net drop over the regenerator might be one half of that total, namely, 0.1. For the r_{h} of Table 15.1 this is achieved at $Ma = 0.035$.

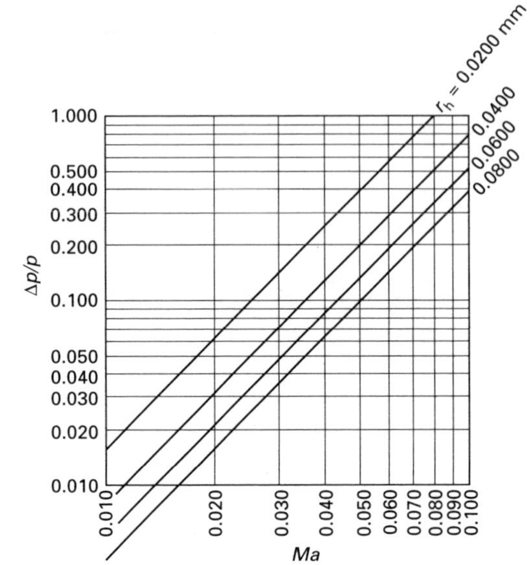

15.5 Fractional pressure drop $\Delta p/p$ vs *Ma* for the data of Table 15.1. Plotted from Eqn 15.9(b).

Table 15.1 Skeleton specification used to illustrate the relationship between *Ma* and *Re* (Eqn 15.8 and Fig. 15.3)

Nominal charge pressure	p_{ref}	10	bar (= 10^6 Pa)
Working fluid	air		
specific gas constant	R	287.0	J/kgK
coefficient dynamic viscosity	μ	0.017×10^{-3}	Pas
Reference temperature	T_C	300	K
Expansion temperature	T_C	900	K
Hydraulic radius	r_h	0.04×10^{-3}	m
Coefficients of $StPr^{2/3}$ – Re correlation $StPr^{2/3} = a/Re^b$			
a	0.588	–	
b	0.385	–	
Coefficients of C_f – Re correlation $C_f = c/Re + d$:			
c	40	–	
d	0.3	–	

Reliance on '*incompressible*' flow data suggests that particle speed may almost double before pressure drop is enhanced to the level now indicated by taking account of the 0.012 Ma_{crit} of Su's results. The prediction is consistent with an emerging picture whereby the 'incompressible' correlation gives an under-estimate of $\Delta p/p$ at Ma above Ma_{crit} for prevailing Re.

15.6 Heat transfer

For incompressible flow through stacked screens, heat transfer parameter $StPr^{2/3}$ is fitted to Re by:

Table 15.2 Gauzes used by Su (1986) in compressible flow tests. Definitions of volume porosity ¶$_v$ and aperture ratio are functions of similarity parameter $d_w m_w$ alone. The close similarity of respective $d_w m_w$ between gauzes tested by Su is the justification offered in this context for discounting geometric variation from test to test

Gauze #	Material	Wire diameter d_w (mm)	Aperture $m_w^{-1} - d_w$(mm)	Similarity parameter $d_w m_w$ (–)
A	Stainless steel	0.032	0.050	0.6400
B	Stainless steel	0.040	0.061	0.6557
C	Stainless steel	0.050	0.080	0.6250
D	Stainless steel	0.100	0.160	0.6250
E	Phosphor bronze	0.040	0.063	0.6349

$$StPr^{2/3} = a/Re^b \qquad 15.11$$

For different values of volume porosity ¶$_v$, a and b take somewhat different respective values. (This author is not aware of a counterpart correlation for compressible flow.)

Substituting for Re in terms of Ma from Eqn 15.8

$$StPr^{2/3} = a(Mar_h p/\tfrac{1}{4}\mu\sqrt{RT})^{-b} \qquad 15.12$$

There is now sufficient in place to allow calculation of work lost due to pumping and to imperfect heat transfer in function of Ma. The detail of the method, which is based on Availability Theory, may be found in Organ (2005a). Values computed for Fig. 15.6 are *proportional* to these losses.

Figure 15.5 suggests a monotonic increase in both losses with Ma. Thus, for the operating conditions of Table 15.1 there is no 'optimum' Ma in the same way as there is an optimum hydraulic radius, r_h. (Alternative operating conditions shift the point of intersection on Fig. 15.5, but over the range investigated by the writer, both losses always increase with increasing Ma.)

Up to $Ma \leq 0.005$ pumping loss is an order of magnitude less than heat transfer loss, which thus determines performance. At the Ma tentatively suggested as a design threshold (0.02), pumping adds 25 percent to the penalty of defective heat exchange. Above that value, pumping loss apparently increases, but cannot be reliably estimated from conventional flow correlations.

15.7 Implications for back-of-envelope design

No design guideline is reliable until it has been comprehensively calibrated against practical experience.

At first sight, Fig. 15.2 amounted to little more than an indifferent correlation

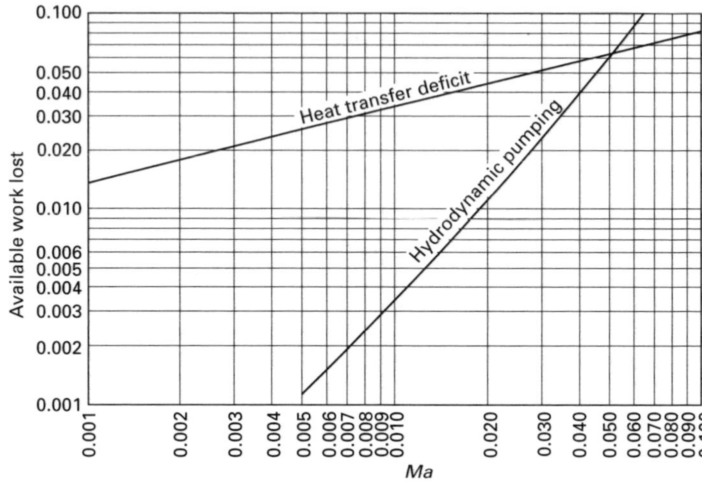

15.6 Penalty of pumping loss and heat transfer deficit as functions of *Ma* for the operating conditions of Table 15.1.

between the left-hand side of Eqn 15.7 and the right. A closer look reveals that two of the air-charged engines, one a massive 103,000cc, the other a mere 61cc, both sit with the Ma_{peak} of their regenerator right on the supposed threshold of compressibility $Ma = 0.02$. The third heavy-gas machine, the CRE, might have fitted better if *rpm actually achieved* (rather than the design value) had been available for use in Eqn 15.7. The six light-gas engines, from diminutive (5cc) to huge (15,000cc), are grouped around a $Ma_{peak} = 0.008$ below the critical value.

Reciprocating engines operate within mechanical as well as fluid flow limits. Inertia forces are proportional to $mr(\text{rpm})^2$. For engines of given configuration $m \propto \rho V_{sw}$ and $r \propto V_{sw}^{1/3}$. If materials inventory is common, then $m \propto V_{sw}$. Inertia load converts to stress on dividing by area – proportional to $V_{sw}^{1/3}$. Making the substitutions, inertia stress within a given class of engine is proportional to $(V_{sw}^{1/3}\ rev/min)^2$.

Table 15.3 lists numerical values of $V_{sw}^{1/3}\ rpm$ (m/min) for the engines of Fig. 15.2. Values calculated at the peak power point of three internal combustion engines are included – a multi-cylinder four-stroke diesel, a single-cylinder four-stroke petrol and a compression ignition two-stroke.

The internal combustion engine (ICE) is highly developed. If not over-designed, it operates at the inertia stress limit divided by design factor of safety. Reciprocating mass in the Stirling engine is a larger fraction of V_{sw} than for the IC engine: multi-cylinder types have cross-heads and piston rods; single-cylinder, coaxial variants have complex crank mechanisms. A Stirling engine operating to a given safety factor may be expected to show lower values of $V_{sw}^{1/3}\ rpm$.

Table 15.3 Inertia stress parameter $V_{sw}^{1/3}$ rpm (m/min) for selected reciprocating internal combustion engines and for the Stirling engines of Fig. 15.2

	Swept volume per cylinder, V_{sw}	*rpm* (min^{-1})	Inertia stress parameter $V_{sw}^{1/3}$ *rpm* (m/min)
Reciprocating IC engines			
Peugeot automotive diesel (205D)	$450 \times 10^{-6} m^3$	4500	***345***
ED 'Racer' model aero engine	$2.5 \times 10^{-6} m^3$	14000	***190***
Ryobi brush-cutter engine	$25 \times 10^{-6} m^3$	7000	***204***
Stirling engines			
Air-charged			
(Stirling 1818	$103{,}000 \times 10^{-6} m^3$	28	13)
MP1002CA	$51 \times 10^{-6} m^3$	1500	***59***
CRE	$5{,}150 \times 10^{-6} m^3$	500	***86***
Light gas			
MAN/MWM	$2{,}070 \times 10^{-6} m^3$	900	***114***
USS V-160	$225 \times 10^{-6} m^3$	1500	***91***
400hp/cylinder	$15{,}300 \times 10^{-6} m^3$	452	***112***
GPU-3	$118 \times 10^{-6} m^3$	3600	***176***
USS P-40	$134 \times 10^{-6} m^3$	4000	***204***
Alison PD-46	$77 \times 10^{-6} m^3$	3000	***128***
(Clapham	$5 \times 10^{-6} m^3$	2000	34)

From Table 15.3 inertia stress limits for the air-charged engines fall below those for the light-gas types, and considerably below those for the ICE. Inertia stress limit does not appear to be the performance limitation. Walker (1980) reports that raising expansion-end temperature of the MP1002CA to 900°C coaxes only another 300 rpm from the engine. There is potential for increased rate of heat transfer at this temperature – if the gas processes are able to respond. The reality is suggested in Fig. 15.2 by the off-design performance point in parentheses – operation at *Ma* above the 0.02 threshold results in an increase in pumping loss out of proportion to increased thermal input.

It has been determined that the operating speed of the original air engine of 1818 was limited to about 28/min by flywheel integrity. But the same forensic study (Finkelstein and Organ 2001) found that pressure drop in the narrow, annular regenerator held operating speed to 28 rpm anyway. The location of this engine on the map is consistent with a Mach number limitation – and at $Ma = 0.02$.

Values of inertia stress parameter for the light-gas engines are not far short of those for the ICE, despite higher reciprocating mass. If $Ma \geq 0.02$ represents a performance boundary, then $Ma = 0.008$ by definition does not. Performance of the light-gas engines is limited by inertia stress.

Opportunities abound for *ad hoc* arithmetic to improve on the achievements of 'modern conceptions in regard to heat transfer, flow resistance etc.' (Rinia and duPré 1946). Regenerator matrix temperature swing relates directly to the ratio of two thermal capacities: that of the matrix and that of the gas per pass. This ratio in turn is proportional to the product N_{TCR} with the factor V_{dr}/V_{sw}. N_{TCR} is nominal thermal capacity ratio, $\rho_w c_w T_{ref}/p_{ref}$ (Organ 1997) and $V_{dr}/V_{sw} = \delta_r$ is the ratio of regenerator dead volume to swept volume. Numerical values of the defining dimensionless product are:

Philips' MP1002CA: 276 (–)

United Stirling V-160: 26 (–)

Matrix temperature swing in the V-160 is a factor of 10 greater than that of the MP1002CA. Volume porosity is of similar order in the two cases, so if the thermal capacity ratio built into the V-160 is appropriate, then regenerator thermal capacity – and thus dead space – of the MP1002CA is excessive by a factor of ten.

15.8 A 'screening' test

Archibald (2006) reports a simple test which may well turn out to offer more information per unit of capital and time expenditure than anything in the repertoire.

A pressure gauge is connected to a thin-walled, metal cylinder containing air at ambient pressure and temperature, and the assembly sealed. If the pressure gauge is of the bourdon-tube type, it is important that its dead volume should be a small fraction of that of the cylinder. Otherwise the experiment can be on a small scale, involving, say, an aluminium cannister for 35 mm film and a pan of water at ambient temperature.

The cylinder is heated until the pressure gauge indicates 1 atm. above ambient. The cylinder is plunged into water and the time noted for pressure to fall to within a percent or two of the starting value. The reader may wish to guess the time taken, carry out the experiment and then prepare for a surprise!

For the fixed mass M of air involved, the rate of fall of the temperature T of the cylinder contents, dT/dt, is given in terms of rate of pressure fall via the ideal gas law $dT/dt = (T_0/p_0)dp/dt$ or, equivalently:

$$d/dt(T/T_0 - 1) = d/dt(p/p_0 - 1) \qquad 15.13$$

Assuming the inside wall temperature to be that of the external water bath T_0, Newton's law of cooling may be applied:

$$hA_wT_0(T/T_0 - 1) = -Mc_p dT/dt$$

Making further use of the ideal gas law $p_0V = MRT_0$, substituting Eqn 15.13 and integrating gives a first estimate for heat transfer coefficient h. If pressure takes Δt seconds to fall to within 1 percent of the original 1 atm:

$$e^{-NTU\Delta t} = p/p_0 - 1 \qquad 15.14$$

$$NTU = (hT_0/p_0r_h)[\gamma/(\gamma - 1)]$$

The Stirling engine has two nominal 'dwell' periods per cycle at each location, to which heat transfer data to traditional $StPr^{2/3}$ vs Re format are of questionable relevance. Archibald's test has self-evident relevance to the 'relaxation' or 'dwell' conditions in the thermal lag engine.

By using an air container having the dimensions of the pulse-tube section the test can be made immediately relevant to a specific design of thermal lag engine. On the other hand, NTU as evaluated via Eqn 15.14 absorbs isentropic index γ and hydraulic radius r_h, suggesting a far more valuable role as a generalized 'screening' test, in the spirit of the Charpy and Izod tests for metals (measure of sensitivity to brittle fracture).

For sheer elegance it belongs in the category of Annand's approach to the acquisition of dimensionless heat transfer coefficient in the cylinder of the motored internal combustion engine (Annand 1963).

15.9 The wider role of scaling

In the 1940s the Philips company of Holland raised the specific power of the Stirling engine by a factor of 100 relative to the norm. By anybody's standards, an improvement of two orders of magnitude is a lot – and it was achieved *without* computer simulation, *without* scaling and (according to contemporary perspective), *without* the benefit of 'modern conceptions of heat transfer and flow resistance'.

On the other hand, the new level of performance had been latent in the concept since 1816: it had merely lain dormant pending improved matching of *concept* to *embodiment*. Depending on how one chooses to do the calculations, the gap between predicted and measured performance remains substantial. This raises the question as to how much of that gap can be closed. Systematic build-and-test is one option – but only to an organization with resources comparable to those of the pre-war Philips.

Simulation thus remains indispensable, and scaling can shape the intuitive processes: it has taken the author some years to appreciate that it is not the properties of any particular gas which determine the internal thermodynamic processes, but the *interaction* between gas properties, flow passage geometry and operating conditions (charge pressure, rpm, temperature limits). The cycle of gas processes when the Philips MP1002CA engine is charged with helium and running at 1500 rpm and 15 bar charge pressure are not those which take place when it operates at the identical pressure and rpm with its rated working fluid, air. An alternative point of view is that substitution of the lighter working fluid shifts operation to a different point on the performance map.

16
'How to make a business out of Stirling engines today'

16.1 Tribal wisdom

The 11th International Stirling Engine Conference (ISEC) was convened (Naso 2003) under the slogan which forms the chapter title above.

If the '*Stirling engines for sale*' columns are an indication, then manufacturing industry has yet to embrace the magic formula.

Nine years plus a few days from the time of writing (early December 2006), Hogmanay festivities will be ushering in the landmark year 2016. It will be embarrassing to say the least if, two centuries on from the famous patent of 1816, we are still trying to make a business out of Stirling engines – *still* celebrating a brilliant invention with a great commercial future behind it.

An independent consultant browsing the balance sheet would be duty-bound to query the rationale behind Stirling development to date – and might well conclude that pursuit of that rationale had amounted to flogging a dead horse. According to ancient tribal wisdom, this should be the cue to dismount and to saddle up a steed showing some sign of life. (Appendix IV offers alternative business-school options. If anyone knows the source of these, please notify, as it merits acknowledgement.)

An autopsy of the dead animal would reveal:

- chronic obsession with the mirage of Carnot efficiency
- futile focus on irrelevancies such as drive mechanisms for replicating the 'ideal' discontinuous Stirling cycle
- distractions such as the relative thermodynamic merits of 'alpha', 'beta' and 'gamma' configurations.
- proprietary exclusivity: understandable, but at the current state of the technology, crippling.

16.2 From alchemy to appropriate technology

Consider some realities:

- the blades of Parson's first steam turbine were cut by hand with hacksaw. The result was a prime mover generating 7.5 kW
- the design calculations which put the gas turbine into military aircraft service extended no further than the velocity triangle and the steady-flow energy equation
- achievement of the high specific power of the petrol engine was accelerated by availability of the Ricardo 'Hydra' research engine. This allowed compression ratio, valve timing, air/fuel ratio and ignition timing to be varied during operation: practical, irrefutable, 'real-time' optimization!

It must by now be obvious to all concerned that the Stirling engine has not evolved – and cannot evolve – via an analogous route. On a casual* parts count it is beguilingly basic: cams, valves and spark plugs have gone. Conversely, the visibility and tangibility of the timing mechanism have given way to the inscrutability of subtle gas process interactions. The effect of an increase in valve overlap in the petrol engine can be guessed at: that of increasing the length of the regenerator stack cannot. At first sight, help is at hand in the guise of the 'virtual' research engine – the computer simulation. However, not until suitable flow and heat transfer data are available for a wide range of flow passage geometries can the resource be effectively mobilized.

Stirling engine technology is challenging – but is scarcely rocket science. Even if it were, four decades have passed since man set foot† on the moon, so rocket science the reality is no longer 'rocket science' the metaphor. All that is wrong with the technology of Stirling engines is (a) that know-how and expertise are spread over a vast number of individuals and organizations and (b) that it is unproductively competitive.

Since the Philips program was wound up, there has been a cataclysmic development which offers to change all this: the *internet* has arrived. Megabytes of data can be interchanged between continents in seconds. This, however, is not a proposal for yet another internet chat-room for disputing the merits of the Schmidt analysis. Neither is it a proposal for a reincarnation of the UK Consortium, which proposed, amongst other things, to harness academic research (including that of this author) and to offer it for sale.

It is about constructing a creative entity, the whole of which exceeds the sum of the parts: It calls for academic engineers (including myself) to make their gas process modelling software freely available in user-friendly form to all who care to log on. Equally it is about involving the likes of Geoff Vaisey

---

* Dismantling a 'serious' pressurized coaxial engine such as the MP1002CA (as opposed to the table-top toy or chalk-board diagram) reveals a disconcertingly large parts count.

† In the interests of objectivity visit, for example, *www.ufos-aliens.co.uk/cosmicapollo.html* and draw your own conclusions as to whether the original landing actually took place.

in Norfolk, designer and developer of micro-gas turbines (180,000 rpm!), sub-contractor to F-1 industry, who sends machining files from his workstations at one end of his machine shop to multi-axis machines at the other, creating engines such as that shown in Fig. 16.1. It is about involving Peter Neal in Derby, ex-Rolls-Royce engineer, also exploiting NC facilities, and undaunted by the challenge of building an exact replica of the Philips MP1002CA air engine. There is potential for harnessing the talents of every Stirling analyst, designer, experimenter and machinist worldwide who is prepared to participate and to benefit.

The combination adds up to a Virtual Development Facility (VDF) – a fusion of resources and manpower which a purpose-built, conventional research centre *could not hire*. The need to register as a commercial corporation does not arise. The venture is simple – it is even free. It could take its first steps tomorrow. With no special effort it would, by its very existence, *raise the threshold* from which any future engine project might start.

In summary, it is about *making insight communicable.* All that is needed is consensus – but the concept is so obvious and elementary that it will

16.1 Rhombic drive engine, synchronized by external double-sided timing belt. With permission of G. Vaisey.

probably never happen. Just in case, however, some basic features might be listed:

- access to every cycle simulation under the sun: key in your data, compare performance predictions, identify thermal bottle-necks, flow restrictions etc.
- a comprehensive specification of a high-performance, single cylinder (or single gas path) engine. The launch specification would be 'elastic', evolving in response to diverse specialist input. At any instant it would be the focus of all stress calculations, solid modelling, performance predictions and materials evaluation.
- with the specification sufficiently firmed-up, all involved (and, indeed, any not involved) could download drawings, solid models, machining files, performance predictions and gas process depictions. Detail drawings would be fully toleranced, so that parts machined to hard copy output would be interchangeable with those from machining files.
- links to every report, paper, book and manufacturer's product catalogue having a bearing on Stirling engine design and development.

The rest of this chapter will take the first steps towards realizing the VDF by proposing an elastic specification for a one-horsepower (746 W), air/N_2-charged engine, the VDF-750(aS). The aS suffix emphasises the status of the specification at the time of writing: it stands for *aunt Sally*. Once constructed, a Stirling engine does not lend itself to ready modification. A priority is therefore to avoid committing to any feature likely to turn into an Achilles heel. To that end critical comment is essential – and welcome - at the formative stage.

This author proposes to set up a pilot VDF web-site as soon as the manuscript of this text is handed to the publisher. In the spirit of the undertaking, those interested in communicating in the meantime are welcome to do so via the author's private *e*-mail address: allan.j.o@btinternet.com

16.3 What else has changed?

Further advantageous developments can be identified, the exact number depending on where the chronological datum is set. Arbitrarily taking the launch of the WhisperGen unit:

- *'Solid oil' bearing technology*: Rolling-element bearings are now routinely available incorporating 'solid oil' technology. Solid oil is a polymer matrix saturated with a synthetic oil. It occupies the internal volume of the bearing, encapsulating cage and rolling elements. The manufacturer's technical literature (SKF Ltd 2002) summarizes the technical features and operational advantages, which include elimination of seals, exclusion of contaminants and maintenance-free operation (lubrication for life). Maximum continuous operating temperature quoted is 85°C, with intermittent peaks allowable to 95°C.

- *Sliding seals*: The relevant technology is that of the oil-free compressor. Nitrogen is one of many gases processed commercially on a large scale. In the 'bone dry' condition it is particularly demanding of seal performance. The brittle thermosets originally used for compression and bearer rings gave lives which were inconsistent above 1000 hours. Lives of *tens of years* are now claimed (Wilson 2000) using rings of polymer alloy running against surfaces of suitable hardness and finish. The figure derives from extrapolation, but is reassuringly based on monitoring of the performance of compressors in service world-wide. Low coefficient of friction would be consistent with the extended lives claimed.
- *Water jet cutting*: The machining revolution introduced by numerically-controlled (NC) wire erosion has been followed up by NC water jet machining. At the current stage of technical development the latter operates with somewhat lower precision than the former, but is markedly quicker and cheaper. Rigid, non-conducting materials can be cut as readily as metals.
- *Friction welding*: The technology is particularly suited joining axi-symmetric components, which are readily spun up to speed with the necessary concentricity. The present project has in mind an expansion cylinder stacked from annular elements. The blank for each element is a short length of thick-walled stainless-steel tube into which axial fins, internal and external, are cut by water jet.
- *Regenerator thermal and flow design:* The 'regenerator problem', identified early last century (Hausen 1929), has been solved for the operating conditions of the Stirling engine. Availability of the solution strategy now highlights a dearth of flow and heat transfer data appropriate to matrix materials suitable for the context. Rectifying this situation is largely a matter of making the commitment to do so.
- *Burner technology*: Use of metal foam as combined flame-holder and regenerative mixture pre-heater in combination with a Spirex exhaust heat recuperator has demonstrated, on a small scale, a combustion chamber capable of extracting the higher calorific value (HCV) of the gaseous fuel.
- *Electro-mechanical energy conversion*: The rare-earth permanent magnet makes possible the impressively compact '*out-runner*' motor, as exemplified by the proprietary Torcman range. In 'axial-flux' configuration (Bumby and Martin 2005) power density encourages consideration for use as an internal flywheel. The price for eliminating the shaft seal by this means would be additional internal windage under (variable) crank-case charge pressure.

16.4 The VDF-750(aS)

16.4.1 Configuration and basic sizing

Philips' MP1002CA air engine was rated at 200W(e) at 1500 rpm and 15 bar charge pressure. The figure for net electrical power suggests some 250W mechanical shaft power from the V_{sw} of 61 cm^3. This converts to 4.1 W/cm^3 – and the value may still be conservative:

- The forerunner of the MP1002CA had been the Type 10, also of V_{sw} = 61 cm^3. Hargreaves (2001 p. 40) quotes 500W at 1500 rpm and rated pressure p_{ref} (given as 5 atm – surely a misprint). At an unspecified value of elevated p_{ref}, one horsepower (746W) was achieved.
- Brake power of the MP1002CA measured by the author (Organ 1962) was 235 W at rpm = 1375 and p_{ref} = 15 bar. The V-belt driven generator remained connected on open circuit, and the engine was driving the cooling fan. Net brake output therefore exceeded the measured 235W by the amount of generator windage and blower power, and by the extent of an unknown loss due to the fact that the centrifugal friction governor was incorrectly set to cut in at 1400 rather than 1500 rpm. 300W might be closer to true shaft power.
- By raising expansion exchanger temperature to 800°C (not a risk this writer would knowingly have taken) Ward (1972) coaxed 430W mechanical from the MP1002CA at 12.4 bar and 1400 rpm.

Crankshaft rpm of 1500 and 3000 are convenient for a.c. electrical power generation. The former offers no scope for pushing the performance envelope of an N_2-charged engine; the latter is ambitious. The specification will settle provisionally on 1500 rpm. When gas process modelling and mechanical stressing stages are reached, the figure will be revised upwards to 2000 rpm.

If the crank-case is oil-lubricated, as in the internal combustion engine, a function of the buffer space and its seals is to prevent entry of the lubricant into the gas circuit and consequent contamination of the heat exchange surfaces. If the working fluid is pressurized air, hydrocarbon lubricants in contact with the internal hot surfaces are an explosion hazard (Hargreaves 2001).

If conventional oil lubrication can be eliminated, so can bulk, complication and cost of the buffer space. Going for combined buffer-space and crank-case does not prevent these functions being separated at some future stage of development, but does call *either* for a rotary shaft seal *or* for the generator to be enclosed within the crank-case. The rotary shaft seal is an easier proposition than the reciprocating seal of the separate buffer space.

16.4.2 Provisional specification

- **Heating provision:** 5kW thermal rating bought-out proprietary recuperative burner. Eventually custom-designed burner based on 'Spirex' exhaust recuperator.
  It is understood that Bekaert Combustion Technology are supplying components and technology to manufacturers of combustion systems for Stirling engines:
  Bekaert Combustion Technology B.V.
  J.C. van Markenstraat 19
  NL-9403 AR Assen
  Postbus 123
  NL-9400 AC Assen
  Netherlands
  Tel: +31/592/345 145
  Telefax: +31/592/345 483
  Email : bct@bekaert.com
- **Fuel**: gaseous butane, methane or mixture
- **Configuration**: single cylinder co-axial 'beta'; crank-case pressurized to minimum cycle pressure and serving as buffer space
- **Drive:** conventional rotary crankshaft driving four-bar linkage (Appendices I and II) formed by lever/bell-crank combination. Near-linear motion to both piston and displacer. Displacer driven from bell-crank, piston from lever
- **Lubrication/sealing:** quasi-dry – oil-free compressor technology for reciprocating seals. SKF/Laval non-migrating 'solid oil' technology for rolling bearings.
  Off-the-shelf oil-impregnated polymer bushes for plain bearings.
  Proprietary pressure-balance crank-shaft rotary seal unilaterally cooled/ oil-lubricated from downstream/low-pressure side
- **Internal exchangers:** multiple interrupted slots parallel to axis
- **External exchangers:** expansion exchanger – multiple interrupted slots parallel to axis compression exchanger – water jacket
- **Regenerator**: annular – matrix specification flexible to allow experimental/ theoretical determination of optimum flow passage geometry and axial gradation thereof
- **Cooling:** water
- **Drive mechanism geometry (Appendix II):** the arrangement drawing (Fig. 16.4) is constructed to slightly different values:

X/r	4.0
Y/r	2.667
h/r	2.667
e/r	3.733
RQ/r	4.266
Angle ζ	95° (the arrangement drawing of Fig. 16.4 indicates 90°)

16.4.3 Swept volume V_{sw}

With P for power (W) and f for cycle frequency (s^{-1}), specific cycle work Z is defined:

$$Z = \frac{P(W)}{p_{ref}\,(\text{Pa})\, f\,(\text{s}^{-1})\, V_{sw}\,(\text{m}^3)} \qquad 16.1$$

In Eqn 16.1 frequency f = rpm/60.
Using the nominal shaft value (250W) of P for the MP1002CA:

$$Z = \frac{250\,(W)}{15.0\text{E}+05\,(\text{Pa}) \times 1500\,(\text{rpm})/60 \times 61.0\text{E}-06\,(\text{m}^3)}$$

$$= 0.10929$$

It is the magnitude of Z, being a 'per cycle' value, which one seeks to maximize by improved thermodynamic design.

Equation 16.1 may be inverted so as to express V_{sw} in terms of the design options available for p_{ref}.

$$V_{sw}\,(\text{cm}^3) = Z^{-1}\,\frac{P\,(W)}{p_{ref}\,(\text{atm}) \times 10^5 \times \text{rpm}\,(\text{s}^{-1})/60}\,10^6\,(\text{cm}^3) \qquad 16.2$$

For given target Z, a modest reduction in V_{sw} may be traded for a proportional *increase* in p_{ref}. For example, V_{sw} may be reduced to 80 percent of the datum value in reasonable expectation that a 25 percent increase in p_{ref} (*and adjustment to gas path design accordingly*) will compensate. Rpm do *not* trade so conveniently – possibly because pumping loss per cycle is linked to the square of particle speed.

In Fig. 16.2 swept volume V_{sw} (cm^3) is plotted according to Eqn 16.2 against p_{ref} (atm.) for rpm = 1500, P = 746 W, and with Z varying over a small range as parameter. At the 'safe' design condition of Z = 0.1, swept volume indicated to achieve 1 HP at the original charge pressure of the MP1002CA (15 bar) is 198.95 cm^3 – or essentially 200 cm^3.

This account proceeds on the basis that a Z of 0.10929 lies well short of the ultimate for this basic concept. Further grounds for supposing a higher potential Z:

- Expansion and compression exchangers of the MP1002CA and Type 10 were symmetrical, while the gas processes are heavily asymmetrical.
- Regenerator hydraulic radius r_h and volume porosity $¶_v$ were uniform, while fluid properties and flow conditions vary substantially end-to-end.
- The later and more sophisticated GPU-3 of General Motors and the engines of United Stirling have asymmetrical heat exchange provision,

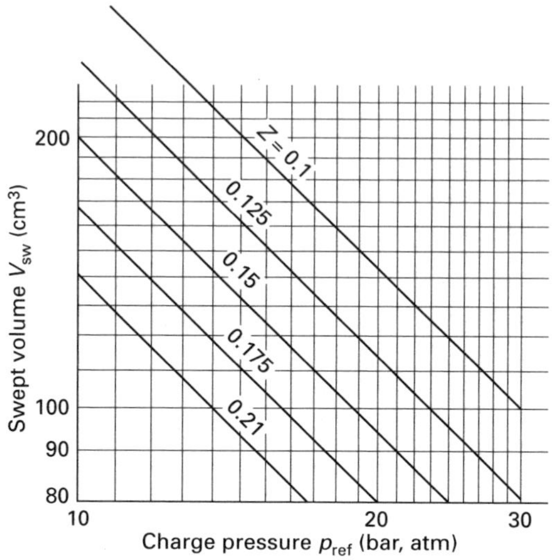

16.2 Graphical equivalent of Eqn 16.2: V_{sw} (cm^3) plotted vs p_{ref} (atm) for rpm = 1500, P = 746 W, and with Z varying over a small range as parameter.

and for this and/or other reasons achieve more than twice the Z of the MP1002CA.

- If performance was explained in terms of 'modern conceptions of heat transfer and flow resistance' (Rinia and duPré 1946), the account as to how is unrevealing, since the only analysis quoted is that of Schmidt – which says nothing about heat transfer or flow resistance. Improvements due to general modernization alone were, nevertheless, dramatic.
- Thermodynamic design will benefit from gas process modelling not available to designers of the MP1002CA
- The drive mechanism to be used offers substantial kinematic flexibility and corresponding scope for tailoring volume variations.

The foregoing support a return to Fig. 16.2 with Z adjusted upwards to 0.175 (the value remains conservative relative to those for the GPU-3 and for the United Stirling machines). Swept volume V_{sw} indicated at 15 bar and 1500 rpm is now 113.7 cm^3. The figure actually settled for the provisional specification is arrived at by adopting rational dimensions for cylinder bore D, crank-pin offset (semi-throw) r, and the coordinates of the linkage. With r = 10.0 mm, the numerical values, in mm, of X, Y, h, e, etc., are simply the earlier tabulated values of the dimensionless ratios X/r, etc. multiplied by 10 (mm). Resulting swept volume, with allowance for a provisional 10 mm diameter of the displacer rod, is 99.78 cm^3 – or, in sensible figures, 100 cm^3.

The evolving specification now embodies the following data:

Bore:	60 mm	
Strokes:	S_p = 36.30 mm (piston)	S_d = 34.25 mm (displacer)
Bore/stroke ratios:	1.65 (piston)	1.75 (displacer)
Swept volume V_{sw}:	100 cm^3	
rpm:	1500	
p_{ref}:	15 atm (15×10^5 Pa)	
hot end T_E:	650°C (923K)	
cold-end T_C:	60°C (333K)	
power P:	750 W (mechanical)	

16.5 Drive mechanism/kinematics

A proven mechanism for the single cylinder coaxial configuration is the bell-crank drive of Philips' MP1002CA air engine, shown in Fig. 16.3(a) to the conventions of *mechanism synthesis* (Johnson 1971), and denoted *mechanism a*. The work piston is subject to the effects of connecting-rod angularity. The displacer, on the other hand, is driven by near-straight-line motion.

An alternative is the linkage previously specified for an experimental hair drier engine (Organ 2001). This is shown in 'equivalent linkage' form at Fig. 16.3(b), and described in some detail in Appendices I and II. Here it becomes *mechanism b*. Near-straight line inputs to both piston and displacer minimize side loads on both. The feature is of greater consequence in the context of dry-running seals – as in the engine now proposed – than in the oil-lubricated MP1002CA.

No designer willingly takes an arbitrary decision. Invoking the formal study of linkages helps to make such a course unnecessary: Figs 16.3(a) and (b) are the 'equivalent linkage' forms. Both have eight links ($L = 8$) and ten joints ($J = 10$). Grübler's equation (Johnson 1971) gives number of degrees of freedom F:

$$F = 3(L - 1) - 2J \qquad 16.3$$

The arithmetic confirms the expectation of a single degree of freedom in both cases: $F = 3(8 - 1) - 2 \times 10 = 1$, but suggests at first sight that there is nothing to choose between the two. Conversely, *mechanism b* has two ternary links vs the single ternary of *mechanism a*. The feature calls for a greater number of numerical dimensions for a complete geometric specification. The larger number translates into increased scope for tailoring the variations of volume with crank angle.

Mechanism a (that of the MP1002CA) does not as it stands encourage interchange of the points of attachment of piston and displacer. In other words, driving the piston via the bell crank and the displacer directly from the connecting rod promises to exceed the capacity of the latter for coping

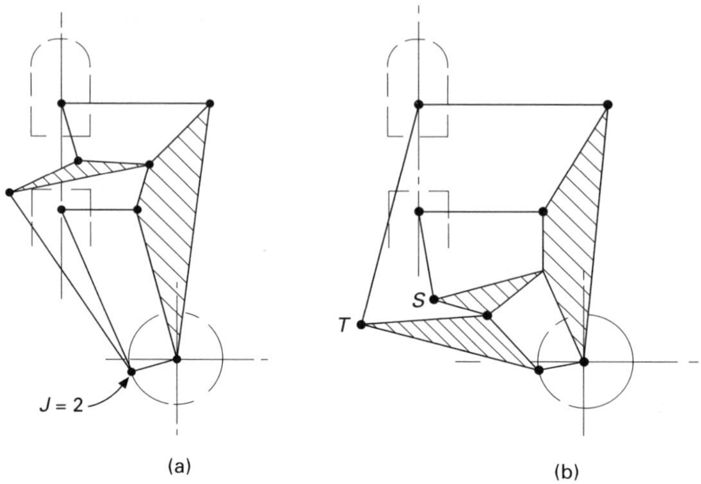

16.3 'Equivalent' linkages for (a) the drive mechanism of the MP1002CA and (b) the bell-crank/lever combination proposed for the VDF-750(aS).

with side loads. *Mechanism b* does not suffer this limitation, so the piston could drive either joint *S* or joint *T*, and either joint *T* or *S* could drive the displacer. The feature amounts to further flexibility (if needed) in tailoring the volume variations. Moreover, *mechanism b* can sit *below* the cylinder unit, as in Fig. 16.3(b), or above, in which case the respective vertical excursions y_S and y_T are both reversed in algebraic sign, and the corresponding variations of volume with crank angle ϕ changed, affording yet further kinematic flexibility.

16.6 General mechanical construction

Figure 16.4 is part of the author's pencil drawing to reduced scale. The two principal orthographic views are in part-section. The projection is 'first-angle', and the aim has been to conform to the drafting conventions of BS 308, 1972 *Engineering Drawing Practice*. A reader interested in pursuing the concept on the basis of Fig. 16.4 should take into consideration that:

- this is not the 'official' start-point of the design, which will be a version generated by ProEngineer, UniGraphics, Solid Works or similar solid-modelling package. An essential purpose of the CAD exercise will be to verify clearances throughout the full range of motion of the mechanism.
- the right-hand section is a composite of more than one section plane, involving some liberties with BS 308. (To include the orthographic view which specifies the section planes would involve scaling-down and further loss of detail from the principal view.)

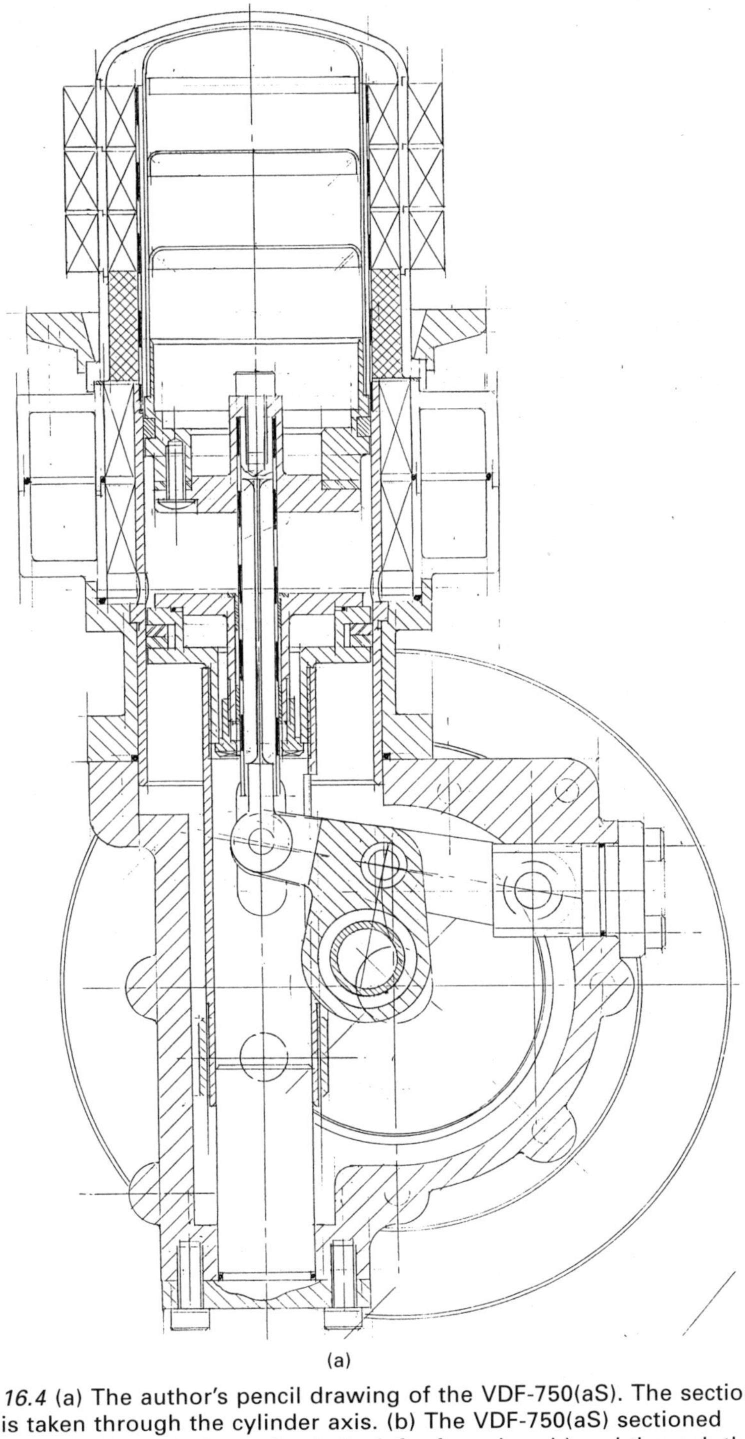

(a)

16.4 (a) The author's pencil drawing of the VDF-750(aS). The section is taken through the cylinder axis. (b) The VDF-750(aS) sectioned through the main journal axis (to left of crank web) and through the cylinder axis (to right of crank web).

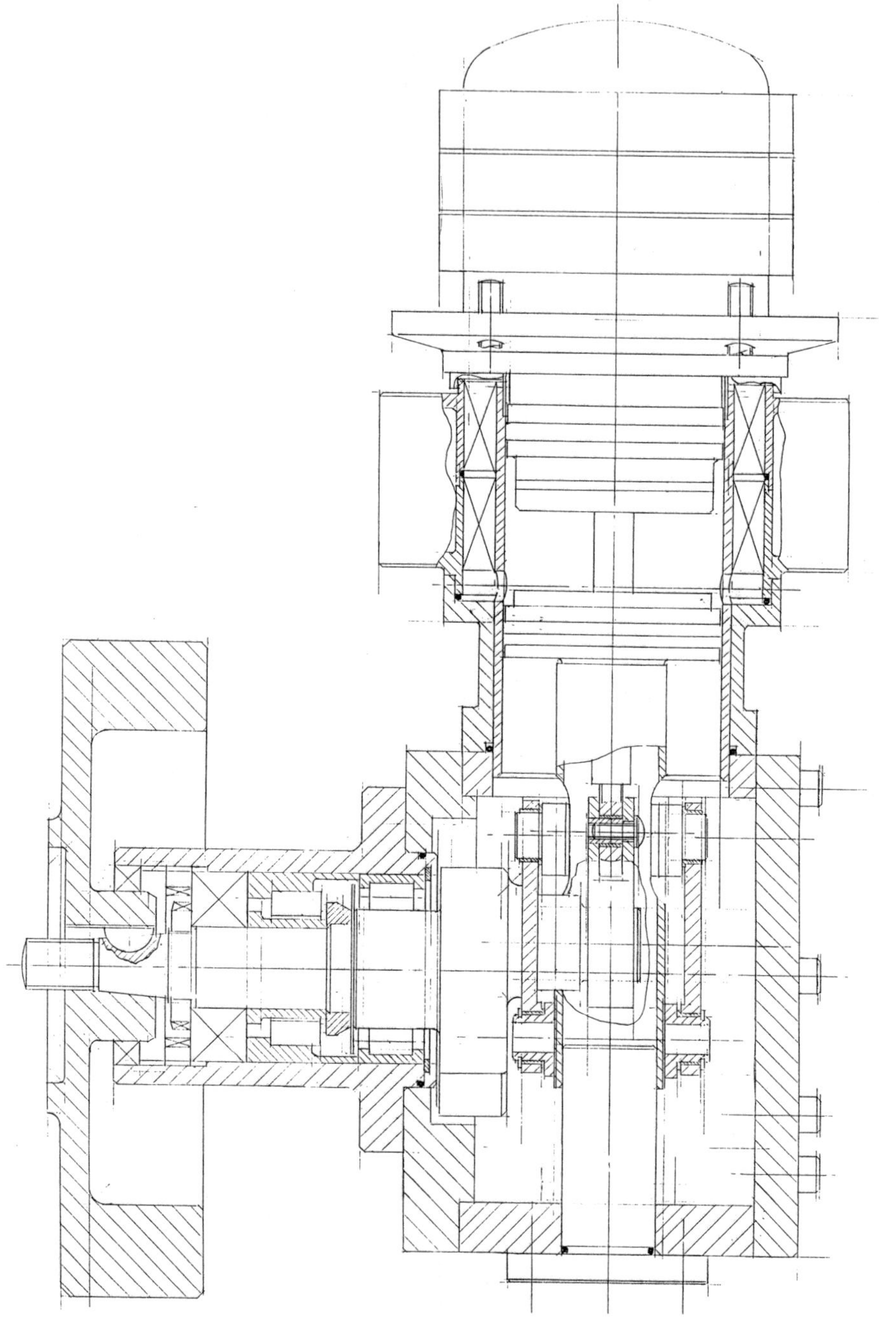

(b)

16.4 Continued

16.6.1 Regenerator

Stirling's original engine appeared in the 1816 patents as a specimen application of the essential invention, the regenerator. If relevant heat transfer and flow data were available today, there would be a strong case for tackling the present design in the same sequence (1) regenerator and (2) engine. While we wait for the matter of design data to receive due priority, the status of the regenerator can be acknowledged by addressing it first.

Stirling Engine Systems (SES) of London are pursuing laser remelt technology (Hislop 2006) for generating microminiature heat transfer surfaces to any geometry which can be fed to a micro-processor controller. Ultimately it offers the possibility of generating a flow-passage system to a pre-determined relationship between heat transfer and pumping loss. While this technology matures it is likely that regenerators will continue to be made by adapting materials in current mass-production for other – usually quite unrelated – purposes.

The annular regenerator of the MP1002CA was 'generated' *in situ* by filament winding. The only significant difference from the multi-layer electrical coil was that the wire was pre-crimped. The technology was no-doubt routine to Philips the mass-manufacturer of electrical components, but in the context of a one-off design for experimental evaluation does not appeal.

The performance characteristics of the regenerators of the GPU-3, V-160, P-40, etc, could no doubt be mimicked by forming stacks of 'washers' punched from woven wire screens. The high percentage wastage of an already expensive stock material cannot be contemplated and could be substantially reduced if the washers could be built up by butting short arcs edge-to-edge. This replaces one problem with another – that of packing the arc elements without overlap or excessive gap. Either way, the annular matrix formed by punching from wire gauze does not appear compatible with the eventual batch- or mass-produced engine.

A 'bandage' of wire gauze wound in place, as in the WhisperGen engine, achieves more promising utilization. On the other hand:

- For given gauze geometry (volume porosity, hydraulic radius) pressure drop/unit length (which in this case is the plane of the weave) exceeds that in the 'conventional' direction by a substantial factor (Magara *et al.* 2000).
- An individual wire is a conduction path between cycle temperature limits. Corresponding thermal short is that of a solid annulus of parent wire material having a cross-sectional area equal to $(1 - \P_v)$ of that of the annular housing, where $\P_v$ is volume porosity of the gauze bandage.
- Heat transfer data for first-principles evaluation of this otherwise straightforward method of production are not available.

Alternatively, the bandage can be formed from alternate layers of plain and corrugated stainless-steel foil (the combination is equivalent to the commercial flame-trap). Flow and heat transfer characteristics of the resulting individual flow passage are nominally those for which data are available. There are hearsay accounts that the MP1002CA ran with increased power when the filament-wound regenerator was substituted in this way – but with reduced efficiency. The change in performance is consistent with the substantially lower friction factor C_f for the parallel flow passage, and the larger hydraulic radius r_h of available flame-trap materials.

The characteristics and disadvantages of the flame trap extend to the dimpled-foil bandage:

- uninterrupted conduction path and thermal short
- anisotropy: there is no radial component of flow – non-uniformities of flow at inlet persist through the regenerator. The favourable 'slab-flow' feature of the more isotropic gauze bandage is not possible. Temperature gradients perpendicular to the flow direction are therefore inevitable, as are associated irreversibilities and efficiency penalties.

A candidate matrix material with some attractions is the metal foam. An annulus of, say, 10 mm height can be produced quickly and accurately by water-jet cutting. Foams from a given manufacturer tend to share common volume porosity $¶_v$ – and thus common free-flow area A_{ff} for given frontal area, but are available in a range of pore sizes, and thus a range of hydraulic radii r_h. The possibility arises of stacking annuli of different pore size as an approximation to continuous grading. Wastage of stock material is as for the wire gauze option. On the other hand, the raw material cost is lower.

Practical criteria for an annular regenerator acceptable for present purposes now emerge:

- must permit flow perpendicular to the main flow direction (recall Chapter 11)
- low cost of parent matrix material
- ease of assembly into the annular housing
- availability in a range of hydraulic radii r_h
- interrupted thermal conduction path (in axial direction)
- r_h (preferably) independent of volume porosity $¶_v$ and of free-flow area A_{ff}
- potential for grading of r_h, $¶_v$ and A_{ff} along flow direction
- until more is known, flow-path elements should probably not have the convergent/divergent characteristic of the wire gauze (Chapters 12 and 14)

The conflicting 'desirable characteristics' listed by Walker in 1980 (Walker 1980, pp. 150–1) makes an interesting comparison:

- for maximum heat capacity – a large, solid matrix
- for minimum flow losses – a small, highly porous matrix
- for minimum dead space – a small, dense matrix
- for maximum heat transfer – a large, finely-divided matrix
- for minimum contamination* – a matrix with no obstruction.

That 'small' might minimize flow losses raises an interesting topic for discussion, but the issue here is how far things have moved away from armchair design to confront the realities of regenerator design.

A commercial product with potential to satisfy all the above criteria (including those of Walker) is *expanded metal mesh* – Expamet™. A commercial producer is the Expanded Metal Co. Ltd, whose product range includes filters for the automotive and chemical industries. Application to the regenerator probably calls for the minimum sheet thickness which can be processed. To see the potential benefit of exploring this it is necessary to return to the concept of *thermal capacity ratio*, N_{TCR}:

Back-of-the-envelope calculation suggests that N_{TCR} for the VDF-750(aS) will be about 900, and thus of similar order to that of the MP1002CA (876). These figures compare with 193 for the GPU-3; 87.8 for the USS P-40; 88 for the USS- V-160; 129 for the Allison PD-46 (and 166 for the remarkable Clapham 5cc engine). The high values for the air-charged engines is no mystery: the charge pressure, and thus density, and thus thermal capacity of the working fluid charge are a small fraction of those of the light-gas engines, while the fractional dead space occupied by the matrix is comparable in all cases.

The N_{TCR} value determines the minimum achievable cyclic temperature swing. If the N_{TCR} of the V-160 is optimal, then that of the VDF-750(aS) is excessive. A moment's thought reveals a number of 'knock-on' effects which would result from aiming for a reduction by reducing matrix envelope volume. An option with a minimum of side-effects would be to make the expanded metal from sheet stock formed by cladding an unsophisticated core material – possibly a material of poor conductivity and/or low specific heat – with a skin of high thermal conductivity and specific heat.

There are similarities with the 'Alclad' process for improving the corrosion resistance of aluminium-alloy sheet. Alternatively, a film of high-conductivity material could be deposited on the base material subsequent to its being expanded to mesh form.

---

*Not a consideration under this heading: there is no scope within regenerator design to remedy an engine producing internal debris.

16.6.2 Materials – preliminary considerations

Loaded ptfe materials, on which the legendary GPU-3 engine depended, appear to have been superseded in general commercial application by other polymers, some oil-impregnated, some not. Of these, some are available in the form of extruded bar for machining to specific requirement, others are supplied as flanged or parallel bushes to a range of stock sizes. For any chance of commercial potential it is necessary to take maximum advantage of the latter.

Many components – possibly the majority – will call for use of one of the many alloys of aluminium. This is evidently an area of confusion, compounded possibly by American usage of *aluminum* (*–num*) for both the element and the structural alloy, and the further quirk whereby usage in Canada (at least as explained to the author in Toronto) is that *–num* is the element, while *–nium* can mean any one of the engineering alloys.

Figure 16.5 shows stress–strain curves for aluminium alloys to a selection of BS/International specifications. With a yield point a factor of 20 below that of alloy 2014A in the TF condition (solution and precipitation treated), the properties of aluminium the element are of no interest for applications where any structural strength is called for. An anecdote illustrates the prevalence of the confusion:

> At one of the universities where the author had the privilege of working, an admirably painstaking system of setting, checking (first and second checking!) and marking engineering examination papers was followed up by a further monitored stage: separate teaching groups, individual members of which had not necessarily been examiners, would hold forensic discussions on the grades achieved by students against the background of individual subject papers, question-by-question.
>
> The opening assertion on a materials paper '*An aircraft is constructed of aluminium* . . .' cried out for comment. The author drew attention to *aluminium* (*–ium*) as a strange choice for aircraft construction, and was rounded on by members of the materials teaching group. It was of no avail to point out that this was the material of saucepans and of electrical conductors, and that no aircraft had *ever* been constructed of aluminium, but of *aluminium alloys*. The author lost the debate to the specialists, who have no doubt moved on to setting questions on space vehicles constructed of aluminium and 'plastic'.

The proposed detail drawings and machining files will specify material, condition (heat treatment etc) for *every* component – metallic or otherwise – as appropriate. To this stage general principles (only) apply. Thus, where components are specified as being machined from aluminium alloy, the material grade may be expected to be BS (British Standard) specifications L-168

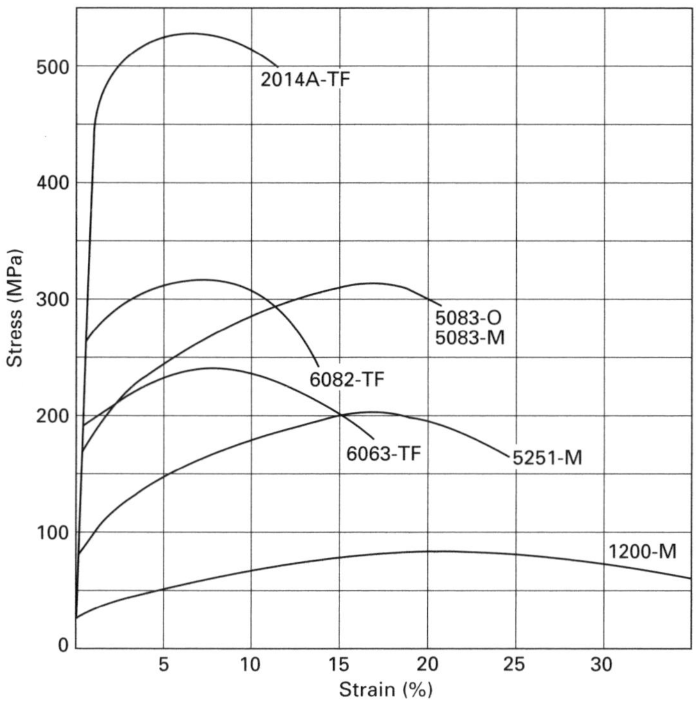

16.5 Stress–strain curves for a selection of engineering alloys of aluminium, revealing a factor of 20 in yield point between commercially-pure element Al (1200-M) and a high-duty grade (2014A in the TF condition). From ALFED (1981) with permission of the Aluminium Federation.

(replacement for L-65), HE-15 or 2014A. Achieving acceptable accuracy and surface finish in cheaper and more readily available alloys such as HE-30 (BS 6082) can compound rather than reduce net part cost. Different materials selection criteria may be expected to apply where a design advances from the one-off, experimental or prototype stage to quantity manufacture.

The kinematics of the drive linkage are implemented differently from the patent description and accompanying illustration (Appendix I): a shallow work piston carries a tubular extension running on, and guided at its far end by, a close-fitting cylindrical plug of oil-impregnated polymer material fixed rigidly to the crankcase with its axis coincident with that of the cylinder. Attaching the 'virtual gudgeon pin' to the piston/tube assembly as far as possible from the piston ensures that the latter takes a minimum of the already light side thrust arising from the small angularity of the connecting rods. The feature means that the piston does not require a skirt – and offers a saving of perhaps 65 mm in overall height of the engine. Bearer rings may now be redundant – or a single pair of rings may take both sealing and bearing functions, although the matter can be resolved only by experiment.

Elastic moduli of candidate polymer materials have yet to be checked, and the guide pillar may eventually be a polymer sleeve around a metal support post.

16.6.3 Crankshaft

The drawing shows the crankshaft as a 'one-piece' construction with integral semi-circular balance web, the axial thickness of which is, to this point, arbitrary. Achieving adequate strength in the crankshaft will almost certainly require use of a through-hardening steel. En-24 is a candidate (in the UK the En, or Emergency Number, designations remain more familiar than the code established to replace them – in this instance, Grade 817M40). It then becomes necessary to consider a two-piece* crankshaft with 'bolt-on' balance mass of, say, low-carbon steel.

A computer package is available which accepts values for the masses of all moving components of the lever/bell-crank combination, including a tentative value for balance mass-moment $m_b r_b$ (kgm) and displays as a polar plot the vector of the out-of-balance force referred to the plane of the cylinder axis. Figure 16.6(a) is a specimen polar. Minimization of the radius of the circle circumscribing the polar by trial-and-error variation of $m_b r_b$ gives the 'optimum' $m_b r_b$. 'Optimum' dynamic balance is then achieved by specifying the dimensions of the crank web outline (Fig. 16.6(b)) which, taken in conjunction with the density ρ (kg) of the web material, has precisely the optimum value of $m_b r_b$. The author's 1991 data book for mechanical design gives alternative crank web having readily-calculated mass-moment.

A hollow crank-pin eases the balancing problem slightly.

The bell-crank drives the crank-pin through an SKF BK2012 needle-roller bearing having basic load rating 8420 N and 'solid oil' lubrication. The outside diameter of crank-pin serves as inner race. The manufacturer specifies diametral tolerance and surface finish, together with surface hardness, which has to be achieved in supplementary heat treatment. The same considerations apply to the inboard main bearing, the unit shown being an SKF NU 2205 EC. Basic dynamic load rating (for life calculations) is 34100 N. The unit will be acquired with 'solid oil' lubrication.

The main journal arrangement owes much to the successful SM-1 test

* The Ryobi™ four-stroke petrol engine fitted to some strimmers (brush-cutters) offers an object lesson in crankshaft design: the web is a stack of laminations punched from sheet material and riveted together. Main journal and crank-pin are pressed into place under heavy interference.

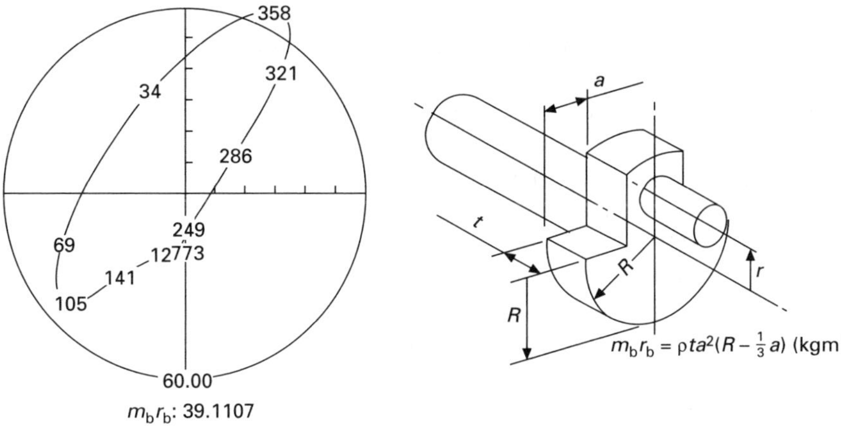

16.6 (a) Out-of-balance force vector referred to the cylinder axis. Polar plot produced by computer model of the entire assembly of moving components; (b) notation for defining balance mass moment $m_b r_b$ (kg/m) of crank web.

engine, detail design of which was in the hands of Rix (1984). An internal circlip provides axial location for the roller bearing outer race, and allows a simple, parallel bore for the housing. An SKF 6304 deep-groove ball race (again with 'solid oil' lubrication) at the out-board position provides axial location for the crankshaft. Basic dynamic load rating of the 6304 is 15,900N. Both inner races are locked up axially by a castellated nut inboard of the flywheel. The pinch force is transmitted through a spacer, through the inner rotating member of the pressure-balance seal and a flinger plate. The latter intercepts any leakage of oil across the seal against the pressure gradient. The flinger discharges into an annular groove, for which drainage provision will be required, bearing in mind that any accumulation of oil is under crankcase pressure.

A further, externally-threaded, internally castellated nut clamps the two outer races either side of a spacer and provides positive location for both against the circlip. The spacer serves to locate the fixed member of the pressure-balance seal.

Flywheel location is by a combination of taper and key (Woodruff). A conventional lip seal runs against the boss of the flywheel. As shown the keyway provides a leakage path – a matter which remains to be addressed.

16.6.4 Displacer assembly

The displacer shell is a length of stock, precision-drawn stainless-steel tube in one of the AISI grades – 304, 314, 316, etc., the choice probably being determined by availability in the seamless specification. The domed cap is of 'tori-spherical' form, and hydro-forming may be a suitable forming method. Subject to appropriate calculation, wall thickness is tentatively 1 mm for both shell and dome. The 4 mm deep, close-fitting ring shown should serve as a jig to hold dome and sleeve in alignment for electron-beam (e-b) welding, and remains in place after that process. The two radiation baffles may be secured by tacking with e-b welds, or by spot welding.

In the MP1002CA the mounting flange is of cast iron – possibly one of the many 'Meehanite' grades, and secured and sealed to the shell by brazing. In that engine the outer diameter of the flange ran against the iron cylinder liner, an arrangement relying on the migration of a small amount of lubricant past the work piston. The VDF-750(aS) is intended to run completely dry, so a bearer ring is necessary. In view of the small pressure differences across the displacer, the bearer ring doubles as seal. The flange should ideally be of the lightest possible material, but it remains to check candidate methods of joining dissimilar metals to the stainless-steel shell.

On completion of fabrication the internal diameter of the flange and its lower face are finish-machined with the shell running true.

A spigoted flange carries the displacer guide rod, and is attached to the lower flange by three equi-spaced, M5 button-head screws. A washer of insulating material (to be specified) is clamped between the flange faces and seals the displacer against cyclic entry and exit of working fluid.

The bell-crank drives the displacer via deep-groove bearing SKF 626 (basic dynamic load rating 1720N and 'solid oil' lubrication). The bearing is mounted in the bell-crank, while the clevis, integral with the displacer drive rod, secures the 6 mm diameter pin. The drive rod is nominally coaxial with the 10 mm-diameter support/seal tube, and has flats machined along the majority of its length, allowing the small lateral excursions of the bell-crank attachment point to be taken up by flexure. The expedient has a precedent in the corresponding component of the Philips MP1002CA.

16.6.5 Work piston assembly

Without practical experience of polymer alloy piston rings it is not possible to know whether satisfactory operation will result if they are sprung into the ring groove following practice for metal rings. Moreover, the possibility of a dual role – seal plus bearer – has still to be discussed with the ring manufacturer. Provisionally the piston is a two-part assembly allowing a two-ring stack to be located without the need for deformation. Subsequent to installation the two components of the piston are clamped together by tightening a slotted nut carried on the externally-threaded boss of the inner member.

The drawing shows the displacer rod guided on Igus™ proprietary flanged polymer bushes. Relying on the guide bushes to double as seals amounts to following Philips' practice for the MP1002CA. Somewhat better results might be expected now, since cyclic departures from straight-line motion at the displacer drive attachment point are less, and rocking forces on the piston are now essentially zero.

It is possible that sealing and/or bush wear may nevertheless determine the service interval. The 10 mm diameter proposed for the displacer guide rod is below the norm for oil-free compressor practice, but the suppliers of the polymer alloy rings to the compressor industry have expressed interest in machining trial sets of 'tangent' or 'penguin' seals. It will be necessary to revise the design of the piston to accommodate these.

With the exception of the rings, all piston components are machined from aluminium alloy BS 2014A (or equivalent), anodized and water-sealed or ptfe-sealed.

The guide tube is machined from extruded tube, and BS HE-30 or equivalent may have to be accepted. The external diameter of 30 mm and 2 mm wall thickness are stock sizes. For the purposes of getting the engine up and running the polymer guide post will run against the unmachined bore of the tube. It will eventually be necessary to explore hard anodizing and ptfe sealing the tube bore.

16.6.6 Crank case

External surfaces and internal cut-out are ruled outlines, permitting roughing-out by water jet cutting. A 100 mm thick plate provides a suitable blank. The water-jet process is virtually indifferent to material specification, but side and top faces have to be skimmed, the piece has to be bored to take the cylinder liner and there are internal threads to be tapped. The choice of material is therefore BS 2014A (TF condition).

The fulcrum for the lever pair which drive the piston is formed by a plug secured to the crank-case by bolting through an external flange. Shims (not shown) beneath allow precise horizontal location of piston and displacer attachment points.

16.6.7 Cylinder head

A preliminary feel for the required wall thickness can be had from the expression for the hoop (circumferential) stress σ_h induced in the wall of radial thickness t, of an infinitely-long cylinder of mean diameter D by constant internal/external pressure difference Δp, namely, $\sigma_h = \Delta pD/2t$. Inverting to make explicit thickness-to diameter ratio t/D:

$$t/D = \Delta p/\sigma_h \qquad 16.4$$

The numerical value to be used for σ_h is that for the creep-to-rupture value for the candidate cylinder material. Figure 16.7 reproduces relevant data published by International Nickel Ltd (Anon. 1966). The curves, originally acquired by Simmons and Cross (1952), apply to AISI grade 304. The data are in imperial units, but both sides of Eqn 16.4 are dimensionless, so the the LHS may be evaluated as a ratio of lengths in mm, the right as a ratio of stresses in lbf/in^2. The proposed charge pressure of 15 atm. converts to Δp of $(15 - 1) \times 14.7$ lbf/in^2. Shell diameter D is estimated as cylinder bore plus twice the radial extent of the internal fins – say $60 + (2 \times 10) = 80$ mm. Provisionally allowing 2,000 hours for creep-to-rupture at 730°C, σ_h is read off as 6500 lbf/in^2, giving t/D from Eqn 16.4 as $14.0 \times 14.7/6500 \approx 0.03166$. The 80 mm diameter suggests a wall thickness t of $0.03166 \times 80.0 \approx 2.533$ mm.

Half-a-century on, stainless steel to AISI 304 remains commonly-available, and can be machined using conventional facilities. On the other hand, it is no longer 'state-of-the-art' for high temperature duty. At 760°C, the 30,000-hour stress-to-rupture of Incoloy™ 802 is about 1.6-times the 2000-hour value for AISI 304, offering scope for a reduction in wall thickness in the same ratio. Whether this offers a net advantage depends on the relative thermal conductivities, raw material costs and machining costs.

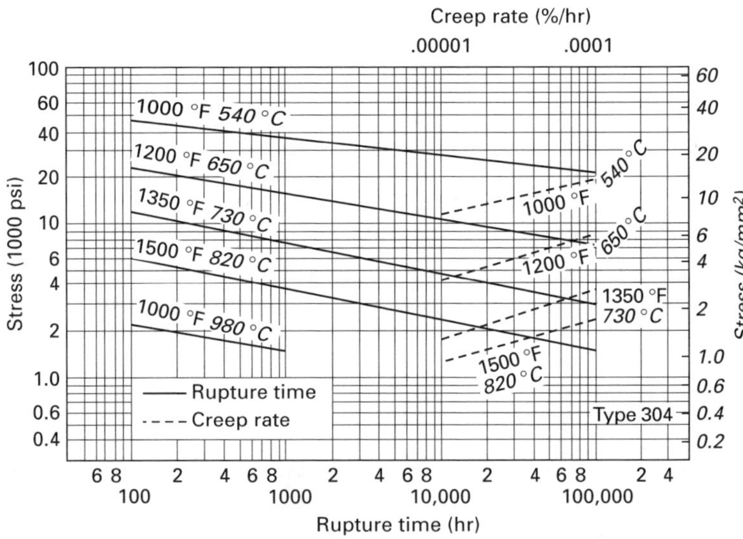

16.7 Creep-to-rupture characteristic of stainless steel to AISI specification 304. Reproduced with permission of ASTM and with approval of Inco Ltd from data acquired by Simmons and Cross (1952) and subsequently incorporated (with acknowledgement) in Inco Ltd (1966).

16.7 Pressure-balanced seal

The Philips MP1002CA and the SM-1 research machine (Rix 1984, Organ 1992) incorporated pressure-balance seals, the former integrated into the crankshaft design, the latter a self-contained cartridge of proprietary manufacture.

Figure 16.8 shows the balance seal schematically. Sleeve 1 is pressed under nominal axial force F_a by a spring against face 2 perpendicular to the axis of the shaft and rotating with it. The seal between the bearing housing and sleeve 1 allows axial movement without significantly affecting the axial load. Upstream pressure p_u acting on effective area A_1 results in net axial closing force of $p_u A_1 + F_a$.

Any flow is down the pressure gradient and radially *inwards*. It is possible to imagine the mating faces sufficiently plane and smooth that the microscopic leakage flow is in the form of a film sufficient to prevent solid-to-solid contact during relative rotation. Discounting centrifugal and coriolis effects within the fluid layer, and using p_d to denote downstream pressure, mean pressure at the seal face A_2 is $\frac{1}{2}(p_u + p_d)$. A force balance for the sliding sleeve is:

$$p_u A_1 + F_a = \tfrac{1}{2} A_2 (p_u + p_d) + p_d (A_1 - A_2) \qquad 16.5$$

Below the value of F_a for which Eqn 16.5 balances, zero net closing force is predicted between the seal faces. A higher value may be expected to keep the seal closed. The service value of F_a is chosen by experiment, and is the minimum at which satisfactory sealing is demonstrated.

The centrifugal effect promises to enhance seal effectiveness, since acceleration of the gas in the gap is radially inwards, requiring a radially-outwards pressure gradient component opposed to the leakage direction.

The pressure-balanced seal of the MP1002CA was lubricated and cooled

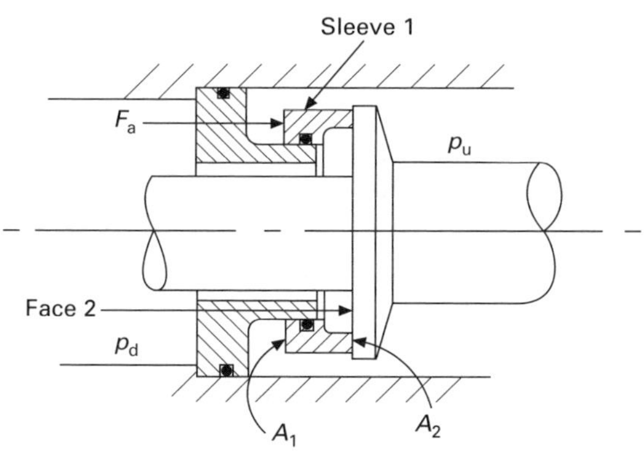

16.8 Notation for analysis of balanced seal.

by oil supplied upstream (on the high-pressure side). Certain proprietary units are rated for dry operation – i.e., for gas lubrication. On the basis of consultation with the manufacturer (Sealol Ltd, now part of John Crane Ltd), the seal cartridge of the SM-1, which otherwise ran dry, was oil-lubricated from the downstream side. The issue is cooling, and on trouble-free experience of the SM-1 it is proposed to lubricate the VDF-750(aS) by a downstream supply.

16.8 Beyond 2006

Two centuries of start-stop development have demonstrated convincingly that there is a small fortune to be made out of Stirling engines. And how was it done? You started with a large one. But things have changed: never has there been a market opportunity on the scale of that now offered by the need for widespread implementation of domestic CHP. The internet removes all constraints on access to relevant design expertise, to data, and to manufacturing facilities.

Dr James Hansen is director of the NASA Goddard Institute for Space Studies. In 1988, in testimony to the US Congress, he was one of the first scientists to warn of climate change. In an interview with *The Independent* newspaper on 1st January 2007, Hansen gives less than 10 years for CO_2 emissions to be curbed if global warming is not to run out of control and to change the landscape for ever: 'We just cannot burn all the fossil fuels in the ground. If we do, we will end up with a different planet. I mean a planet with no ice in the Arctic, and a planet where warming is so large that it's going to have a large effect in terms of sea level rises and the extinction of species.'

The scientific community is not in unanimous agreement that the phenomenon is due to CO_2 emissions. It is not yet conclusively proven that the problem is man-made. But man is, *de facto*, custodian of the planet, and to fail to explore man-made solutions is hardly a logical recipe for survival.

This book has drawn attention to recent developments having potential to speed the introduction of Stirling-powered domestic CHP. These do not add up to a magic formula for instantly raising engine reliability and efficiency. When, during the Second World War, Marshall of Cambridge received battle-damaged DC-3 transport aircraft of an evening and returned them to service two days later (Marshall (1994)), they were not following a formula prepared by the Air Ministry in anticipation of conflict. The unprecedented rate of turn-round was achieved by doing it – and each time doing it more slickly. The response was, in fact, no different from that called for by any engineering challenge when the chips are down.

The chips are now down, the stakes have never been higher, and the

Stirling engine has emerged as a valuable card. Not since the mid-1960s have conditions been so favourable for obtaining financial support for energy-related research. What has changed since that optimistic era of Prime Minister Wilson's 'white heat of technology' is that the means to co-ordinating expertise and know-how which the Stirling engine uniquely requires is now waiting on every desk and in every workshop.

Appendix I

Draft Patent Specification – Improvements in or relating to Stirling engines and 'hot-air' engines

Inventor: Allan John Organ

This invention relates to Stirling, hot-air and similar energy converters in which synchronization of work piston and displacer motions is achieved by means of a planar kinematic linkage.

The Stirling cycle machine is a thermodynamic device capable of functioning as a prime mover by converting heat to work, or as a refrigerator or heat pump by applying work to the lifting of heat from one temperature and delivering it at a generally different temperature. This account will proceed by reference to the prime mover variant as an example, but will be understood to imply the heat pump variant equally.

In order for the Stirling engine to offer commercial advantage over established prime movers such as the internal combustion engine the costs of its construction must be as low as possible and the power which it affords per unit of its size and mass should be as great as possible.

Many energy converters working on the Stirling principle are of so-called single-cylinder, or single gas-circuit type, having both principal moving components, namely, work piston and displacer, operating coaxially in a common cylinder. The arrangement offers the possibility of minimizing unswept volume, and correspondingly of maximizing compression ratio and, ultimately, of maximizing thermodynamic work converted per unit mass of working fluid. There remains, however, the problem of converting that thermodynamic work efficiently to mechanical work at the output shaft. This problem is most critical in the so-called 'hot-air' variant, which tends to operate at atmospheric pressure, but whose constructional simplicity is otherwise a commercial attraction.

At a conceptual level, the problem is minimized if the lateral loads on members subject to sliding are minimized. In practical terms, this means that gas pressure and inertia forces arising at the work piston should be transmitted via a connection making the minimum angle with the direction of motion of that piston. A similar criterion applies to displacer actuation.

Previous proposals for achieving this end have included a bell-crank mechanism, which relieves the side load on the displacer, but not that on the piston. They also include the patented 'rhombic drive', which eliminates side loads on both sliding members, but at the penalty of intricacy, bulk and high manufacturing cost. The patented Ross drive mechanism reduces side loads except those due to friction torque of the residual connecting rod angularity at the gudgeon pins, but is restricted to Stirling machines of 'opposed-piston' configuration. 'Free-piston' variants dispense completely with the offending lateral load components, but introduce substitute problems which have so far militated against their making a commercial impact.

It is accordingly an object of the present invention to provide a combination of cylinder, work piston, displacer and arrangement of linkages whereby actuation of both work piston and displacer along their respective axes, or along their common axis, is achieved through connecting links subject to minimal angularity relative to the axes or axis in question.

According to the present invention a Stirling engine or hot-air engine embodies a crankshaft conventional to the extent of having a crank-pin offset from the main journal axis by a radial distance. The axis of that crankshaft is however disposed to one side of the cylinder axis. The crank-pin drives a lever via one arm of a coupler, which may conveniently have the form of a bell-crank. The lever is constrained to arcuate motion by being linked near one of its extremities – the distal end – to the engine frame. Those skilled in the art of planar mechanism analysis will recognize the combination of crank throw, bell-crank arm, lever and engine frame as a four-bar linkage and will thus anticipate the possibility of identifying a point on the coupler the locus of whose motion as the crank-pin executes a complete revolution is approximately a straight line. The distal end of the lever moves through an arc, but by an amplitude in the chordal direction which is small in relation to the radial distance, its motion thus deviating only slightly from a straight line. By expeditious choice of proportions of the mechanism the near-linear locus of the point on the coupler is arranged to superimpose substantially over the near-linear arc of motion of the point on the lever. The second arm of the coupler or bell-crank being so angled relative to that of the first driving arm as to turn the input motion substantially through 90° or so, this angular phase difference is reflected in the two superimposed motions, which are accordingly suitable for actuation of the piston and displacer of a Stirling cycle machine.

The invention will now be described in greater detail by way of an example with reference to the accompanying drawing.

Figure AI.1 is a sectional elevation view through a hot-air engine representing a preferred embodiment of the invention, which view includes displacer, work piston, part of an enclosing cylinder and assembly of connecting rods and coupler which together comprise the invention. The flywheel and usual

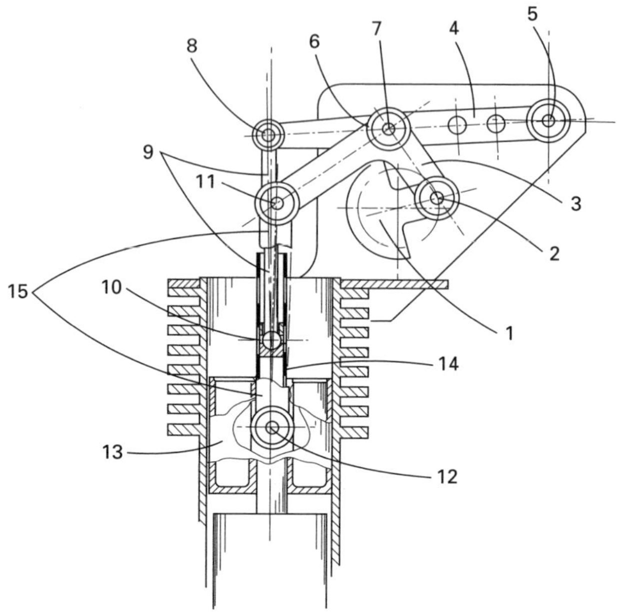

AI.1 Drawing submitted with the patent application.

heating and cooling provisions essential for operation as a prime mover are omitted in the interests of clarity. For the same purpose, some components are partially cut away.

Referring to the drawing, the crank shaft (1) lies to one side of the cylinder axis, carrying the crank-pin (2) on which the coupler (3) is free to rotate in or parallel to the plane of motion of the crank-pin. The lever (4) attaches to the engine frame at 5 in such a way as to permit rotation in or parallel to the plane of motion of crank-pin. Crank throw, coupler, part of the lever and engine frame form, in combination, the four-bar linkage anticipated above. The heel (6) of the coupler (3) is pivoted at some intermediate point (7) of the lever (4). Rotary motion of the crank causes attachment at point 8 on the lever to execute arcuate motion about the mounting point (5). Mounting point 8 on the lever lies substantially over the cylinder axis, and thus substantially over the axis of motion of piston and displacer. A connecting rod (9) links pivot point 8 of the lever to a pivot point 10 on or within the displacer drive plunger (14), which pivot point is conveniently a spherical joint. A connecting rod or link (15) connects attachment point 11 on the coupler pivotally to the gudgeon pin (12) on the work piston (13). A complete revolution of the crankshaft (1) in anti-clockwise direction causes piston and displacer to induce a cycle of volume variations within the cylinder consistent with operation of the complete device as a prime mover.

The preferred embodiment allows the flexibility of interchanging the attachment points of work piston and displacer, the former instead now driving point 8, the latter being driven by point 11, in which case functioning as a prime mover calls for clockwise rotation of the crankshaft.

I do not claim the operating principle of the Stirling or hot-air engine when separately considered; neither do I claim the using of a four-bar linkage when separately considered, but my invention relates to the combined using of Stirling or hot-air engine with inversions of the four-bar linkage as described above.

What I claim is...

1. A Stirling or hot-air machine comprising the usual cylinders work piston and displacer coaxially or substantially coaxially disposed together with mechanism for actuating and synchronizing said work piston and displacer and comprising a crankshaft with axis lying to one side of the cylinder axis, a coupler and a lever. A point on the coupler is pivotally connected to the crank-pin while the heel of the coupler is pivotally connected to a point on the lever intermediate between its extremities. Adjacent to one of its extremities, the lever is pivotally attached to the engine frame. The proportions of the various members and the points of connection are so chosen that a point on the coupler and a point on the lever execute respective motions lying substantially along the axes of motion of the reciprocating engine components with which they cooperate, namely, of work piston and displacer, and so chosen also that the points in question attain their extremes of motion with an angular separation referred to crank-pin rotation corresponding to the angular amount required for satisfactory actuation of the combined system as a Stirling cycle or hot-air machine.
2. A machine according to Claim 1 above, in which the lever actuates the displacer, and the piston drives through the coupler.
3. A machine according to Claim 1 above, in which the coupler actuates the displacer, and the work piston drives through the lever in the converse sense to that of Claim 2.
4. A machine according to Claim 2 above, in which connection between piston and coupler or between displacer rod and lever is by a connecting rod or where both connections are achieved by connecting rods.
5. A machine according to Claim 3 above, in which connection between piston and lever or between displacer rod and coupler is by connecting rod or where both connections are achieved by connecting rods.
6. A machine according to all or any of the foregoing Claims in which piston and displacer are disposed on the side of the crankshaft opposite to the lever.

7. A machine according to all or any of the foregoing Claims in which piston and displacer are disposed on the same side of the crankshaft as the lever.
8. A machine according to all or any of the foregoing claims in which the coupler is in the form of a bell-crank.
9. A machine according to all or any of the foregoing claims in which the point of connection between the cylindrical drive rod which directly actuates the displacer body and its associated connecting rod is made internally to said cylindrical drive rod.
10. A machine according to Claim 8 above in which the internal connection is in the form of a joint capable of spherical motion.

Abstract

A coaxial Stirling or 'hot-air' engine has a crankshaft (1) whose axis lies to one side of the axes of translation of the usual work piston and displacer. A lever (4) is pivoted to the engine frame adjacent one of its extremities. Point 8 is chosen adjacent the opposite extremity so that its motion resulting from rotation of the crankshaft (1) superimposes substantially over the axis or axes of translation of piston and displacer. Lever 4 is actuated from the crank-pin (2) via one arm of a coupler or bell crank (3). The complementary arm is angled relative to the first, so that a point (11) adjacent its end may be found whose motion resulting from rotation of the crankshaft (1) also superimposes substantially over the axis or axes of translation of piston and displacer. A connecting rod links point 8 on the lever to a point 10 on or within displacer drive plunger (14), and a further connecting rod links point 11 on the coupler to the piston at its gudgeon pin (12). The motions of piston and displacer on actuation of the crankshaft (1) are consistent with operation of the device as a prime mover or heat pump, depending upon the direction of crankshaft rotation.

Appendix II

Crank mechanism kinematics

In Fig. AII.1 the crankshaft rotates clockwise about point *O*, driving crank-pin at point *P* which is offset from the crankshaft axis by amount *r*. Crank-angle ϕ is measured from the upper vertical axis. Lever *S–R–Q* is pivoted at ground-point *Q*. Bell crank *P–R–T* pivots at crank-pin *P* and at point *R* on the lever. Point *S* may drive the piston and *T* the displacer – or *T* the piston and *S* the displacer, reversing the direction in which power-production is positive.

By application of Pythagoras' theorem:

$$c/r = \sqrt{[(X/r)^2 + (Y/r)^2]}$$

Fixed angle ξ is expressed in terms of the horizontal and vertical coordinates of *Q* relative to *O*:

$$\xi = \text{atan}(X/Y)$$

Point *S* is uppermost at the value of ϕ for which *O–P–R* (in that order) form a straight line, and lowermost when *P–O–R* (in that order) are aligned. These conditions determine the dead-centre (inner/outer) positions of the member (piston or displacer) connected at *S*.

By the cosine formula applied to triangle *O–R–Q* with *O–P–R* in alignment:

$$\psi_{\text{TDCs}} = \text{acos}\{\tfrac{1}{2}[RQ^2 + C^2 - (r + H)^2]/(RQ + C)\}$$

$$= \text{acos}\{\tfrac{1}{2}[(RQ/r)^2 + (C/r)^2 - (1 + H/r)^2]/(RQ/r + C/r)\}$$

Similarly:

$$\psi_{\text{BDCs}} = \text{acos}\{\tfrac{1}{2}[(RQ/r)^2 + (C/r)^2 - (1 - H/r)^2]/(RQ/r + C/r)\}$$

The dead-centre values of angle γ follow by simple subtraction:

$$\gamma_{\text{TDCs}} = \psi_{\text{TDCs}} - \xi$$

$$\gamma_{\text{BDCs}} = \psi_{\text{BDCs}} - \xi$$

With angle ϕ measured from the uppermost position in Fig. AII.1

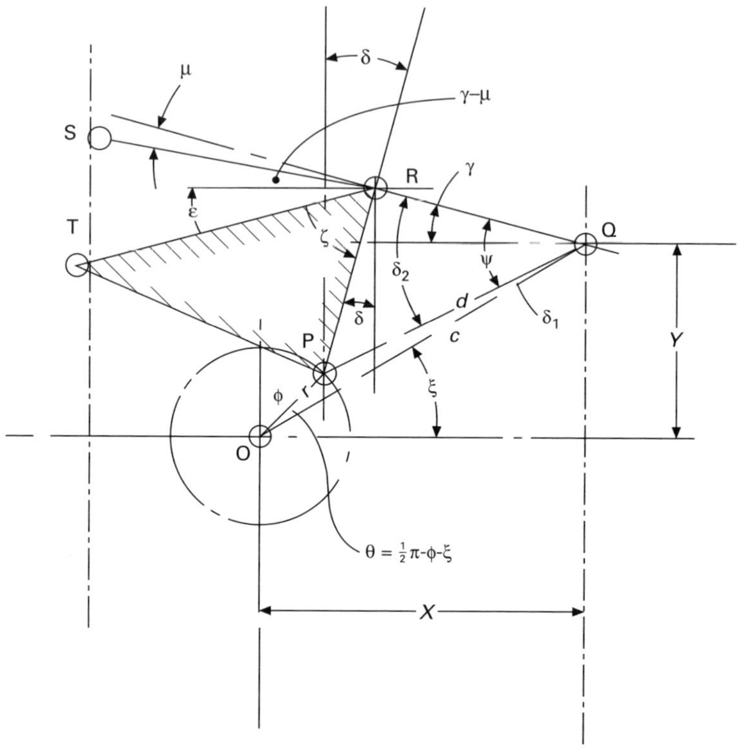

AII.1 Notation for drive mechanism kinematics.

$$\theta_{anticlock} = \pi/2 + \phi - \xi$$

$$\theta_{clockwise} = \pi/2 - \phi - \xi$$

By a further application of the cosine formula:

$$d^2 = c^2 + r^2 - 2cr\cos\theta$$

$$d/r = \sqrt{(c/r + 1 - 2c\cos\theta)}$$

An expression for auxiliary angle δ_1 may be found by considering the right-angled triangle formed by sides P–Q, $r\sin\theta$ and $c - r\cos\theta$:

$$\delta_1 = \text{atan}[r\sin\theta/(c - r\cos\theta)]$$

$$= \text{atan}[\sin\theta/(c/r - \cos\theta)]$$

Variable angle ψ is the sum of angles δ_1 and δ_2. The latter follows from a further application of the cosine formula to triangle formed by sides H, R–Q and d:

$$H^2 = RQ^2 + d^2 - 2dRQ\cos\delta_2$$

$$\delta_2 = \text{acos}\{[\tfrac{1}{2}(RQ/r)^2 + (d/r)^2 - (H/r)^2]/(d/r \cdot RQ/r)\}$$

$$\psi = \delta_1 + \delta_2$$

Vertical and horizontal excursions of point S are available in terms of angle γ:

$$\gamma = \psi - \xi$$

$$x_R/r = e/r + X/r - Q\cos\gamma$$

$$y_R/r = Y/r - Q\sin\gamma$$

$$x_S/r = x_R/r - RS/r\cos(\gamma - \eta)$$

$$y_S/r = y_R/r + RS/r\sin(\gamma - \eta)$$

The considerable kinematic versatility of the mechanism is appreciated on noting that rotation may be clockwise or anticlockwise (with different volume variations in each case); that the piston may be connected at T, the displacer at S – or the other way around, again with different respective variations of volume with ϕ; and, moreover, that the cylinder may be situated *below* the mechanism ('inverted' relative to the usual orientation) or *above*.

Let $z = X - RQ\cos\gamma + r\sin\phi$ (anticlockwise rotation)

$= r\sin\phi - (X - RQ\cos\gamma)$ (clockwise rotation)

$$\delta = \text{asin}(X \cdot z/H)$$

$$\zeta - \varepsilon + \delta + \pi/2 = \pi$$

Angular inclination ε of side R–T to the horizontal is

$\varepsilon = \zeta + \delta - \pi/2$ (anti-clockwise rotation)

$= \pi/2 - \zeta - \delta_I$ (clockwise rotation)

Recalling that horizontal distances originate at point O

$$x_T/r = x_R/r - (RT/r)\cos\varepsilon$$

$$y_T/r = y_R/r - (RT/r)\sin\varepsilon$$

Appendix III

Equilibrium or 'temperature-determined' picture of thermal lag engine

If the time-dependence of the gas processes is overlooked, the pulse-tube cycle degenerates into compression and expansion curves which superimpose and which yield zero net indicated work. The algebra is essentially Finkelstein's reworking (1960) of the original approach by Schmidt (1871).

The total mass content M of the gas path is assumed constant, so that $M = p/R\Sigma(V/T)$:

$$M = p/R\{V_{dxe}/T_E + V_{dr}\,f(N_T)/T_C + V_{pt}/f(N_T)/T_C$$
$$+ V_{dxc}/T_C + V_C(\omega t)/T_C\}$$

$$f(N_T) = (\log_e N_T)/(N_T - 1)$$

$$N_T = T_E/T_C$$

$$M = pV_{sw}/RT_C\{\nu + V_C(\omega t)/V_{sw}\} \qquad \text{AIII.1}$$

$$\nu = \{(V_{dr} + V_{pt})f(N_T) + V_{dxe}/N_T + V_{dxc}\}/V_{sw}$$

The equilibrium cycle is not time-dependent, so crank angle ϕ may replace the independent variable ωt. Eqn AIII.1 is valid at $t = 0$ when $V_C(0) = V_{sw}$, so that M is also given by $M = p_{min}V_{sw}(\nu + 1)/RT_C$, where p_{min} is the outer dead-centre value of p. Dividing the two expressions for M:

$$p/p_{min} = (\nu + 1)/(\nu + \mu(\phi)) \qquad \text{AIII.2}$$

$$\mu(\phi) = V_C(\phi)/V_{sw}$$

Differentiating with respect to crank angle ϕ:

$$D(p/p_{min})/(p/p_{min}) = Dp/p = -(\nu + 1)d\mu(\phi)/(\nu + \mu(\phi))^2 \qquad \text{AIII.3}$$

Appendix IV

Tribal wisdom

Wisdom of the Dakota Indians (so legend has it), passed down from generation to generation, says that, 'When you discover that you are riding a dead horse, the best strategy is to dismount.'

However, in modern government, education and the corporate world, more advanced strategies are often employed, e.g.:

1. Buying a stronger whip.
2. Restructuring the dead horse's reward scale to contain a performance-related element.
3. Scrutinizing and clawing back a percentage of the horse's past 12 months' expenses payments.
4. Setting up a working party to study the horse.
5. Arranging to visit other countries to see how other cultures ride dead horses.
6. Convening a dead horse productivity improvement workshop.
7. Lowering standards so that dead horses can be included.
8. Reclassifying the dead horse as living-impaired.
9. Hiring outside contractors to ride the dead horse.
10. Outsourcing the management of the dead horse.
11. Harnessing several dead horses together to increase speed.
12. Providing additional funding and/or training to increase dead horse's performance.
13. Carrying out a productivity study to see if lighter riders would improve the dead horse's performance.
14. Declaring that, as the dead horse does not have to be fed it is less costly, carries lower overhead and therefore contributes substantially more to the bottom line of the economy than do live horses.
15. Rewriting the expected performance requirements for all horses.

Most effective of all:

16. Promoting the dead horse to a supervisory position.

References and bibliography

Adler A A (1946) Variation with Mach number of static and total pressures through various screens. National Advisory Committee for Aeronautics (NACA) Wartime Report # L5F28 (Feb.).

Alfed (1981) *The Properties of Aluminium and its Alloys*, 8th edn (amended) Aluminium Federation (ALFED) Broadway House, Birmingham B15 1TM (address at 1981).

Annand W J D (1963) Heat transfer in the cylinders of reciprocating internal combustion engines. *Proceedings Institution of Mechanical Engineers.* **177** (36) pp. 973–990.

Anon. (undated, a) NHS Estates Health Technical Memorandum HTM 2025 Ventilation in health care premises. Pt 2 Design considerations, ISBN 0113217528.

Anon. (undated, b) SIMQX – appears in documentation for IBM 1130 Scientific Subroutine Library.

Anon. (1966) *Mechanical and Physical Properties of the Austenitic Chromium–Nickel Stainless Steels at Elevated Temperatures.* International Nickel Ltd, Thames House, Millbank, London (address at 1966).

Archibald J (2006) Communication with the author by e-mail.

Barree R D and Conway M W (2004) Beyond beta factors: a complete model for Darcy, Forchheimer and trans-Forchheimer flow in porous media. Paper 89325. Proceedings Annual Technical Conf., Society Petroleum Engineers.

Bird R B, Stewart W E and Lightfoot E N (1960) *Transport Phenomena.* John Wiley and Sons, New York.

Bomford N (2004) Personal communication (3 Jan).

Boomsma K, *et al.* (2003) Simulations of flow through open cell metal foams using an idealized periodic cell structure. *International Journal of Heat and Fluid Flow.* **24** pp. 825–834.

Bopp G and Co. (undated) Woven wire cloth. Document G 0100/02, G. Bopp AG, Zurich.

Bowden C E (1947) *Diesel Model Engines.* Percival Marshall, London.

Bryson W (2003) *A Short History of Nearly Everything* Doubleday, London.

Bumby J R and Martin R (2005) Axial flux permanent magnet gernators for engine integration. Proceedings 12th International Stirling Engine Conference, University of Durham, pp. 463–471.

Cambell Tools Co. (2005) Website of Campbell Tools Company, Springfield, OH 45504 USA.

Chen Z *et al.* (2001) Derivation of the Forchheimer law *via* homgenization. *Transport in Porous Media.* **44** (2) (Aug.) pp. 325–335.

CNW (undated) Precision engineering of stainless steel small tubes. Product catalogue of Coopers Needle Works Ltd, 261–5 Aston Lane, Birmingham B20 3HS.

Coppage J E (1952) The heat transfer and flow friction characteristics of porous media. PhD diss., Department of Mechanical Engineering, Stanford University.

Coppage J E and London A L (1956) Heat transfer and flow friction characteristics of porous media. *Chemical Engineering Progress*, **52** (2) pp. 57F–63F.

CSC (undated) Illustrated catalogue of ceramic products. Ceramic Substrates and Components Ltd, Lukely Works, Carisbrooke Road, Newport, Isle of Wight.

Dando R (1997) Selected details of CRE Stirling engine supplied by personal communication.

Darcy H (1856) *Les Fontaines Publiques de la Ville de Dijon*, Dalmont, Paris.

de Brey H, Rinia H and van Weenen F L (1947) Fundamentals for the development of the Philips air engine. *Philips Technical Review* **9** (4) pp. 97–124.

Denney D (2005) Beyond beta factors: a model for Darcy, Forchheimer and trans-Forchheimer flow in porous media. Highlights for *JPT-online* of paper of similar title by Barree R D and Conway M W for Society Petroleum Engineers (SPE) annual technical conference and exhibition, Houston, 26–29 September 2004.

Devois J F and Durastanti J F (1996) Modelling of a spiral plate heat exchanger by a finite difference method. Proceeding of Conference: *New Developments in Heat Exchangers*, Eds. N Afgan *et al.* ISBN 905699512X Part 5 (Plate-type exchangers) Gordon and Breach, Amsterdam pp. 377–86.

Dunlop Co. plc (undated) Retimet® metal foam. Manufacturer's sales brochure, Dunlop Ltd. Aviation Division – Equipment, Holbrook Lane, Coventry CV8 4QY.

Dunn A R (1980) Selection of wire cloth for filtration/separation. Reprinted from *Filtration and Separation*, Sept./Oct. issue (pages of re-print un-numbered.)

Ergun S (1952) Flow through packed columns. *Chemical Engineering Progress* **48**, pp. 89–94.

Faires V M (1969) *Design of Machine Elements*. Collier-MacMillan, New York.

Feulner P (2003, 2005) Personal communications with author.

Finkelstein T (1960) Optimization of phase angle and volume ratio for Stirling engines. Paper 118C Proceedings, SAE Annual Meeting, Detroit, Michigan, 11–15 Jan.

Finkelstein T and Organ A J (2001) *Air Engines*. Originally published by Professional Engineering Publishing, London and Bury St Edmunds. (2005 onwards: John Wiley and Sons, Chichester).

Galloway E (1830) *History and Progress of the Steam Engine (with a Practical Investigation of the Structure and Applications)*, p. 667, (printed for) Thomas Kelly, London.

Gifford W E and Longsworth R C (1964) Pulse-tube refrigeration. *Transactions of the American Society of Mechanical Engineers (ASME)* Aug. pp. 264–268.

Granta Design (2006) Web site http://www/grantadesign.com.solutions.

Gunston W (1998) *The Life of Sir Roy Fedden*. Rolls Royce Heritage Trust.

Hargis A M and Beckmann A T (1967) Applications of spiral plate heat exchangers. *Chemical Engineering Progress*. **63** (7) pp. 62–67.

Hargreaves C M (2001) *The Philips Stirling Engine*. Elsevier Science, Amsterdam.

Harness J B and Newmann P E L (1972) A theoretical solution of the shuttle heat transfer problem. *Proceedings of International Conference Cryogenic Engineering*. **16** pp. 97–100.

Harvey J P (2003) Oscillatory compressible flow and heat transfer in porous media – application to cryo-cooler regenerators. PhD diss. Georgia Institute of Technology.

Hausen H (1929) Waermeaustausch in Regeneratoren. *Zeitschrift des Vereins deutscher Ingenieure* **73** p. 432.

Haywood R W (1990) *Thermodynamic Tables in SI (Metric) Units*, 3rd edn. Cambridge University Press, Cambridge.

Heaton H S, Reynolds W C and Kays W M (1964) Heat transfer in annular passages. Simultaneous development of velocity and temperature fields in laminar flow. *International Journal of Heat Mass Transfer*. **7** pp. 763–781.

Hislop D (2006) Personal communications (e-mail and telephone) throughout the year.

Hodgman C D (ed.) (1954) *Mathematical Tables from Handbook of Chemistry and Physics*. Chemical Rubber Publishing Co., Cleveland, Ohio (5th printing 1957).

Igus UK (2006) Technical sales information on polymer bearing bushes and related products (web-site readily located under Igus).

IMI/Amal (undated) Flame arresters. Illustrated trade catalogue published by Amal, a division of IMI, Holdford Rd, Witton, Birmingham B6 7ES.

Inco Ltd (1966) Mechanical and physical properties of the austenitic chromium–nickel stainless steels at elevated temperatures. Inco Alloys International, Hereford UK.

Jakob M (1957) *Heat Transfer II*. Wiley, New York (also Chapman and Hall, London).

Jenkin H C F (1885) Gas and caloric engines. Lectures delivered to the Institution of Civil Engineers on 21 February 1884. Published in *Heat and its Mechanical Applications*, Institution of Civil Engineers, London.

Johnson R C (1971) *Mechanical Design Synthesis with Optimization Applications*. Van Nostrand Reinhold, New York.

Kays W M and London A L (1964) *Compact Heat Exchangers*. McGraw Hill, New York.

Kays W M and London A L (1998) *Compact Heat Exchangers*, 3rd edn. Krieger, Malabar, Florida.

Kennedy R (1904-5) *Modern Engines and Power Generators*. Vol II (of six) Caxton, London.

Khayargoli P *et al*. (2004) The impact of microstructure on the permeability of metal foams. Proceedings of CSME Forum (Canadian Society for Mechanical Engineering) pp. 220–228.

Kolin I (1999) *Motive Power of Fire: three centuries of heat engines*. Technical Museum, Zagreb.

Laidlaw-Dickson D J (1946–7) *Model Diesels*. Harborough Publishers, Leicester.

Larque I (1998) Personal communication.

Larque I (2002) Design and test of a self-pressurized high compression-ratio Stirling air engine. Proc. European StirlingForum, Osnabrueck, Published on CD-rom by ECOS Gesellschaft für Entwicklung und Consulting Osnabrück GmbH, Westerbreite 7 D49084 Osnabrück, Germany.

Laws E M and Livesey J L (1978) Flow through screens. *Annual Review of Fluid Mechanics* 10 pp. 247–66 (Jan.).

Lo C-B (1999) Appraisal of a 25cc Stirling cycle engine. MPhil dissertation, University of Cambridge.

Ludwig H (1964) Experimentelle Nachprüfung der Stabilitätstheorien fur reibungsfreie Stromungen mit schraubenlinienformigen Stromlinien. *Zeitschrift Flugwissenschat* **12** Heft 8.

Magara Y *et al*. (2000) Effects of stacking method on heat transfer characteristics in stacked wire gauze. Proceedings 4th Symposium on Stirling Cycle, Japan Society of Mechanical Engineers, pp. 43–6.

Marshall A (1994) *The Marshall Story: a Century of Wheels and Wings*. P. Stephens, Sparkford.

Martini W M (1978) *Stirling Engine Design Manual*. US DOE/NASA report.

Miyakoshi R *et al.* (2000) Study of heat transfer characteristics of regenerator matrix (2nd report). Proc. 4th Symposium on Stirling Cycle, Japan Society of Mechanical Engineers.

Model Engineering Company (1974) Correspondence ref. SWI/ABM from the company to the author.

Morgan P G (1959) High speed flow through wire gauzes. *Journal of the Royal Aeronautical Society* **63** (Aug.).

Naso V (2003) Chairman's introduction, call for papers, 11th International Stirling engine conference, Rome, 19–21 Nov.

Ochoa F, Eastwood C, Ronney P D and Dunn B (2003) Thermal transpiration based on micro-scale propulsion and power generation devices. Proceedings 7th International Microgravity Combustion Workshop, Cleveland, Ohio (June – 4pp un-numbered).

Organ A J (1962) An appraisal of the hot-air engine. Final year dissertation, Department of Mechanical Engineering, University of Birmingham.

Organ A J (1992a) *Thermodynamics and Gas Dynamics of the Stirling Cycle Machine.* Cambridge University Press, Cambridge.

Organ A J (1992b) 'Natural' coordinates for analysis of the practical Stirling cycle. *Proceedings Institution Mechanical Engineers* Pt. C 206 pp. 407–15.

Organ A J (1995) The hot-air engine – a very un-scale model. Paper read at the annual Model Engineering Exhibition, Olympia.

Organ A J (1997) *The Regenerator and the Stirling Engine.* PEP (Professional Engineering Publishing), London and Bury St Edmunds. (2005 onwards: John Wiley and Sons, Chichester.)

Organ A J (2000) Regenerator analysis simplified. Proc. European Stirling Forum Osnabrück, 22–24 Feb. pp. 27–37.

Organ A J (2004) Solution of the 'shuttle heat transfer' problem. Proceedings Int. Stirling Forum, Osnabrück. Published on CD-rom by ECOS Gesellschaft für Entwicklung und Consulting Osnabrück mbH, Westerbreite 7 D49084 Osnabrück, Germany.

Organ A J (2005a) *Stirling and Pulse-tube Cryo-coolers.* John Wiley and Sons, Chichester.

Organ A J (2005b) The flow friction correlation – an obstacle to regenerator design. Conditionally accepted for publication in *Cryogenics*, Elsevier.

Organ A J (2005c) Regenerator thermal analysis – unfinished business. Under consideration for possible publication in *Proceedings Japan Society Mechanical Engineers.*

Organ A J (2005d) Regenerator design and the neglected art of back-of-the-envelope calculation. Under review for publication in *Proceedings Institution of Mechanical Engineers. Pt. C.*

Organ A J and Larque I (2004a) A recuperative combustion chamber for a small air engine. Proceedings European Stirling Forum, Osnabrück, Published on CD-rom by ECOS Gesellschaft für Entwicklung und Consulting Osnabrück mbH, Westerbreite 7 D49084 Osnabrück, Germany.

Organ A J and Larque I (2004b) The strange case of the self-regulating Stirling engine. Proceedings European Stirling Forum, Osnabrück, Published on CD-rom by ECOS Gesellschaft für Entwicklung und Consulting Osnabrück mbH, Westerbreite 7 D49084 Osnabrück, Germany.

Organ A J and Lo C-b (2000) Thermodynamic personality of a 25cc Stirling engine. Pt. II: Hydrodynamic pumping loss and thermal short. Proceedings European Stirling Forum, Osnabrück, pp. 22–24 Feb. 49–55.

Organ A J and Maeckel P (1996) Connectivity and regenerator thermal shorting. Proceedings European Stirling Forum, Fachhochschule Osnabrück pp. 229–43.

Organ A J *et al.* (2001) A shot in the dark. Proceedings 10th International Stirling Engine Conference, Osnabrück, 24–26 Sept., Verein deutscher Ingenieure, pp. 47–54.

O'Sullivan M (1953) *Twenty Years A-Growing*. Oxford University Press, London (in World's Classics Series).

Oswatitsch K (1956) *Gas Dynamics*. English version by Gustav Kuerti, Academic Press, New York p. 60.

Pinker R A and Herbert M V (1967) Pressure loss associated with compressible flow through square-mesh wire gauzes. Proceedings Institution of Mechanical Engineers Pt. C (*Jnl. Mech. Eng'g Sci.*) **9**, pp. 11–23.

Rinia H and duPré F K (1946) Air engines. *Philips Technical Review* **8** (5), (May) pp. 129–160.

Rios P A (1971) An approximate solution to the shuttle heat transfer losses in a reciprocating machine. *ASME Journal of Engineering for Power*. Apr., pp. 177–182.

Rix D H (1984) An enquiry into gas process asymmetry in Stirling cycle machines. PhD diss., University of Cambridge.

Rogers G and Mayhew Y (1992) *Engineering Thermodynamics Work and Heat Transfer*, 4th edn. Adison Wesley Longman, Harlow.

Rollet H & Co. (2005) Web site of manufacturer, postal address Dawley Trading Estate Stallings Lane, Kingswinford, West Midlands DY6 7BW.

Ross M A (1989) States-side hot air. *Model Engineer* **162**, (3842) pp. 155–6.

Schmidt G (1871) Theorie der Lehmann'schen calorischen Maschine. *Zeitschrift des Vereins deutscher Ingenieure*. Band XV, Heft 1 (January).

Shapiro A H (1954) *The Dynamics and Thermodynamics of Compressible Fluid Flow*. Ronald, New York, **I** p. 184, **II** p. 1131.

Sier R (1995) *The Revd Robert Stirling DD*. L A Mair, Chelmsford.

Sier R (1999) *Hot Air, Caloric and Stirling Engines* – I A History. L A Mair, Chelmsford.

Simmons W F and Cross H C (1952) Report on the elevated-temperature properties of stainless steels. ASTM Special Technical Publication # 124.

SIMQX (undated) FORTRAN code. Originally part of the IBM 1130 Scientific Subroutine Library, but subsequently available as part of other library packages. Original contribution to the 1130 library possibly made by the late Dr Wiley Gregory of the University Engineering Department., Cambridge, who once claimed authorship in conversation with the author.

Stirling J (1845) Description of Stirling's improved air angine. *Proceedings Institution Civil Engineers* **IV** pp. 348–61.

Stirling J (1853) Contribution to discussion 'Heated Air Engine'. Proceedings Institution Civil Engineers **XII** pp. 558–600.

Stirling R (1816a) London patent 4081 – Improvements for diminishing the consumption of fuel, and in particular an engine capable of being applied to the moving of machinery on a principle entirely new (accompanied by orthographic drawings).

Stirling R (1816b) Edinburgh patent 4081 (as above, but superior illustrations).

SKF Ltd (2002) SKF bearings with *solid oil* – the third lubrication choice. Publication 4827/1 E, SKF Corporation, Sweden.

Su C-C (1986) An enquiry into the mechanism of pressure drop in the regenerator of the Stirling cycle machine. PhD diss., Engineering Department, Cambridge University.

Tailer P L (1995) Thermal lag machine. US Patent # US5414997.

Taylor G I (1935) Distribution of velocity and temperature between concentric rotating cylinders. *Proceedings of the Royal Society of London series* A, **135** pp. 494–512.

Taylor G I (1936) Fluid friction between rotating cylinders. I – Torque measurement. *Proceedings of the Royal Society of London* series A, **157** pp. 546–564.

Thompson B E (undated) Evaluation of advanced heat recovery systems. Web page of Foam Application Technologies Inc.

Thwaites B (1949) Approximate calculation of laminar boundary layer. *Aeronautical Quarterly*. **1**, pp. 245–80.

Urieli I and Berchowitz D M (1984) *Stirling Cycle Engine Analysis*. Adam Hilger, Bristol.

van Rijn C (2005) Report on work carried out at Cambridge University Engineering Department (CUED) as part of engineering degree programme at the University of Eindhoven, Netherlands.

van Rijn C (2006) Interim account of further work at CUED towards engineering degree programme at the University of Eindhoven, Netherlands.

Walker G (1980) *Stirling Engines*. Oxford University Press, Oxford.

Warbrooke E T (2005) *Building Stirling-1*. Camden Miniature Steam Services, Rode, Somerset.

Ward G (1972) Performance characteristics of the Stirling engine. MSc diss. University of Bath.

Weisstein E W (2005) Cubic Close Packing. From MathWorld – A Wolfram Web Resource. http://mathworld.wolfram.com/CubicClosePacking.html

Wellington Engineering (2006) Web site of Parkside House, Rigby Lane, Hayes, Middlesex, UB3 1ET UK, www.welleng.co.uk

West C D (1993) Some single-piston closed-cycle machines and Peter Tailer's thermal lag engine. Paper 93105, Proceedings 28th. Inter-Societies Energy Conversion Engineering Conference (IECEC), American Chemical Society, pp. 2.673–2.679.

White F M (1974) *Viscous Fluid Flow*. McGraw-Hill, New York, p. 206.

White F M (1979) *Fluid Mechanics*. McGraw-Hill Kogakusha, Tokyo.

Wicks F and Caminero C (1994) The Peter Tailer external combustion thermal lag piston/cylinder engine analysis and potential applications. Paper AIAA-94-3987-CP, Proceedings of Inter-Societies Energy Conversion Engineering Conference (IECEC) pp. 951–954.

Wilson R S (2000) Advances in piston and packing ring materials for oil-free compressors. Presented to 4th Workshop on Piston Compressors, Kötter Consulting Enginers, Rheine, Germany (25–26 Oct.).

Wirtz R A *et al.* (2003) Thermal/fluid characteristics of 3-D woven mesh structures as heat exchange surfaces. *Transactions of the Institution of Electrical and Electronic Engineers IEEE – Components and Packaging Technologies* **26** (1), pp. 40–7.

Wünnung J G (1996) Flammlose Oxidation als Verbrennungsverfahren für Stirling-Maschinen. Proceedings, European Stirling Forum, 26–28 Feb., Osnabrück, pp. 273–285.

Zimmerman F J and Longsworth R C (1970) Shuttle heat transfer. *Advances in Cryogenic Engineering*. **16** Paper H-5, pp. 342–51.

Index